Symposium on aesthetic surgery of the nose, ears, and chin

Volume six

Symposium on aesthetic surgery of the nose, ears, and chin

Editors

FRANK W. MASTERS, M.D.

Professor and Chief, Section of Plastic Surgery;
Vice-Chairman, Department of Surgery, University of Kansas
Medical Center, Kansas City, Kansas

JOHN R. LEWIS, Jr., M.D.

Chief, Department of Plastic Surgery, Crawford Long Hospital
of Emory University; Chief, Department of Plastic Surgery,
Doctors' Memorial Hospital, Atlanta, Georgia

Proceedings of the Symposium of the Educational
Foundation of the American Society of Plastic and Reconstructive Surgeons, Inc.,
and the American Society for Aesthetic Plastic Surgery, Inc.,
held at Phoenix, Arizona, November 10-13, 1971

With 514 illustrations, including 3 full-color plates

The C. V. Mosby Company

Saint Louis   1973

Contributors

Morrison D. Beers, M.D., F.A.C.S.

Clinical Assistant Professor, Abraham Lincoln School of Medicine, The University of Illinois; Attending Surgeon, Department of Plastic Surgery, Cook County Hospital, Chicago, Illinois; Attending Surgeon, Plastic Surgery, Lake Forest Hospital, Lake Forest, Illinois

James E. Bennett, M.D., F.A.C.S.

Professor of Surgery and Director of Plastic Surgery, Department of Surgery, Indiana University Medical Center, Indianapolis, Indiana

Edward L. Bitseff, M.D.

Major, U.S. Air Force and Chief Resident, Plastic Surgery, Tulane Service, Charity Hospital, New Orleans, Louisiana

Silas Braley

Director, Dow Corning Center for Aid to Medical Research, Midland, Michigan

T. Ray Broadbent, M.D.

Associate Clinical Professor of Surgery (Plastic), College of Medicine, University of Utah; Co-Director, Plastic and Reconstructive Surgery Training Program, Latter-Day Saints Hospital and Primary Children's Hospital, Salt Lake City, Utah

Gene A. Carlin, M.D., F.A.C.S.

Assistant Clinical Professor of Surgery/Plastic Surgery, University of California Center for the Health Sciences, Los Angeles, California

James E. Conklin, M.D.

Clinical Assistant Professor of Plastic Surgery, University of Pittsburgh School of Medicine, Pittsburgh, Pennsylvania

Bard Cosman, M.D.

Associate Clinical Professor of Surgery, College of Physicians and Surgeons, Columbia University, New York, New York; Associate Attending in Surgery, Columbia-Presbyterian Medical Center, New York City; Visiting Attending, Francis Delafield Hospital, New York City; Associate Plastic Surgeon, St. Elizabeth's Hospital, New York City; Consultant in Plastic Surgery, Blythedale Children's Hospital, Valhalla, New York

Eugene H. Courtiss, M.D.

Clinical Instructor in Surgery, Boston University School of Medicine; Assistant Surgeon in Plastic Surgery (III Surgical Service), Boston City Hospital; Associate in Surgery, University Hospital, Boston, Massachusetts; Chief, Plastic and Reconstructive Surgery Division, Newton-Wellesley Hospital, Newton Lower Falls, Massachusetts; Consulting Staff, Waltham Hospital, Waltham, Massachusetts

George T. Craig, M.D.

Surgeon to Outpatients, The New York Hospital–Cornell University Medical Center, New York, New York

George F. Crikelair, M.D.

Professor of Surgery, College of Physicians and Surgeons, Columbia University; Director, Plastic Surgery Division, Columbia-Presbyterian Medical Center, New York, New York

John F. Crosby, Jr., M.D.

Historian, American Society for Aesthetic Plastic Surgery, Inc., Mobile, Alabama

Simon Fredricks, M.D., F.A.C.S.

Clinical Associate Professor of Plastic Surgery, Baylor University College of Medicine; Associate Chief of Plastic Surgery, Hermann Hospital of the University of Texas Medical School, Houston, Texas

Nicholas G. Georgiade, M.D., F.A.C.S.

Professor of Plastic and Maxillofacial Surgery, Duke University Medical Center, Durham, North Carolina

Robert M. Goldwyn, M.D.

Assistant Clinical Professor of Surgery, Harvard Medical School; Senior Associate of Surgery, Peter Bent Brigham Hospital; Chief of Plastic Surgery, Associate Surgeon, Beth Israel Hospital, Boston, Massachusetts

Mark Gorney, M.D., F.A.C.S.

Clinical Instructor in Plastic Surgery, Stanford University School of Medicine, Palo Alto, California; Chief, Plastic Surgery, St. Luke's Hospital; Attending Surgeon, St. Francis Memorial Plastic and Reconstructive Surgery Center, San Francisco, California; Consultant, U.S. Army Letterman General Hospital, U.S. Public Health Service Hospital, San Francisco, California; Veterans Administration Hospital, Palo Alto, California; Chairman, Medico-legal Committee, American Society of Plastic and Reconstructive Surgeons, Inc.

William C. Grabb, M.D.

Clinical Associate Professor of Surgery (Plastic Surgery), University of Michigan Medical School, Ann Arbor, Michigan

Michael M. Gurdin, M.D., F.A.C.S.

Associate Clinical Professor of Surgery/Plastic Surgery, University of California Center for the Health Sciences, Los Angeles, California

James H. Hendrix, Jr., M.D.

Professor of Surgery, Chief, Plastic Surgery Section, Department of Surgery, University of Tennessee, Memphis, Tennessee

Charles H. Hill, M.D.

Former Senior Resident, Section of Plastic Surgery, Indiana University Medical Center, Indianapolis, Indiana

Charles E. Horton, M.D., F.A.C.S.

Chief of Plastic Surgery, Norfolk Medical Center; Professor of Surgery (Plastic), East Virginia Medical School, Norfolk, Virginia

Bernard L. Kaye, M.D., D.M.D., F.A.C.S.

Clinical Assistant Professor of Plastic Surgery, University of Florida; Chairman, Department of Plastic Surgery, Memorial Hospital of Jacksonville; Chief, Section of Plastic Surgery, Hope Haven Children's Hospital, Jacksonville, Florida

Herbert Lipshutz, M.D.

Clinical Professor of Plastic Surgery, University of Pennsylvania School of Medicine; Head, Section of Plastic Surgery, Division of Surgery, Pennsylvania Hospital, Philadelphia, Pennsylvania

Gary H. Manchester, M.D.

San Diego, California

Frank W. Masters, M.D.

Professor and Chief, Section of Plastic Surgery; Vice-Chairman, Department of Surgery, University of Kansas Medical Center, Kansas City, Kansas

Robert M. McCormack, M.D.

Professor of Plastic Surgery, Chairman, Division of Plastic Surgery, Vice-Chairman, Department of Surgery, University of Rochester Medical Center; Plastic Surgeon-in-Chief, Strong Memorial Hospital, Rochester, New York

Frederick J. McCoy, M.D.

Clinical Professor of Surgery, University of Missouri-Kansas City; Chief, Division of Plastic Surgery, Kansas City General Hospital and Medical Center; Chief, Division of Plastic Surgery, The Children's Mercy Hospital, Kansas City, Missouri

Allyn McDowell, M.D.

Assistant Clinical Professor of Surgery/Plastic, University of California at Los Angeles School of Medicine, Los Angeles, California

Mar W. McGregor, M.D.

Clinical Assistant Professor of Surgery (Plastic), Stanford University School of Medicine, Palo Alto, California; Senior Attending Staff, St. Francis Memorial Hospital, Plastic and Reconstructive Surgery Center, San Francisco, California

D. Ralph Millard, Jr., M.D., F.A.C.S.

Professor of Clinical Surgery, Chief, Division of Plastic Surgery, University of Miami School of Medicine, Miami, Florida

Ross H. Musgrave, M.D.

Associate Professor of Surgery (Plastic), University of Pittsburgh School of Medicine, Pittsburgh, Pennsylvania

George C. Peck, M.D.

Associate Professor of Plastic Surgery, Temple University School of Medicine, Philadelphia, Pennsylvania

William J. Pollock, M.D.

Associate Professor of Surgery (Plastic), Tulane University School of Medicine, New Orleans, Louisiana

Robert Pool, M.D.

Chairman, Division of Plastic Surgery, William Beaumont Hospital, Royal Oak, Michigan

Thomas D. Rees, M.D., F.A.C.S.

Clinical Associate Professor of Surgery (Plastic), New York University Medical School, New York, New York

Robert F. Ryan, M.D.

Professor of Surgery (Plastic), Chief of the Section of Plastic Surgery, Tulane University School of Medicine, New Orleans, Louisiana

Eugene C. Sherlock, M.D.

Clinical Assistant Professor, Chief, Plastic Surgery Section, Department of Surgery, University of Alabama Medical Center, Birmingham, Alabama

James W. Smith, M.D.

Associate Professor of Clinical Surgery, The New York Hospital–Cornell University Medical Center, New York, New York

Joseph Still, M.D.

Instructor in Plastic Surgery, Duke University Medical Center, Durham, North Carolina

Lars M. Vistnes, M.D., F.R.C.S.(C), F.A.C.S.

Assistant Professor of Surgery (Plastic), Stanford University School of Medicine; Chief of Plastic Surgery, Veterans Administration Hospital, Palo Alto, California

Richard C. Webster, M.D.

Chief of Plastic Surgery, Melrose-Wakefield Hospital, Melrose, Massachusetts

Malvin F. White, M.D., D.M.D.

Associate Clinical Professor of Surgery (Plastic), Tufts University Medical School; Chief of Plastic Surgery, St. Elizabeth's Hospital, Boston, Massachusetts

Robert M. Woolf, M.D.

Associate Clinical Professor of Surgery, University of Utah College of Medicine, Co-Director, Plastic and Reconstructive Surgery Training Program, Latter-Day Saints Hospital and Primary Children's Hospital, Salt Lake City, Utah

Setrag A. Zacarian, M.D.

Assistant Clinical Professor of Dermatology, Boston University School of Medicine, Boston, Massachusetts; Director of Dermatology Service, Springfield Hospital Medical Center, Springfield, Massachusetts

Preface

Sponsored by the Educational Foundation of the American Society of Plastic and Reconstructive Surgeons with the invaluable cooperation of the American Society for Aesthetic Plastic Surgery, the second Symposium on Aesthetic Surgery was held in Phoenix, Arizona, November 10-13, 1971. The symposium was designed to present a comprehensive review of cosmetic surgery of the nose, ears, and chin, and a variety of therapeutic approaches were discussed in detail, with free interchange between both those on the program and those in the audience. Although somewhat limited by space, the authors have attempted to present in book form the salient features of those problems discussed.

In any field of interest as broad as that contained within the boundaries of aesthetic plastic surgery, individual tricks, variations of therapeutic approach, and specialized methods of postoperative care are as numerous as there are surgeons. It is patently impossible to present all possible variations in technique; therefore, this volume was designed to concentrate primarily upon time-tested principles involved in the surgical care of aesthetic deformities of the nose, ears, and chin. Even though many of the techniques described may seem repetitive to those who have already established personal preferences among the methods of therapy, the combination of established principles and new variations in technique will provide material for all. In this vein, the Educational Foundation of the American Society of Plastic and Reconstructive Surgeons is pleased to present the second volume on cosmetic surgery in the hope that, within these pages, there will be something of value for all.

Frank W. Masters
John R. Lewis, Jr.

Contents

Part III

**Aesthetic surgery of the chin and nonsurgical
considerations of surgery of nose, ears, and chin**

Color plates

Part I

Aesthetic surgery of the nose

Chapter 1

A history of aesthetic rhinoplasty

Frank W. Masters, M.D.
Gary H. Manchester, M.D.

The love of life is next to our love of our face and thus the mutilated cry for help.
Sushruta Samhita, 200 B.C.

The nose is the central and most prominent feature of the face; and on its shape, size and appearance, to a great degree, depends the relative facial beauty of the person.
John O. Roe, 1887

Since the beginning of history, man has considered the nose to be the key to facial appearance and expression. The Papyrus of Edwin Smith (3000 B.C.) and the Nuzu Tablet laws (1500 B.C.) state clearly that nasal mutilation was one of the main forms of punishment for prisoners of war and persons convicted of serious civil offenses. The Ebers Papyrus written about 1600 B.C. devotes a whole section to nasal deformities and their correction.[1]

The first written record of an attempt at nasal reconstruction was not until the year 200 B.C. when Sushruta Samhita wrote his treatise on the "Indian" technique of total rhinoplasty. It is surprising that no attempt at nasal reconstruction had been made before this; however, Galen (138-201 A.D.) states that although the Egyptians performed total rhinoplasty, they did not describe their secret technique.[1]

Paulus Aegineta (625-690) in his seven-volume work *On Medicine* describes the treatment of nasal deformity resulting from blunt trauma. Aegineta clearly described two deformities: dislocation of the septal cartilage and fracture of the nasal bones. He specifically states that fractures should be reduced early, "during the course of the first day, or not long afterwards, because the bones of the nose get consolidated about the tenth day." For collapse of the nasal bones, "the head of a probe" is inserted into the nostrils for elevation of the parts and then "they are to be put into the proper position with the index finger and thumb externally. In order to prevent the bones from changing their position two wedge-like tents formed of twisted linen rag, are to be applied one to each nostril even if but one part of the nose be deranged and these are to be allowed to remain until the bone or cartilage gets consolidated."[1] For deviation of the nasal septum, Paulus Aegineta instructs the reader to place his fingers into the nostrils in order to pull and push the septum into correct alignment. He also discusses the use of external splinting in order to maintain the shape of the corrected nose but "abhors" its use when the splint might depress a good result.

Guy de Chauliac (1300-1368), in his *Chirurgia* in 1363, discusses nasofacial fracture, stating: "And you should reduce the fragments with the finger in such a way that neither bulging nor depression occurs at the place of fracture."

Total nasal reconstruction was lost to the world until 1430 when the Branca family began using arm flaps for skin coverage. The Branca family never wrote about this technique and tried to keep it as a secret part of their family's surgical armamentarium.

Leonardo da Vinci (1452-1519) analyzed the

nose and its relation as a part of the face. "The ear is precisely as long as the nose. The length of the ear should equal the distance from the bottom of the nose to the top of the eyelid. The space from the chin to the beginning of the bottom of the nose is the third part of the face and equal to the nose and to the forehead."

Gaspar Tagliacozzi (1546-1599) gave to the world a detailed description of the "Italian method" of rhinoplasty.[2] This was almost identical to the method of the Branca family, but Tagliacozzi is given credit for disseminating its use throughout the world. It is of interest that this complicated procedure was performed on the patient without anesthesia. Authorities at the time allowed Tagliacozzi to proceed with his work, considering his reconstruction as part of the punishment for the original offense.

In 1455 the Spanish explorer Augustin de Zarate discovered Peruvian coca.[1] Medical use of this discovery was not realized for over 300 years, however, until Koller introduced cocaine as an analgesic in 1884. Gardehe isolated the alkaloid in 1855 and Albert Niemann isolated the active substance in 1860 and called it cocaine.[1] H. Knapp[3] is given credit for being the first to use cocaine as a local block in nasal surgery in 1884, and J. Roe[4] used it effectively in 1887.

The nineteenth century heralded a new era for plastic surgery. With the rediscovery of cocaine and improved means of communication, people became more conscious of their appearance. Daumier's caricatures poke fun at people by distorting their facial features, particularly the nose. With this increased concern over appearance, it is not surprising that physicians began receiving inquiries concerning aesthetic alteration of the nose. J. F. Dieffenbach in his *Die Operative Chirurgie* devotes an entire section to total rhinoplasty and then discusses the subcutaneous correction of the intact nose.[5] He states that he has performed such operations and describes various methods used (Fig. 1-1).

In 1875, William Adams,[6] a dentist, published his report *On the Treatment of Broken Nose by*

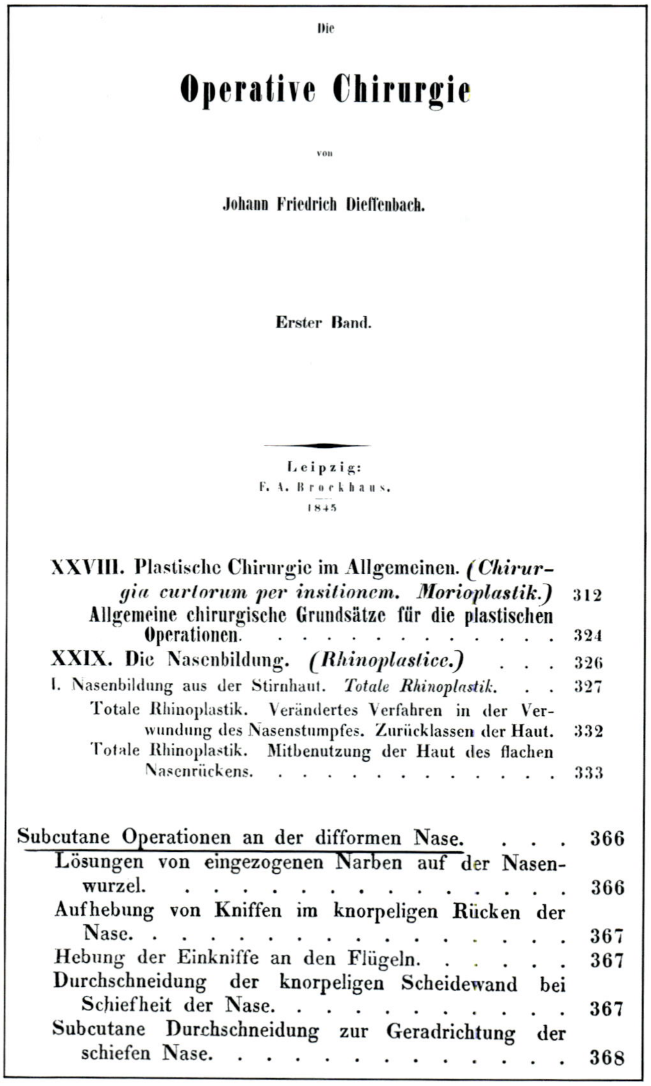

Die

Operative Chirurgie

von

Johann Friedrich Dieffenbach.

Erster Band.

Leipzig:
F. A. Brockhaus.
1845

XXVIII. Plastische Chirurgie im Allgemeinen. *(Chirurgia curtorum per insitionem. Morioplastik.)* 312
Allgemeine chirurgische Grundsätze für die plastischen Operationen. 324
XXIX. Die Nasenbildung. *(Rhinoplastice.)* . . . 326
I. Nasenbildung aus der Stirnhaut. *Totale Rhinoplastik.* . . 327
Totale Rhinoplastik. Verändertes Verfahren in der Verwundung des Nasenstumpfes. Zurücklassen der Haut. 332
Totale Rhinoplastik. Mitbenutzung der Haut des flachen Nasenrückens. 333

Subcutane Operationen an der difformen Nase. . . . 366
Lösungen von eingezogenen Narben auf der Nasenwurzel. 366
Aufhebung von Kniffen im knorpeligen Rücken der Nase. 367
Hebung der Einkniffe an den Flügeln. 367
Durchschneidung der knorpeligen Scheidewand bei Schiefheit der Nase. 367
Subcutane Durchschneidung zur Geradrichtung der schiefen Nase. 368

Fig. 1-1. Table of contents from Dieffenbach's *Die Operative Chirurgie* published in 1845 with an entire section devoted to rhinoplasty.

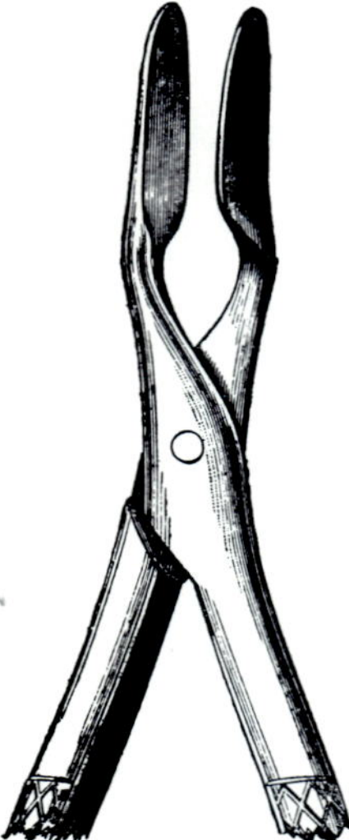

Fig. 1-2. Original forceps described by Adams for reduction of the nasal bones and nasal septum.

Forcible Straightening and Mechanical Retentive Apparatus. Dr. Adams felt that the medical profession had not been aggressive enough in the treatment of nasal fractures and that the resulting deformities could be averted if treated early. His paper subdivides nasal fractures into those related to the cartilaginous and the bony framework. Although he discusses acute fractures, his reported cases are 6 months to 6 years from the date of injury. He created the first nasal forceps (Fig. 1-2) and used them to refracture the nasal bones and replace the septal cartilage in the midline. Dr. Adams' device for external support is similar to those in use today.

Dr. R. F. Weir[7, 8] published two papers in 1880 discussing Dr. Adams' work. In his thesis he modified the Adams forceps by making them thinner, and he describes an "osteoplastic operation" in which a chisel is used to achieve an even reduction.

"The deformity termed pug nose and its correction by a simple operation" was reported by Dr. John Roe of Rochester in 1887.[4] In this article the nasal physiognomy is classified into: (1) the Roman nose, (2) the Greek nose, (3) the Jewish (Semitic) nose, (4) the snub or pug nose, and (5) the celestial nose. The "pug nose" deformity results from increased tissue at the nasal tip either from excess septum or prominent alar cartilages. Roe described

The Medical Record

A Weekly Journal of Medicine and Surgery

Vol. 31, No. 23   NEW YORK, JUNE 4, 1887   Whole No. 865

Original Articles.

THE DEFORMITY TERMED "PUG NOSE" AND ITS CORRECTION, BY A SIMPLE OPERATION.[1]

By JOHN O. ROE, M.D.,
ROCHESTER, N. Y.

THE nose is the central and most prominent feature of the face; and on its shape, size, and appearance, to a great degree, depends the relative facial beauty of the person.

Physiognomists emphasize the importance of the nose in the category of anatomical conformations that are indicative of special traits of character; and regard it as a measure of force in nations and individuals.

Says Wells: " A skilful dissembler may disguise, in a degree, the expression of the mouth; the hat may be slouched over the eyes; the chin may be hidden in an impenetrable thicket of beard; but the nose will stand out 'and make its sign' in spite of all precautions. It utterly refuses to be ignored, and we are, as it were, compelled to give it our attention."

Even in ancient times much attention was given to its shape and appearance. Among the ancient Persians no man who had a crooked or deformed nose was allowed to sit upon the throne. Cyrus, it is said, had an asymmetrical nose, which was made a thing of beauty through the kind assistance of his emasculated attendants. In order to secure symmetrical and handsomely formed noses, in the children of the royal blood, the eunuchs who had charge of the royal offspring were accustomed to mould their noses into perfect shape (Mackenzie).

Considered from the profile point of view alone, noses are classified according to their shape by students of physiognomy into five main classes:

1. The Roman noses;
2. The Greek noses;
3. The Jewish noses;
4. The Snub or Pug noses; and
5. The Celestial noses.

These classes of noses, considered in the light of the characteristics of the race or class to which they are peculiar, are observed to indicate prominent traits of character, as follows:

The Roman indicates executiveness or strength; the Greek, refinement; the Jewish, commercialism or desire for gain; the Snub or Pug, weakness and lack of development; the Celestial, weakness, lack of development, and inquisitiveness.

"Le nez retroussé" of the French is applied to the Celestial nose, which is simply the pug lengthened and turned upward so as to form a gentle curve from the root to the tip.

The fact that the deductions of physiognomists almost completely harmonize with the anatomical and physiological facts in the case of the last two classes becomes striking, when we consider that those deductions have been made from observation alone.

Mr. Warwick says: " A snub-nose is to us a subject of most melancholy interest. We behold in it a proof of the degeneracy of the human race."

Tristram Shandy's father, regretting his son's misfor-

[1] Read before the Medical Society of the State of New York, February 1, 1887.

tunes, remarked: " No family, however high, could stand against a succession of short noses; " and his grandfather " when tendering his hand and heart to the lady who afterward consented to 'make him the happiest of men,' was forced to capitulate to her terms owing to the brevity of his nose."

There are three conditions that may occur during the development of the nose that give it the appearance called snub or pug. They are: (1) Excessive development of the alæ and cartilaginous portions on the end of the nose; (2) a lack of sufficient development, or a sunken or flattened condition of the base and bridge of the nose, while the end of the nose may be but normally developed; 3, the combination, to a greater or less degree, of the conditions just mentioned. This last condition is the one most frequently found.

During development the nose and parts comprising the central portion of the face, as the ethmoid and sphenoid bones, and parts adjacent, are late in developing, and are also the last portions of the face to undergo ossification. At birth the nose, at its base and central portions, is flat and nearly level with the face, but later this depressed line is replaced by a more prominent one as the nose becomes developed. From this it will be seen that anything interfering with the proper development of these parts so as to cause them to remain in their infantile condition, while the end of the nose undergoes due development, will give the nose a snubbed and unsightly shape.

The best developed and most beautiful noses are one-third the length of the face. But noses often vary from this proportion, and in some instances an ill-formed nose is inherited, it being a special family mark. Ribot says, " that of all the features, the nose is the one which heredity preserves the best" (" Hereditary Traits," Richard A. Proctor). But in other instances it is the result of diseased conditions affecting its growth and proper development during infancy and early childhood.

There are many conditions that operate to produce this result. The principal one is obstruction of the nasal passages which cuts off nasal respiration. During the inspiratory act of respiration and deglutition, when the nasal passages are obstructed, a partial vacuum is produced in the naso-pharynx. This suction force, being exerted on the inner side of the yielding cartilaginous nasal tissues, tends thereby to draw them inward, and thus in a corresponding degree retards or prevents their normal expansion and development. This obstruction of the nasal passages may also cause an enlargement or undue development of the portion of the nose below and beyond the obstruction, especially if this obstruction is composed of firm tissues, through interfering with the return circulation. The end of the nose thus becomes engorged, the vessels distended, and a marked thickening of the tissues takes place. The importance of attention to obstructed nostrils in infants, commonly called snuffles, is thus clearly demonstrated.

All chronic affections of the nose, even when unattended by obstruction of the passages, tend to produce by sympathetic irritation more or less congestion of the vessels of the end of the nose, and, by reason of these vessels having less power of resistance, an undue distention of them takes place; the surrounding tissues become thickened, and an enlargement of the end of the nose occurs. This is very commonly observed during the treatment of nasal diseases.

Fig. 1-3. Front page from Roe's original article on rhinoplasty under cocaine anesthetic.

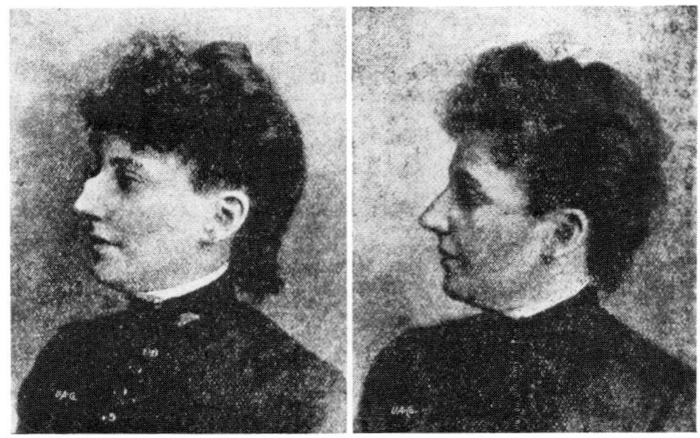

Fig. 1-4. The first pre- and postoperative rhinoplasty photograph published by Roe in 1887.

intranasal incisions and trimmed the dorsum of the septum in its distal third. For prominent cartilages, Dr. Roe molded them after "cutting with a tenotomy knife through those cartilages in different places, sufficiently to destroy their elasticity." The tip of the nose was then molded by means of external splinting in order to maintain the desired shape (Fig. 1-3).

In 1891, Dr. Roe published his work, *The Correction of Angular Deformities of the Nose by a Subcutaneous Operation.*[9] He described the use of cocaine topically and subcutaneously in order to obtain anesthesia, and the incisions for the removal of the nasal hump were made internally between the nasal wall and upper lateral cartilage. After the dorsal nasal skin was dissected free, the hump was removed by means of a bone cutter. Dr. Roe molded his skin "by strapping it down with gentle pressure" and maintained his position by means of an internal spring splint for the nose (Fig. 1-4). The results are documented quite well with similar pre- and postoperative photographs.

The first clear description of a nasal infracture was presented by Dr. R. F. Weir in 1892 in his paper, *On Restoring Sunken Noses without Scarring the Face.*[10] "It is not difficult to overcome the deformity which results from the flattened bony ridge of the nose by division with a chisel of the ossa nasi, first in the median line and subsequently at the line of their attachments to the superior maxilla, or by the use of forceps to fracture their attachments to the same bone." The author also discusses making these cuts from either internal or external incisions.

In this same article Weir describes using a bone graft from a duck's sternum to fill the deformity of a saddle nose. In two additional cases of saddle nose deformity, Dr. Weir used a platinum prosthesis

and a celluloid strut to compensate for the missing cartilage. With the interest created by Roe and Weir, multiple papers began to appear concerning new devices or machines used in correcting nasal deformities.[11-15]

In 1898, Jacques Joseph, an orthopedic surgeon, published his paper *Ueber die operative Verkleinerung einer Nase (Operative Reduction of the Size of a Nose) (Rhinomiosis).*[16] In his first reported case, Joseph made external incisions over the nasal hump in order to remove excess skin; the hump was removed with a chisel and the tip elevated by removing the inferior portion of the septum. He makes no mention of trimming the alar cartilages or infracturing the nasal bones. This was to come later. In his original article, Joseph acknowledges Weir's paper of 1892 and discusses it. He states that a "Dr. Row" (Roe) had performed a similar operation but that he was not familiar with his technique. Seven years before Joseph's report, however, Roe removed the nasal hump through internal nasal incisions. It was Joseph who went on to further develop rhinoplasty as an art. His final method of reduction rhinoplasty exists today as the standard technique for total nasal reduction.

The endonasal approach, however, was first introduced by Dieffenbach (1845) and then by Roe in 1887 and later by Joseph.[17] The removal of the nasal hump was first performed by Roe (1891) by means of a bone cutter, followed by Weir (1892) and Joseph (1898) with a chisel. The lateral osteotomy was performed and described by Weir in 1892. Joseph began performing infracturing after 1900 and developed the nasal saw in order to control his line of fracture. This controversy, chisel versus saw, continues to rage till the present day.[7, 18-23] Aufricht,[24] a pupil of Joseph, was the first to describe the removal of a small wedge of bone from the wide nasal root so that the infracture would be more exact and the nasal pyramid sharp. The use of the nasal forceps to facilitate freeing the nasal bones for infracture belongs to Adams (1875) and Weir (1890). Weir was the first (1892) to excise a wedge from the septum in order to correct a drooping tip, and Joseph reported a similar technique in 1898. Trimming of the alar cartilages was taught by Joseph in the early 1900's through an intercartilaginous incision. It is interesting to note that Miller[25] in 1924 published his report of this same incision.

The alar contouring that is discussed so widely today was first performed by Weir and then

Joseph.[19] Joseph's excisions were mainly placed in the nasal vestibule and inferiorly so as not to be noticed. This technique continues to be used today.[19, 26, 27] Miller in 1924 wrote of excising transverse diamonds from the midportion of the alae in order to narrow a nasal flare.[25]

Many subtle changes in the Joseph technique of rhinoplasty have been suggested over the past 70 years, but it was the orthopedist who made rhinoplasty the art of human sculpture. Aufricht[28] in 1934 reported on the "concert" of the human face and the importance of the chin or nasal profile. Since his initial paper, this step has become increasingly more popular with rhinoplasty. [19, 24, 29-35]

SUMMARY

According to Dieffenbach, the important pioneer of rhinoplasty, "the nose (is) man's most paradoxical organ. It has its root above, its back in front, its wings below, and one likes best of all to poke it into places where it does not belong.[19]

Historically, however, the main emphasis of nasal surgical ingenuity was directed at reconstruction of major defects rather than at the alteration of functionally normal structures for aesthetic reasons. The history of the aesthetic rhinoplasty is indeed brief. It apparently developed on a basis of demand and has progressed so rapidly that the major advances have occurred within the purview of many of the senior plastic surgeons still in active practice today. Although "art is long," the art of aesthetic rhinoplasty is a virtual newborn child in the field of reconstructive surgery.

REFERENCES

1. Maltz, M.: Evolution of plastic surgery, New York, 1946, Forben Press.
2. Tagliacozzi, G.: De curtorum chirurgia per insitionem, libri duo . . . , Venice, 1597, Gaspar Bindonus, Jr.
3. Knapp, H.: Hydrochlorate of cocaine—experiments and application, Med. Record **26**:461, 1884.
4. Roe, J. O.: The deformity termed 'pug nose' and its correction by a simple operation, Med. Record **31**:621, 1887.
5. Dieffenbach, J. F.: Die operative chirurgie, Leipzig, 1845, F. A. Bockhaus.
6. Adams, W.: On the treatment of broken nose by forcible straightening and mechanical retentive apparatus, Brit. Med. J. **2**:421, 1875.
7. Weir, R. F.: On the relief of the deformity of a broken nose by some new methods, Med. Record **17**:279, 1880.
8. Weir, R. F.: On the relief of the deformity of a broken nose by some new methods, Trans. Med. Soc. New York, pp. 273-282, 1880.
9. Roe, J. O.: The correction of angular deformities of the nose by a subcutaneous operation, Med. Record **40**:57, 1891.
10. Weir, R. F.: On restoring sunken noses without scarring the face, New York Med. J. **56**:449, 1892.
11. Daly, W. H.: An eligible method of repairing a broken nose, New York Med. J. **56**:508, 1892.
12. Goodale, J. L.: The correction of old displacements of the nasal bones, Boston Med. Surg. J. **145**:538, 1901.
13. Israel, J.: Two new methods of rhinoplasty (reprinted), Plast. Reconstr. Surg. **46**:80, 1970.
14. Johnson, W. B.: An original device for correcting deformities of the nose resulting from traumatism; with a report of two cases, Med. Record **41**:374, 1892.
15. Monks, G. H.: Correction by operation of some nasal deformities and disfigurements, Boston Med. Surg. J. **139**:262, 1898.
16. Joseph, J.: Ueber die operative Verkleinerung einer nase (Rhinomiosis), Berlin. Klin. Wschr. **35**:882, 1898.
17. Joseph, J.: Nasenplastic und Sonstige Gesichtsplastik, Leipzig, 1931, C. Kabizsch.
18. Byars, L. T.: Surgical correction of nasal deformities, Surg. Gynec. Obstet. **84**:65, 1947.
19. Denecke, H. J., and Meyer, R.: Plastic surgery of head and neck, vol. I: Corrective and reconstructive rhinoplasty, New York, 1967, Springer-Verlag, Inc.
20. Gatewood, W. L.: Substitution of the chisel for the saw in reconstructive surgery, Surgery **2**:149, 1947.
21. Safian, J.: Failures in rhinoplastic surgery, Am. J. Surg. **50**:274, 1940.
22. Safian, J.: Fact and fallacy in rhinoplastic surgery, Brit. J. Plast. Surg. **11**:45, 1958.
23. Webster, G. V.: Rhinoplasty: the saw technique, Plast. Reconstr. Surg. **28**:249, 1961.
24. Aufricht, G.: A few hints and surgical details in rhinoplasty, Laryngoscope **53**:317-335, 1943.
25. Miller, C. C.: Cosmetic surgery—the correction of featural imperfections, Philadelphia, 1924, F. A. Davis Co.
26. Millard, D. R.: External excisions on rhinoplasty, Brit. J. Plast. Surg. **12**:340, 1960.
27. Seeley, R. C.: Reconstruction of the nasal tip: a new technic, Plast. Reconstr. Surg. **3**:594, 1948.
28. Aufricht, G.: Combined nasal plastic and chin plastic, Am. J. Surg. **25**:292, 1934.
29. Aufricht, G.: Combined plastic surgery of the nose and chin, Am. J. Surg. **95**:231, 1958.
30. Aufricht, G.: Rhinoplasty and the face, Plast. Reconstr. Surg. **43**:219, 1969.
31. Millard, D. R.: Adjuncts in augmentation mentoplasty and corrective rhinoplasty, Plast. Reconstr. Surg. **36**:48, 1965.
32. Blair, V. P., and Brown, J. B.: Nasal abnormalities, fancied and real, Surg. Gynec. Obstet. **53**:797, 1931.
33. Millard, D. R.: Alar margin sculpturing, Plast. Reconstr. Surg. **40**:337, 1967.
34. Safian, J.: A critical review of recent literature of rhinoplasty, Plast. Reconstr. Surg. **2**:463, 1947.
35. Straatsma, C. R.: Surgery of the bony nose: comparative evaluation of chisel and saw technique, Plast. Reconstr. Surg. **28**:246, 1961.

Chapter 2

Anatomy of the nose and septum

Eugene C. Sherlock, M.D.

The most prominent and characteristic feature of the face is the nose. Its shape, which is both familial and individual, varies from grecian to aquiline. The external nose is roughly pyramid shaped, thus it is sometimes referred to as the nasal pyramid. The bony and cartilaginous structures are most important in determining the nasal profile and shape.

The basic structures manipulated during rhinoplasty are the nasal bones, frontal process of the maxilla, upper lateral cartilages, alar cartilages, and septum. The best results are obtained after a well-executed rhinoplasty that is blessed with a fine-textured skin drape.

SKIN

The skin of the nose varies from individual to individual in its oiliness, thickness, and porosity. The skin over the bony nose and upper lateral cartilages is movable and generally thinner than its more distal counterpart, which is densely adherent to the alar cartilages and contains sebaceous glands.

The blood supply to the covering tissue of the nose is derived from both the external (facial artery) and internal (ophthalmic and anterior ethmoid arteries) carotid systems. These vessels transverse the area in the superficial layers, requiring that surgical dissections remain close to the osteocartilaginous structures (Fig. 2-1).

The cutaneous innervation of the external nose is derived from the trigeminal nerve through its V_1 and V_2 divisions. The root, bridge, and upper part of the side of nose is innervated by the supratrochlear and infratrochlear nerves, while the lower side is innervated by the infraorbital branch of the

maxillary nerve. The lower dorsum and tip area is innervated by the external nasal branch of the anterior ethmoidal nerve (Fig. 2-2).

EXTERNAL NOSE

Cartilages

Alar cartilages. The alar cartilages are thin plates that are adherent to the skin and that mold the nasal apertures. Each cartilage is made up of a medial and lateral crus. The medial crura of the

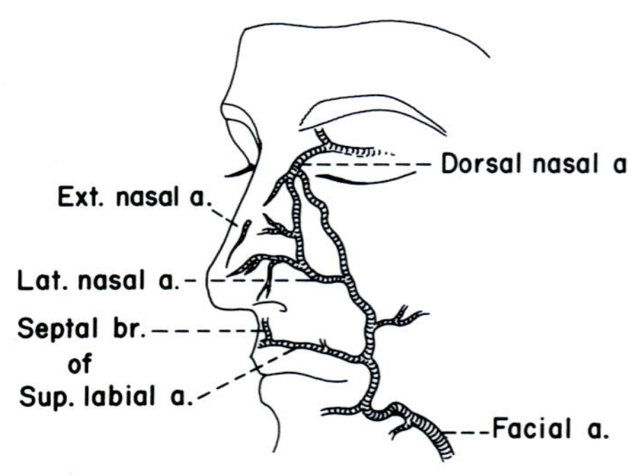

Fig. 2-1. Blood supply of the external nose.

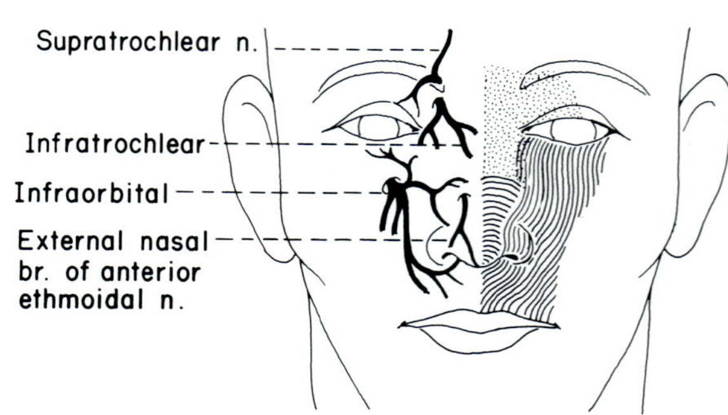

Fig. 2-2. Cutaneous innervation.

two alar cartilages lie within the membranous columella, where they are intimately adherent to the lateral skin and loosely connected to each other medially, and diverge at the base of the columella. At the dome of the nose (tip), the alar cartilages curve backward and diverge, leaving a thin area medially between the alar and upper lateral cartilages known as the weak triangle.[1,2] The alar cartilages are bound to each other and to the upper lateral cartilages by an aponeurotic tissue, which acts as the suspensory ligament to the nasal tip.

Upper lateral cartilages. The upper lateral (or lateral) cartilages meet the dorsum of the septum edge to edge. There may be a mutual perichondrium but there is never continuity. The septum with its adjoining lateral cartilages dips beneath and is attached to the nasal bone and frontal process of the maxilla for perhaps 1 cm. Distally at the septal angle, the upper lateral cartilages are separated from the septum by a narrow cleft (Fig. 2-3).

Bony framework. The bony nose is formed by the nasal bones, which are joined in the midline and which articulate with the upper septum and perpendicular plate of the ethmoid. Posteriorly they are supported by the nasal process of the frontal bone and laterally by the frontal processes of the maxilla. The lateral articulation is especially strong since the nasal bone is beveled—on its inner surface in the upper portion and on its outer surface in the lower portion.

NASAL CAVITY

The nasal cavity begins at the external nares where that portion corresponding to the ala of the nose is called the vestibule. It is lined with skin and contains vibrissae (hair) and sebaceous and sweat glands. The lining of the remainder of the nasal cavity is mucous membrane, which is tightly attached to the periosteum and perichondrium of its walls. Most of the mucous membrane contains mucous and serous glands and is highly vascularized. It is covered with pseudostratified columnar ciliated epithelium constituting respiratory epithelium (schneiderian membrane).

Septum

The septum divides the interior nose into two cavities. It has cartilaginous and bony components.

Cartilaginous. Anteriorly in the columella are the medial crura of the alar cartilages. These give support and shape to the columella. This part is sometimes referred to as the mobile septum.

Quadrangular or septal cartilage. Posteriorly and superiorly septal cartilage articulates in a groove in the perpendicular plate of the ethmoid bone. Its inferior surface is similarly situated in the vomer and nasal crest. The perichondrium of the quadrangular cartilage is continuous with the periosteum of the ethmoid and vomer. However, there is a loose connective tissue actually between the cartilage and bony groove that may contain some fat. This allows mobility of the septum and acts as a cushion.

The dorsal border of the quadrangular cartilage becomes bifid near the nasal bones, resulting in a supraseptal groove that is palpable in most individuals but not visible because of the interposed tissues. The upper lateral cartilages are joined to the septum over a great portion of its external course, as discussed previously (Fig. 2-4).

Innervation

The innervation of the nasal cavity is essentially by the trigeminal nerve, although some authors have

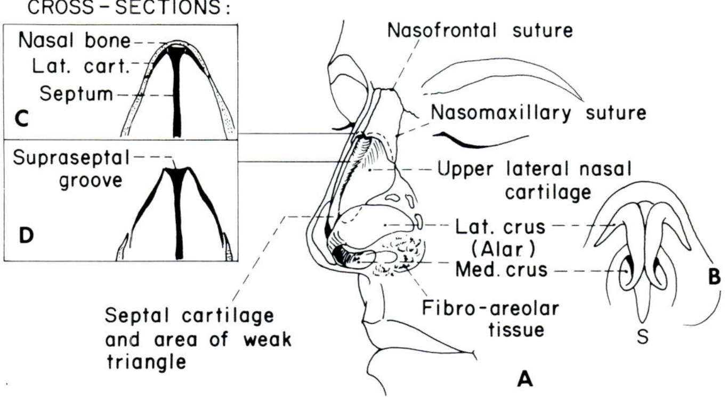

Fig. 2-3. External nose.

demonstrated sensory facial nerve fibers within the sphenopalatine ganglion. Their function and distribution, however, are not clearly understood.[3]

The sphenopalatine (pterygopalatine) ganglion is suspended from the maxillary nerve and lies in the pterygopalatine fossa, located in the posterior lateral nasal cavity. Most of the branches from the sphenopalatine nerve exit through the sphenopalatine foramen, which is just superior and posterior to the middle nasal concha.

The septum is partially innervated by the internal nasal nerve (superiorly and anteriorly), while the larger area draws innervation from the nasopalatine nerve, which is the largest branch

from the sphenopalatine ganglion. It traverses the septum in an anterior-inferior direction, penetrating the incisive foramen along with the anterior superior alveolar nerve that innervates the hard palate in the retroincisor area.

Most of the lateral wall is innervated by the lateral posterior branches (superior and inferior) and the pharyngeal nerve. The superior and anterior portion of the lateral wall is innervated by the internal nasal nerve. Other minor contributors are the terminal branches of the infraorbital nerve and the anterior superior alveolar nerve (Fig. 2-5).

The greater palatine nerve passes downward in the greater palatine canal, giving off small posterior

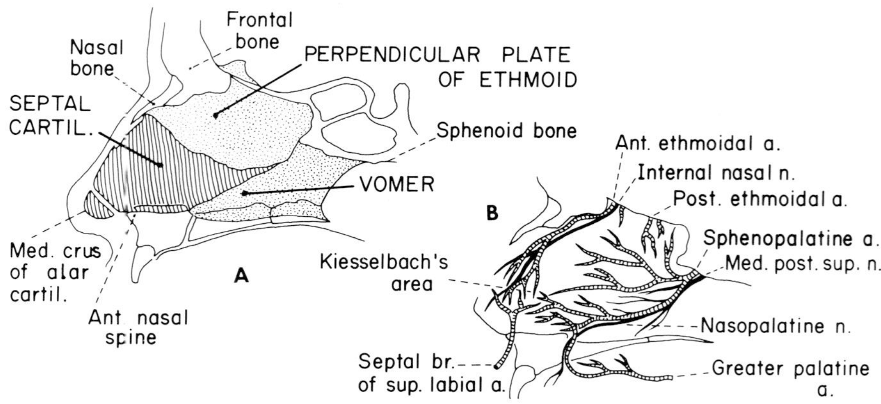

Fig. 2-4. Nasal septum.

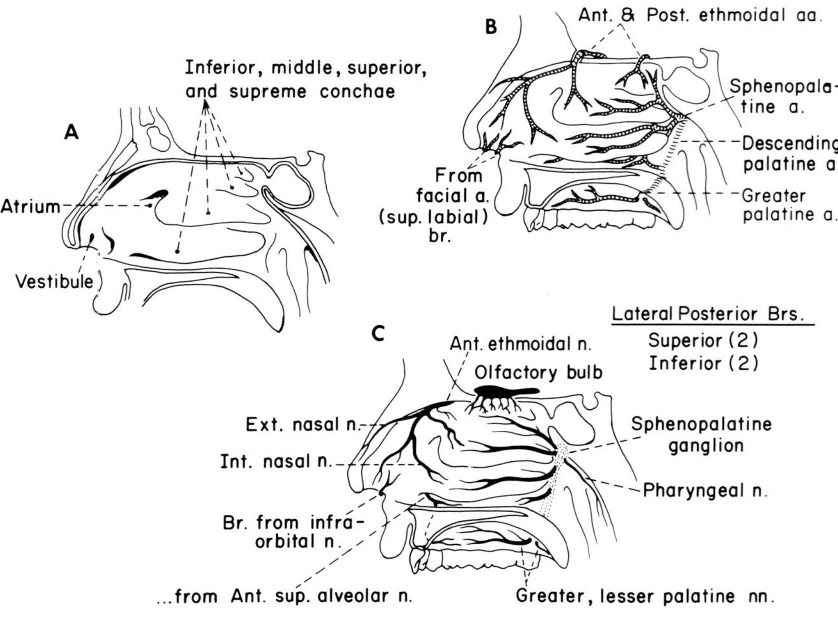

Fig. 2-5. Lateral nasal wall.

inferior branches to the inferior conchal area posteriorly.

The arterial supply of the nasal cavity parallels that of the nerve. The major supply is via the sphenopalatine artery (internal maxillary) and the anterior and posterior ethmoid arteries from the internal carotid system. The anterior nose is supplied by the septal branch of the facial artery. The common meeting ground of these three systems is called Kiesselbach's area.

SUMMARY

The anatomy of the nose is generally in twos. It has two kinds of skin, sebaceous and thin; both carotid systems (internal and external) supply it; it has a dual nerve supply (V and VII); and its structure, both external nose and septal, is made up of bone and cartilage. An intimate knowledge of its structures is of paramount importance to the rhinoplastic surgeon.

REFERENCES

1. Converse, J. M.: Cartilaginous structures of the nose, Ann. Otol. **64:**220-229, 1955.
2. Converse, J. M., and others: Deformities of the nose in reconstructive plastic surgery, Philadelphia, 1964, W. B. Saunders Co., pp. 694-828.
3. Hollinshead, W. H.: The nose and paranasal sinuses in anatomy for surgeons, ed. 2, New York, 1968, Harper and Row, Publishers, pp. 253-305.

Chapter 3

Physiology of the nose and its alteration after rhinoplasty

James E. Bennett, M.D., F.A.C.S.
Charles H. Hill, M.D.

Before the plastic surgeon performs a corrective rhinoplasty, he should note the functions of the nose and be aware of the relationship between nasal anatomy and physiology.

Nasal functions are significant. The nose provides an airway to the lower respiratory tract. Through moistening, heating, and cleansing of inspired air, it protects the lungs. The nose cleanses itself of foreign material that has been filtered from the air. Finally, it is the seat of the sense of smell. Most of these functions are taken for granted, but serious alteration is soon noted by the patient and occasionally can cause significant physiologic impairment.

Proetz[1] has studied the pattern of air flow through the nose. In the normal nose, the nozzle effect of the structures directs the airstream upward, where most of it passes through the olfactory fissure close to the septum and continues in a high arch to the anterior surface of the sphenoid bone and then through the choana to the pharynx. On expiration, the pattern is reversed, this time with a stream of air passing forward in a high arch, entering the meati from the back, and eddying into them from the bottom. Humidification takes place as the inspired air passes over the turbinates. The result is roughly 90% relative humidity of inspired air before it reaches the larynx. Performing this function, the nose emits approximately 1 liter of water every 24 hours. Approximately 70 kilocalories of energy are required to perform this task of humidification. Temperature control is likewise regulated, bringing inspired air to body temperature prior to entering the lungs.

The cilia and mucous glands provide a self-cleansing action to the nose. The human cilia are 7 μ long and less than 0.3 μ in diameter. They beat from three to twelve times per second and always in the same direction. There is an "effective stroke," which is a sharp, stiff, and quick motion, and a "recovery stroke," which is slower and more languid. In the nose the cilia beat toward and through the ostia. Moisture is essential for this activity, since drying is the cilia's only natural enemy. Therefore the surgeon must not produce nonciliated areas in locations where they will impede the cleansing action of the nose and its sinuses. Ill-placed scars may produce hurdles over which the ciliary stream cannot pass. This leaves an accumulation of mucus, often with resultant infection. If the nostril is rendered too small (Weir excisions), hypoventilation may result with metaplasia of the ciliary epithelium and reduction of cleansing action. The nostrils direct the airstream almost vertically into the nose. If the air flow is altered to a backward direction, pharyngitis sicca or, later, laryngitis sicca may occur. Thus, the nose may be thought of as a series of valvular structures constituting the entrance of the respiratory tract and effecting maximum mechanical efficiency of respiration, including air conditioning and resistance to invasion by microorganisms.

Frequently, normal nasal physiology has been altered without the patient being fully aware of it. A careful history will frequently uncover these abnormalities preoperatively. It is important to correct internal anatomic distortion if a good functional result is to be obtained. One of the most commonly encountered abnormalities is that of the deviated nasal septum. By producing partial nasal obstruction, airway resistance is increased. Also, laminar flow rates are changed into eddying currents and directional flow is altered. The convex side of the septum loses ciliary action and exposes mucous membrane in such a way as to produce drying, crusting, and, occasionally, erosion with bleeding. Straightening or relocating the septum usually corrects both anatomy and physiology. Alar collapse (apposition of the alar cartilages to the septum during deep inspiration) may follow overenthusiastic cartilaginous resection. If only the alar cartilages are involved, the central portion of the nose assumes a "pinched" appearance and the nostrils dilate.

Goldman[2] has studied rhinoplasty sequelae causing nasal obstruction. He found that nasal obstruction was caused by (1) lack of appropriate management of the nasal septum or hypertrophied inferior turbinates and/or (2) improper treatment of upper and lower lateral cartilages. Usually the standard intranasal incisions play no significant role in nasal physiology, provided the osseous and cartilaginous vaults are restored to their normal anatomic divisions. However, poor surgical technique may lead to significant physiologic abnormalities by creation of denuded regions, improperly replaced intranasal flaps, poor incisions, or improper use of instruments. Goldman found that the nasal airway could be compromised by (1) excessively long upper lateral cartilages with projecting ends; (2) suturing of turn-up flaps of the upper lateral cartilage and approximating them too close to the septum; (3) indiscreet removal of upper lateral cartilages, leading to pinching on either side of the nose; and (4) scarring or undue thickening of the intercartilaginous incisions, which may cause a funnel-like narrowing of the nostrils. The treatment of alar cartilages may also cause respiratory obstruction. Excessive removal leads to a pinched-in, scarred, and collapsed nostril.

Indiscriminate sacrifice of vestibular skin with inevitable cicatrization of raw surfaces should be avoided since it leads to distortion or stenosis of the areas. Vestibular stenosis may also occur from excessive removal of alae to narrow the size of nostrils in the floor of the nose, especially in thin-skinned individuals.

Champion[3] found that anosmia following rhinoplasty was usually temporary, for 7 to 10 days (packs, clots, edema), or the result of injury to the mucous membrane after the use of cocaine or epinephrine. If anosmia was prolonged, there were usually psychologic problems.

In a review of 1,000 rhinoplasties, Flowers[4] found that the lacrimal apparatus was frequently injured but rarely caused serious disability. The damage was usually produced by the lateral osteotomy and was limited to the lacrimal sac, which healed spontaneously, rapidly, and without stenosis or sequelae.

COMMENTS

One would think from reading the foregoing that rhinoplasty is a procedure fraught with danger as far as permanent, untoward physiologic alterations are concerned. Yet clinical experience has shown that if care is taken, neither subjective nor objective symptoms of significance follow rhinoplasty. Basically, the surgeon is dealing with normal or abnormal physiology preoperatively, the latter resulting in most instances from deviation of the septum. When this situation pertains, correction of septal deviation usually solves the problem. When the septum is not deviated prior to operation and anatomy and physiology are basically normal, the surgeon must take care not to alter this situation to a point where troublesome symptoms occur. Careful placement of incisions to avoid compromising scar contracture, judicious handling of both the upper and lower cartilages, careful trimming and reapproximation of mucous membrane, and proper postoperative packing and splinting should enable the surgeon to correct aesthetically the nasal deformity without unfavorably altering physiology.

REFERENCES

1. Proetz, A. W.: Physiology of the nose from the standpoint of the plastic surgeon, Arch. Otolaryng. **39:**514, 1944.
2. Goldman, I. B.: Rhinoplastic sequelae causing nasal obstruction, Arch. Otolaryng. **83:**151, 1966.
3. Champion, R.: Anosmia associated with corrective rhinoplasty, Brit. J. Plast. Surg. **19:**183, 1966.
4. Flowers, R. S.: Injury to the lacrimal apparatus during rhinoplasty, J. Plast. Reconstr. Surg. **42:**577, 1968.

Chapter 4

Aesthetics: the ideas and ideals of beauty

John F. Crosby, Jr., M.D.

The question of what constitutes beauty is one that mankind has debated and about which there have been vastly different opinions throughout the ages. Generally, it seems easier for people to agree on what is beautiful in the world of nature, such as the oceans, forests, mountains, lakes, and streams, and in the animal world than it is for them to see eye to eye regarding human shapes, proportions, coloring, hair, clothing, and the like. Considering the fact that all normal human beings have basically the same anatomic structures and functions, the natural differences between the races and between individuals in proportion, physical and mental development, hair, and facial features are really astonishing.

GREEK INFLUENCE

The first attempt to establish an ideal of human proportion and beauty was slowly evolved and established by the ancient Greeks. The results of this were inherited by the entire western world, and they still have the power to affect ideas on ideal proportions. Greek influence and Greek ideals of human beauty were not only extended throughout the entire Mediterranean world but, through the conquests of Alexander the Great, also into the Orient, India, Afghanistan, and beyond. Today, the great works of Greek sculptors such as Phidias (Fig. 4-1, *A*) and Praxiteles (Fig. 4-1, *B*), or even copies of them, are still some of the most famous and admired statuary ever produced.

In Athens, Greece, in the 5th century B.C., an unparalleled combination of circumstances, events, men and ideas came together to create a cultural climax for which later generations should be forever grateful. It was a multiple birth of human genius, and the greatest flowering of politics, architecture, literature, sculpture, religion, etc. that the world has ever known. The crowning glory of the Acropolis was the Parthenon, the temple honoring Athene, the patroness of the city of Athens.

The objective thought of the period was exemplified by its architects, Ictinus and Callicrates, and was as logical as a geometric proposition. The columns, appearing to be a single piece of marble, are actually made up of drums pinned together with wooden plugs. They are fluted to correct an optical illusion. Since an ungrooved round column seen in bright sunlight will appear to be flat, the shadows of the fluting preserve the beautiful roundness of the columns.

The corner columns seem dark when viewed against the sky, while others, viewed against the dark cella walls appear lighter. If the diameters of all were equal, those on the end would appear thinner. The architects corrected this optical distortion by increasing the diameter of the corner columns and by spacing them closer at the ends, thus increasing the sense of stability at these points (Fig. 4-2).[1]

The Parthenon goes far beyond technical execution, however, and becomes a work of art through the infinite care in the design as a whole and the minute perfection of the details of its parts. Balance between vertical and horizontal elements was sought. However, on careful analysis, not one straight line can be found anywhere in the entire design! It is psychologically rather than mathematically correct—a system of curved lines is made to compensate for the optic aberrations of the human eye. Paradoxically, the construction is carried out on a subjective basis in order to achieve the ideal of objectivity.

The columns are not straight but curve slightly outward as they rise, reaching their greatest deviation about one third of the way above the base,

Fig. 4-1. A, Apollo of Piambino; **B,** Phyrne by Praxiteles. Early Greek ideals of human proportion and beauty still have the power to affect our ideas of those ideals today.

Fig. 4-2. The Parthenon goes far beyond technical execution and becomes a work of art through the infinite care in the design as a whole and the minute perfection of the details of its parts.

then gradually tapering toward the top. This curving is called *entasis,* and it constitutes a tempering of mathematical correctness in order to give the appearance of life and elasticity to the columns, which thus seem to give slightly under the great weight they carry.

The stylobate on which the columns rest and the entablature above them, while giving the appearance of straight lines, actually have slightly upward and outward curves, in order to counteract the illusion of sagging. The center lines of the columns and the walls of the cella tilt slightly inward to avoid the feeling of top-heaviness.

If we are to learn anything from the marvelous construction of the Parthenon, it is that the geometry of a structure must be a harmonious whole in which the component parts function as subordinate members. It is never a mere assemblage of parts, but an integrated unit.

As far as appearance is concerned, this architectural problem is directly related to our own problems. In facial reconstruction, the partial visual patterns of the separate features exist not separately and not for themselves alone, but they must be correctly related elements forming one particular, unique human face.

In the search for a viable philosophy and the real truth, the more advanced and intellectual people of every age have always had a dichotomy between ideals and the real world, which is full of illusion. Plato,[2] for example, was greatly concerned with such things as balance, clarity, order, simplicity, and standards of excellence in everything. He was disturbed by the illusions and distortions that were sometimes used by the greatest architects and other visual artists and points out that "in works either of sculpture or painting, which are of any magnitude, there is a certain degree of deception; for artists give up the truth in their images and make only proportions which appear to be beautiful, disregarding the real ones!"

DEVIATION FROM MATHEMATICAL REGULARITY

The same necessary deviation from strict size relationships necessary in art and sometimes, for different reasons, in our own profession of plastic surgery appears again in recited poetry and in performed music. If strict mathematical regularity prevails, the result is monotonous, dull, and mechanical. The pitch of musical tones must waver slightly in order to possess life and remain interesting. With his profoundly inquiring mind, this again disturbed Plato, who felt that this was irrational and that "the arts of measuring and numbering and weighing (must) come to the rescue of human understanding."[2]

In spite of the worries of Plato over the liberties that men of art felt they were obliged to take at the expense of mathematics and pure science, Greek architecture turned out to be the most rational solution to building and design problems. The orderly placement of structural members in solving both the artistic and the construction problems proved finally to be just as logical and well reasoned as Euclid's mathematics or Plato's dialogues.

The geometric principles of design, which play so large a part in drawing and painting, are perhaps even more important in sculpture. Because it is a three-dimensional figure, the ratios and proportions are subtle and are concerned with depth and volume and balance, all of which must stay in correct relationship as the roving eye of the observer moves to different positions.[3] The sculptor also has a problem in common with the architect in answering the question: "From what position, in what lighting conditions, and from what distance will my creation be most often viewed?"

The sculptor has an advantage over the surgeon, since he can control the entire concept and design. The plastic surgeon is confronted with a living face and must perform his work, whatever it might be, in such a way that it blends in with, and hopefully improves, the physiognomy by becoming a harmonious element of the whole human face.

RACIAL CHARACTERISTICS

The entire question of race and racial differences is, by its very nature, so fraught with controversy and emotionalism that it is fortunate for us, as plastic surgeons, that we can leave most of these questions to the anthropologists, the ethnologists, and even the politicians!

The plastic surgeon, however, must be aware of racial characteristics as evidenced in an individual patient, and particularly as this may affect the harmony and relationship of his individual features. This area of thought is of vital importance, in fact the sine qua non in arriving at a satisfactory solution to any readjustment or reconstruction of an entire human face or its minor constituents.

A particular reconstruction or reshaping of a feature that might be very desirable and well proportioned in one individual patient might be entirely inappropriate or even ridiculous in another, simply because it would not fit in with, or contribute to, the harmony and logical balance of the face considered as a whole!

Recall, if you can, the American Indian whose classic profile graced the nickel for a quarter of a century. The relative position, size, and shape of nearly every one of that individual's features violated, or did not agree with, western concepts of physical beauty.

It is extremely difficult for any human to look at any object in nature, especially a human countenance, without having impressions and reactions clouded and distorted by subconsciously acquired and deeply ingrained prejudices and preconceived ideas.

Nothing in this world means very much, nor can it be correctly judged, except in its entirety and in proper context. Judged in this light, the Indian's features, no matter how alien to our usual ideas, are balanced, in proper proportion to each other, and therefore beautiful and a classic example of nature's ability to form and mold individual elements into a harmonious whole. If for any reason it became necessary to operate on or change such an individual's facial features, it would be almost criminal to do it in such a way that the inherent beauty and harmony of the whole countenance were upset or destroyed.

ART RESTORATION

Since valuable and irreplaceable works of art do deteriorate and decay, the esoteric skill of the art restorer has become as important in the art world as the plastic surgeon has become important in the field of medicine. A tradition has grown up in the field of art restoration that the restorer has best performed his function when the vital work he has done is not obvious. This is a lesson and a goal that we should keep uppermost in our minds in every operation!

VISION AND ILLUSION

Vision is one of the most marvelous gifts to man, but we cannot have vision without illusion. The world of illusion is fascinating. For example, the study of why certain curves or combinations of curves are extremely pleasing to the eye, whereas other curves will be disturbing, is a field that deserves much more scientific study. It has been theorized that no curve is pleasing unless it is based on some regular, rhythmic progression related to some mathematical formula. A brief listing would include the parabola, hyperbola, ellipse, curves showing gradation, and French curves, which are usually based on a logarithmic or exponential basis (Fig. 4-3). These beautiful curves are everywhere in nature: leaves and flowers, shells, and the markings of birds and animals.

The double reversed curve (Fig. 4-4, *A*) occurs often in nature and is much admired, being especially pleasing to the eye. The name of the English painter, William Hogarth, is attached perhaps forever to this curve, because he employed it so often and found it so beautiful. Hogarth had a theory that beauty is ultimately contained in "the precise

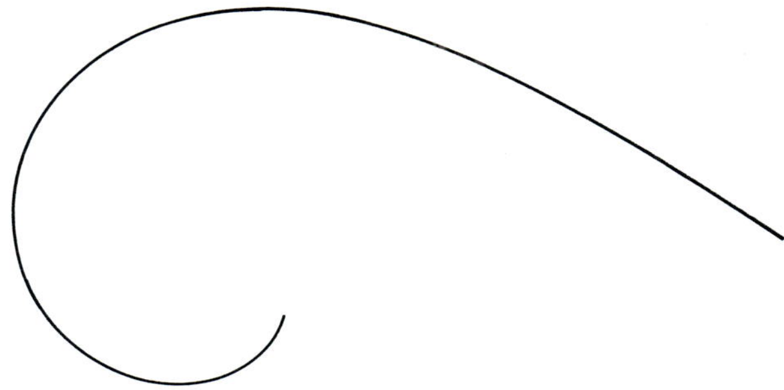

Fig. 4-3. An exponential curve—very pleasing to the eye because the curve is continuously changing its radius. (Illustration by Bernard B. Timon.)

Fig. 4-4. The double-reversed curve of Hogarth used in flower arranging. (Illustration by Bernard B. Timon.)

serpentine line." Today, ladies studying western world flower arranging are urged to follow in their designs, wherever possible, some variation on "the Hogarth curve" (Fig. 4-4, *B*).

Why consider anything about curves in a talk to plastic surgeons? What is the connection? There are two reasons. First, the very center of our professional attention is a curving, rounded, three-dimensional head and face. Great care must be taken to keep all incisions curved. As we all know, the general rule is to have the curve of the incision follow or harmonize with existing lines or the outlines of the facial feature nearest it. Second, certain lines and curves of a face, particularly in an adult, are essential for expression and, therefore, communication. It is essential that we distinguish not only between these expression lines and mere lines of aging but also between good and bad expression lines.

There is a phenomenon, or perhaps a mathematical and artistic concept, called "radiation in natural forms" (Fig. 4-5). The dotted lines in these diagrams indicate the sources of this radiation into lines that are visible, and hopefully pleasing. They are the unseen origins that give beauty, balance, and meaning to the visible lines. Primarily found on the face, these lines occur about the mouth, forehead, and eyelids.

When facial muscles contract to express such emotions as pleasure or laughter, anger or concern, perpendicular to these contracting muscles, lines must appear. They are the immediately visible evidence of emotional reaction and therefore are an important means of direct communication.

An infant can produce a fleeting rudimentary kind of smile, can screw up his eyelids and squeeze out tears to communicate some form of distress, and that is about all. But an adult, and particularly an older and more mature one, is expected to be able to do much more if he or she is to have an alive, intelligent responsive face.

Sometimes these lines are quite pleasing and at other times considerably less pleasing. At times, they may be so unpleasing that surgery can help little or none, and in still other cases there is a disappointing lack of lines, as in certain very inexpressive faces.

The only real justification of aesthetic surgery is that the repaired or reconstructed part must blend harmoniously with the whole; whatever the

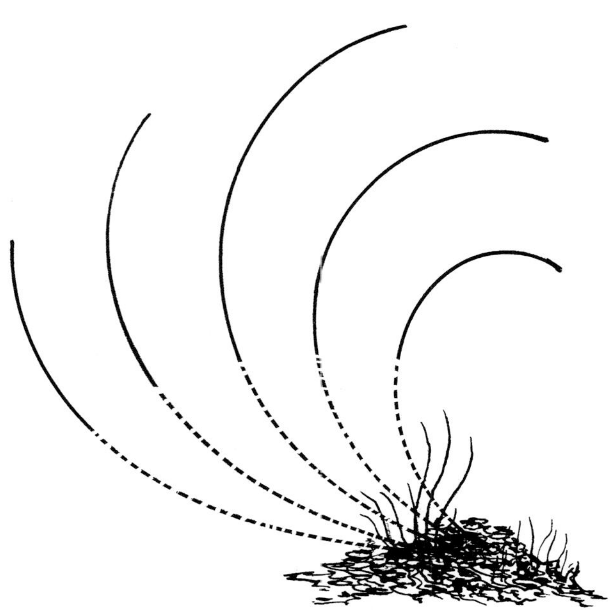

Fig. 4-5. The beauty of curves in natural materials radiating from a common point. (Illustration by Bernard B. Timon.)

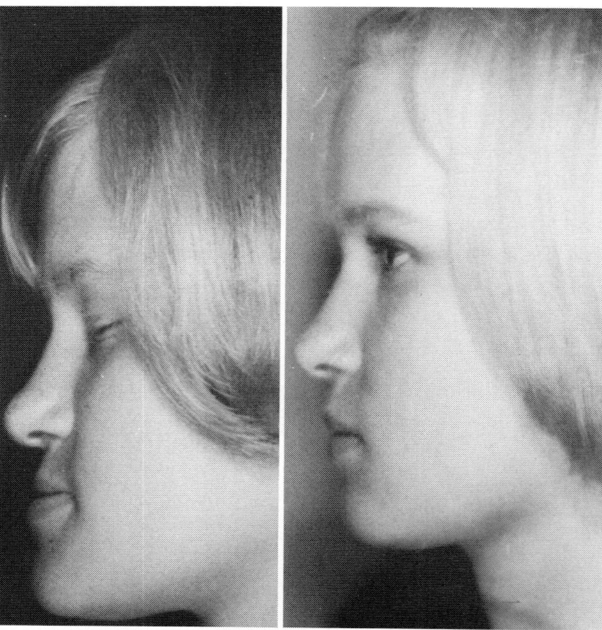

Fig. 4-6. The only justification of aesthetic surgery is that the reconstructed part must blend in harmoniously with the whole. Whatever the surgery done, the hand of the restorer, once his vital labor is completed, must be withdrawn and no longer be in evidence.

surgery done, the hand of the restorer, once his vital labor is completed, must be withdrawn and be no longer in evidence (Fig. 4-6). It is obvious that in no case will a slavish application of proportion or measurements produce a work of art, because a product of skill is never based upon an assemblage of ready-made parts.

A knowledge of space-relations, of distance-vision, of color effects, and of geometric proportions all must be at the command of the artist. Without them, he suffers a deprivation as great as when he lacks skill in the use of his tools. But neither should his knowledge of design nor his technical skill be allowed to gain the upper hand in his work. If he does give undue emphasis to either rationalization or technique, his product will be as clearly an abnormality or perversion of insight and talent as it is in any other product of behavior where a partial pattern is allowed an inappropriate degree of dominance.

Over-emphasis on arithmetical or geometrical measurements will render a work of art lifeless and mechanical. Over-emphasis on technique is likewise abnormal or perverse. The ultra refinement of lines, the minute care with which a surface is colored or shaped may result only in setting it off in isolation, thus destroying its relationship to the rest of the work.[3]

In order for a work of art to be fully integrated, no detail of its technique should be dominant! In fact, in art it is axiomatic that its necessary limitations should be made to add to its effectiveness. James McNeill Whistler, the famous American painter, had some thoughts on this that apply directly, with only a little translation to our own work: "A picture is finished when all trace of the means used to bring about the end has disappeared. . . . To say of a picture, as is often said in its praise, that it shows great and earnest labor, is to say that it is incomplete and unfit to view. . . . Industry in art is a necessity—not a virtue—and any evidence of the same, in the production, is a blemish, not a quality; a proof, not of achievement, but of absolutely insufficient work, for work alone will *efface the footsteps of work!*"[4]

CONCLUSION

Aesthetic surgery is not actually in its infancy but in its early maturity. Much of what we do is really quite elementary and often not terribly imaginative. We tighten skin to reduce sagging and wrinkling; we standardize noses and ears and chins to bring them within average and inconspicuous limitations. Perhaps this suffices for the age in which we live, where the demands for security are in youthfulness, sameness, conformity, and a retreat from individualization.

Hopefully this will change. How many of us have ever had a patient come to us and say: "My face is sad and uninteresting; it just isn't expressive, and I wish I had more lines of character"? Can we visualize where those lines should be, and if we can, do we know how to create them? This, I believe, is the challenge of our future: to create not merely a younger appearance or a good feature, but a better face.

REFERENCES

1. Fleming, W.: Arts and ideas, New York, 1955, Holt, Rinehart and Winston, Inc., pp. 3, 12, 13, 14.
2. Plato: Republic, collected dialogues, New York, 1961, Pantheon Books, Inc.
3. Ogden, R. M.: The psychology of art, New York, 1938, Charles Scribners Sons, p. 233.
4. Whistler, J. M.: The gentle art of making enemies, New York, 1967, Dover Publications, Inc.

Chapter 5

Anatomy of a rhinoplasty—saw technique

T. Ray Broadbent, M.D.
Robert M. Woolf, M.D.

The request for functional and cosmetic nasal reconstruction is an ever-increasing one in our society. Facial symmetry, better appearance, and good function are concomitant desires of the patient and goals of the surgeon. To achieve these goals regularly, it is necessary that the surgeon understand the anatomy of the nose and the contribution of individual parts to nasal form. Success will be evident when the nose complements other well-balanced facial features; and it is especially gratifying when the nose does not have an "operated look" or a "bobbed-nose" appearance.

ANESTHESIA

An important part of anesthesia is preoperative medication. Euphoric sedation (barbiturates and diazepam [Valium]), modest drying of the mucosa (atropine), and freedom from apprehension regarding pain, nasal packing, and the administration of local anesthesia (meperidine [Demerol]) are desired.

A topical anesthetic agent is placed on cotton applicators or gauze packing and inserted in appropriate areas in the nose to anesthetize the anterior ethmoid nerve and branches from the sphenopalatine ganglion, since they supply both the turbinates, the lateral and medial walls, and the floor of the nasal cavity. Xylocaine with epinephrine is used for local anesthesia and hemostasis. Branches of the supratrochlear nerves, the infratrochlear nerves, and the infraorbital nerve are blocked, as are the septum and columella (Fig. 5-1). From 8 to 10 ml. of anesthetic agent are required. One should wait 5 to 7 minutes following injection, until the needle puncture wounds stop oozing, before beginning the

operation if one is to appreciate good hemostasis. Proper premedication and a good anesthetic block allows the patient to enjoy the operation.

BASIC ANATOMY

The mobile lower lateral cartilage forms the lateral wall and the dome of the nasal tip before turning down into the columella where it supports itself as a foot on either side of the nasal spine. The upper lateral cartilage is a hooded structure, is semi-mobile, and extends to and over the dorsum of the cartilaginous septum to the opposite side. It articulates firmly with the rigid nasal arch and is aligned loosely with the lower lateral cartilage by fibroareolar tissue and mucosa (Fig. 5-2, *A*). The anterior arm of the inner canthal ligament runs from the medial canthus to the periosteum of the nose. The posterior arm of the ligament inserts posterior to the lacrimal sac. Deep to the anterior arm of the ligament and attached to the bone is the orbicularis oculi (Fig. 5-2, *C*). The procerus muscle streams down over the root of the nose, and the qaudratus labii superioris runs along the side of the nose with its angular head reflecting into the transverse nasal musculature. The bony and cartilaginous humps terminate in the nasal tip as the septum surfaces between the lower lateral cartilages (Fig. 5-2, *B*).

INCISIONS

How incisions are closed is secondary only to why they are made. Proper alignment and correct position of anatomic parts can be expected only when they are closed exactly and carefully.

The first incision is made intranasally between

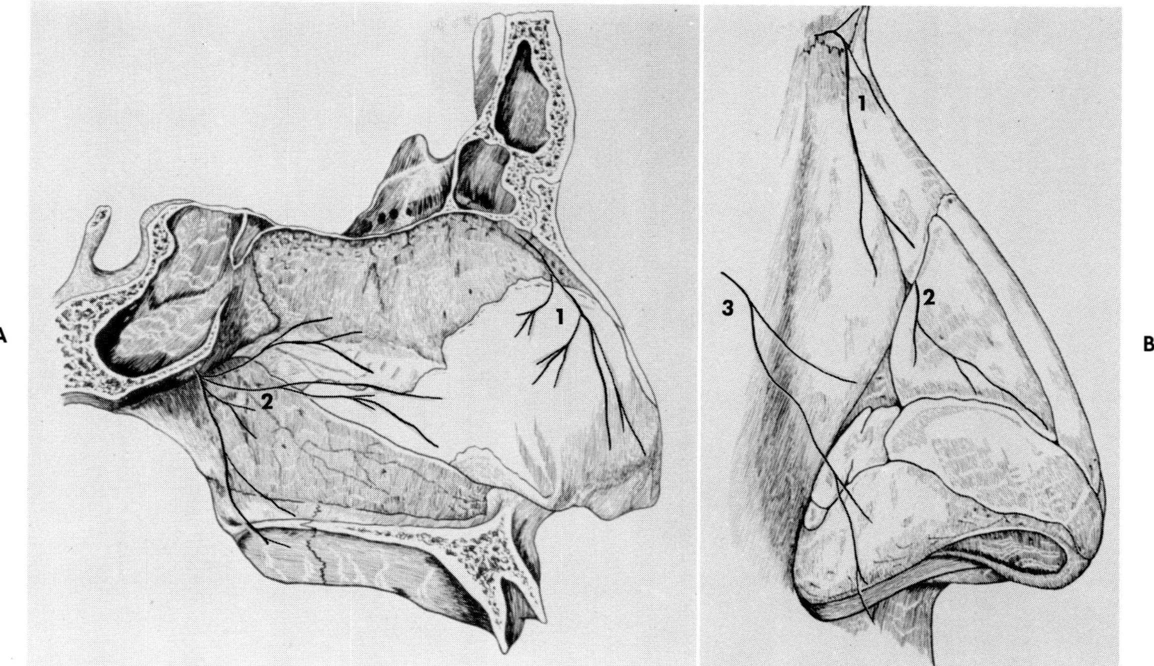

Fig. 5-1. A, Anterior ethmoid (*1*). Branches from sphenopalatine ganglion (*2*) to the turbinates; lateral and medial walls and floor of nose. **B,** Supratrochlear (*1*), infratrochlear (*2*), and inferior orbital nerves (*3*).

the upper and lower lateral cartilages. This incision is carried into the dome of the nostril. The columella is separated from the septum through the membranous septum. This incision is connected to the first one by simply snipping the remaining bit of mucosa in the dome of the nostril. Second, a small stab wound is made in the floor of the nose for access to the lateral nasal wall. Last, an incision is made through the mucosa along the inferior edge of the lower lateral (alar) cartilage. This incision is not on the nostril rim. Separation of these incisions will give access to the framework of the nose and flexibility to impose on oneself the rule of seeing all that is done during the rhinoplasty (Fig. 5-3, *A* and *B*).

MOBILIZATION OF SOFT TISSUES

The periosteum, skin, and procerus muscle over the dorsum of the nose are elevated as a blanket in preparation for removal of the hump. The periosteum can be elevated, but not as a repairable sheet of tissue. It simply falls away and rolls up as fragments lifted from the bone (Fig. 5-3, *C*).

Elevation of the periosteum from the nasal process of the maxilla is delineated along a line drawn from the inner canthal ligament to the nasofacial groove (Fig. 5-3, *D*) and accomplished through a stab wound in the floor of the nostril. The inner canthal ligament can be torn with ease by passing the elevator deep to its insertion. One should keep the elevator superficial to this ligament.

The skin is separated from the alar cartilage through the incision along the lower border of the lower lateral cartilage.

TRIMMING

The upper lateral cartilage is cut free from the septum in preparation for the reduction of the nasal hump. This leaf of cartilage and nasal mucosa should be cut clean from the septum, leaving no reflection laterally (Fig. 5-4, *A*). When the dorsal septum is to be trimmed, as in the hump nose, fragments of the upper lateral cartilage along the septal ridge will be removed with the hump; in other instances, any remaining fragments will give irregularities in width in the central third of the nose. Removal of the profile prominence of the upper lateral cartilage is the first stage in reduction of the nasal hump, since it produces part of the cartilaginous hump (Fig. 5-4, *B*). This trimming

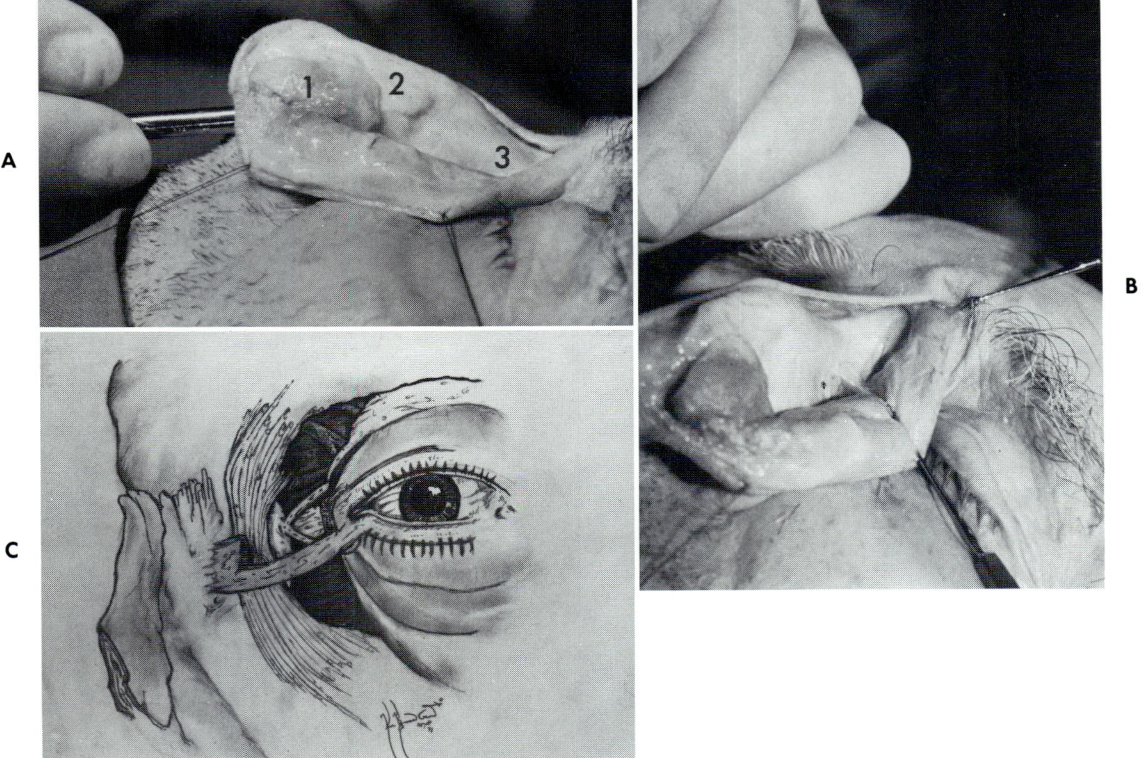

Fig. 5-2. A, Mobile lower lateral cartilages (*1*). Semimobile upper lateral cartilages (*2*). Rigid bony nasal arch (*3*). **B,** Upper hook in procerus and lower hook in quadratus labii superioris muscle. The latter overlies the inner canthal ligament and orbicularis oculi insertions. **C,** Anterior arm of inner canthal ligament (section cut out), crossing in front of orbicularis oculi. Deep to orbital ridge are tear ducts and sac and deep to this, the posterior arm of the inner canthal ligament.

shortens the upper lateral cartilage and narrows the central third of the nasal arch also. A cut is made in the cartilaginous septum for removal of a portion of the hump contributed by that structure (Fig. 5-4, *C*). A saw is inserted through the original incision between the upper and lower cartilages and placed in line with the cut made in the cartilaginous septum for removal of the bony hump (Fig. 5-4, *D*). One can completely cut across the hump with the saw and remove it intact with the nasal bones extending from the sides of the nasal septum. If this is not an equal removal, it is compensated for by rasping or further trimming on the high side. After removing the hump, a great deal of correction of underlying bony irregularities can be managed by simple rasping (Fig. 5-4, *E*). It should be carried well into the nasal root to avoid a step off in that region. After the hump has been removed, it is usually necessary to further trim the

upper lateral cartilage and the slight dorsal projection of nasal bone near the septum, to avoid a persistently high position of the upper lateral cartilage when infracturing and narrowing have been completed. One may additionally trim the cartilaginous septum to form a straight or saucerized profile line.

To trim the tip, the lower lateral cartilage and mucosa are everted as a bipedicle flap (Fig. 5-5, *A*). The overlying fat and areolar tissue is trimmed (Fig. 5-5, *B*) from the cartilage, and if fatty tissue exists in excess on the undersurface of the skin, it is trimmed also in an effort to avoid lateral thickening of the tip. The cartilage is trimmed and the nasal tip narrowed under direct visualization. From 2 to 2.5 mm. of cartilage at the inferior rim of the lower lateral cartilage is preserved (Fig. 5-5, *C*). Trimming is carried well into the dome and down the reflection in the columella.

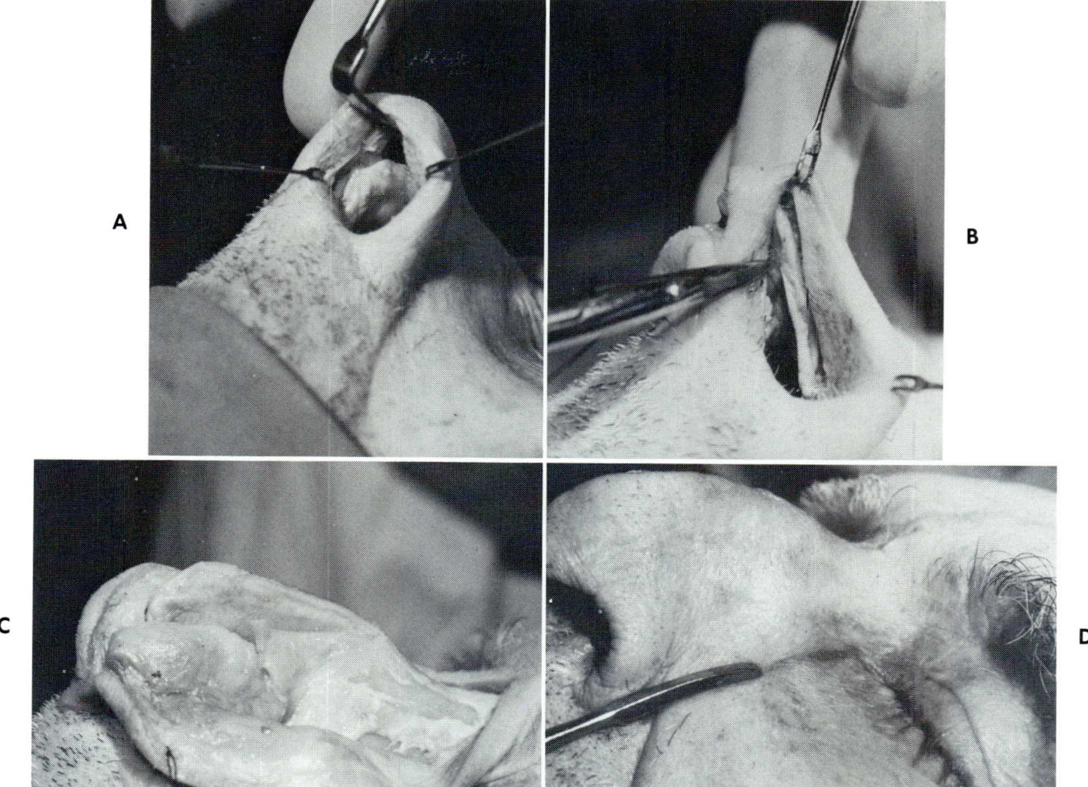

Fig. 5-3. A, Incision between columella and septum, continuous with one between upper and lower lateral cartilage. **B,** Incision along inferior edge of lower lateral cartilage. *Note:* This is not on the nostril rim. **C,** Periosteum lifted from bone, in line for hump removal and lateral osteotomy, does not persist as a sheet of tissue. **D,** Line for bony cut to narrow nasal arch, inner canthal ligament to alar wing groove.

Failure to do this well into the tip will result in small excrescences and irregularities in the tip proper. To assist in the narrowing of the tip, trimming of the lower lateral cartilage is tapered laterally. An excessively wide excision, or one carried too far laterally, may contribute to nasal wall collapse on inspiration. The lower lateral cartilage rim (2 to 2.5 mm.) is preserved for continuity of the rim and nostril form. It may be serrated, however, if further narrowing or lowering of the dome is desired. This should be done with great care, for it may also produce irregularities. The full mucosa (with attached alar rim) is saved and replaced in the nostril for proper lining.

Shortening of the nasal septum and upper lateral cartilages is necessary in long noses. Even though the nose may tilt back after removal of a hump, it will end up as a long nose unless the upper lateral cartilage and the septum are appro-

priately shortened. Removal of a rectangular or triangular segment of the distal septum will shorten the nose and widen the nasolabial angle (Fig. 5-6, *A*). In addition, removal of a triangular segment near the tip of the septum will tilt the tip (Fig. 5-6, *B*). When the nose has been shortened, there will be an excess of nasal mucosa. This mucosa is trimmed so that it fits properly and does not allow the tip to fall following corrective rhinoplasty (Fig. 5-6, *C* and *D*).

Section of the nasal reflection of the maxilla with a saw is done through the previously made tunnel along the line from the nasofacial groove to the canthal ligament. This cut will almost always tear the anterior arm of the inner canthal ligament free from its periosteal attachment (Fig. 5-7, *A* and *B*). Hypertelorism does not develop, however, unless there has been gross disruption of the insertion of the orbicularis oculi immediately deep to the

Text continued on p. 29.

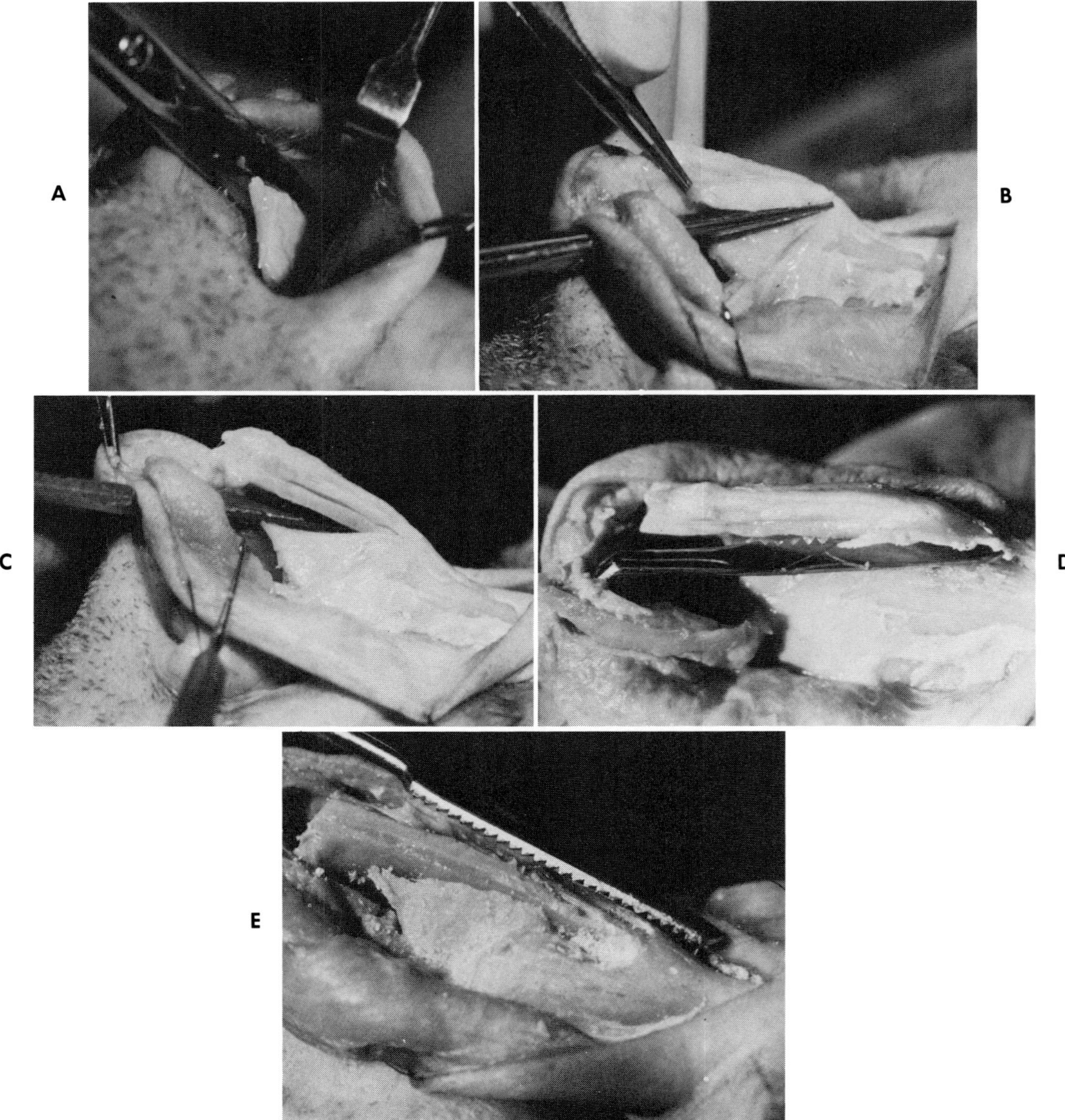

Fig. 5-4. A, Upper lateral cartilage cut cleanly from septum. B, Removal of profile (hump) prominence of upper lateral cartilage. C, Incising the cartilaginous septum for hump removal. D, Cutting the bony hump and remaining cartilage with saw. E, Rasping smooth, bony irregularities on sides and in nasal root.

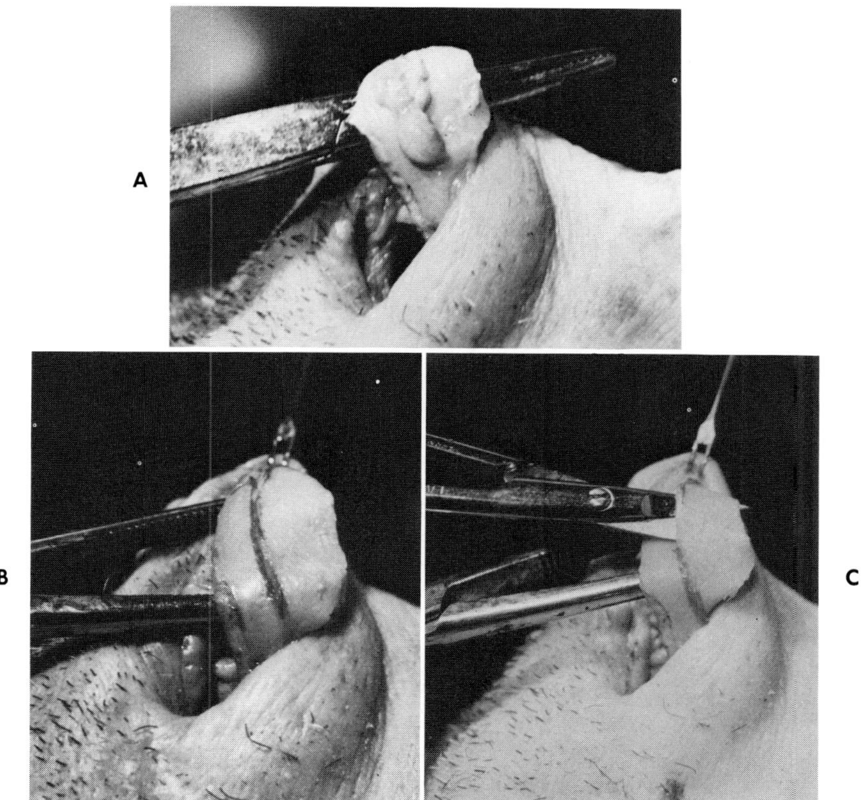

Fig. 5-5. A, Lower lateral cartilage everted as a bipedicle flap. Overlying fat on cartilage and underside of skin is to be removed. **B,** Cartilage marked for removal of upper part. **C,** Mucosa separated from cartilage. Inferior border rim of cartilage is saved along with all of the mucosa.

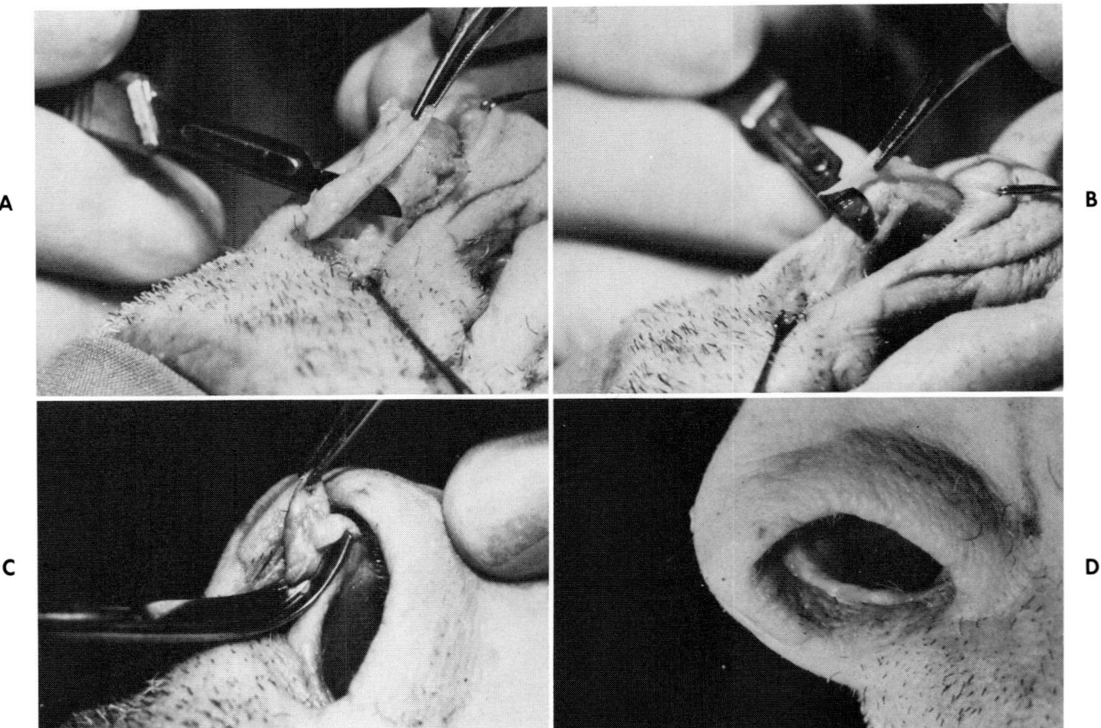

Fig. 5-6. A, Shortening of septum (in long noses) and change in the nasolabial angle. **B,** Cartilaginous triangle removed from distal septum to tilt the tip. **C** and **D,** Excessive mucosa, after shortening septum, trimmed to fit and hold new position.

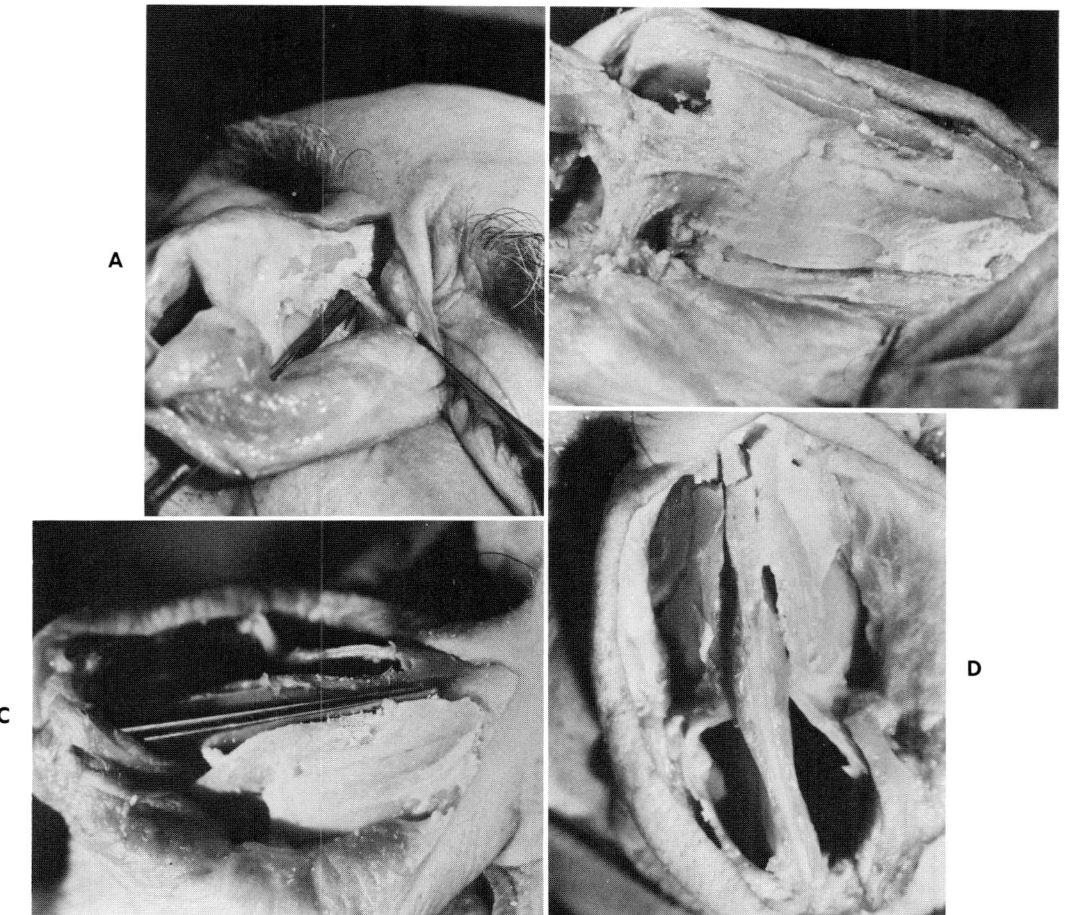

Fig. 5-7. A, Elevator under canthal ligament. Saw cut for lateral osteotomy, even though on top of ligament, usually tears ligament free from bone. **B,** Ligament absent; saw cut completely through bone. **C,** Osteotome between septum and nasal bone; outfracturing of bone and attached upper lateral cartilage. **D,** Nasal bones lying against septum following digital infracturing.

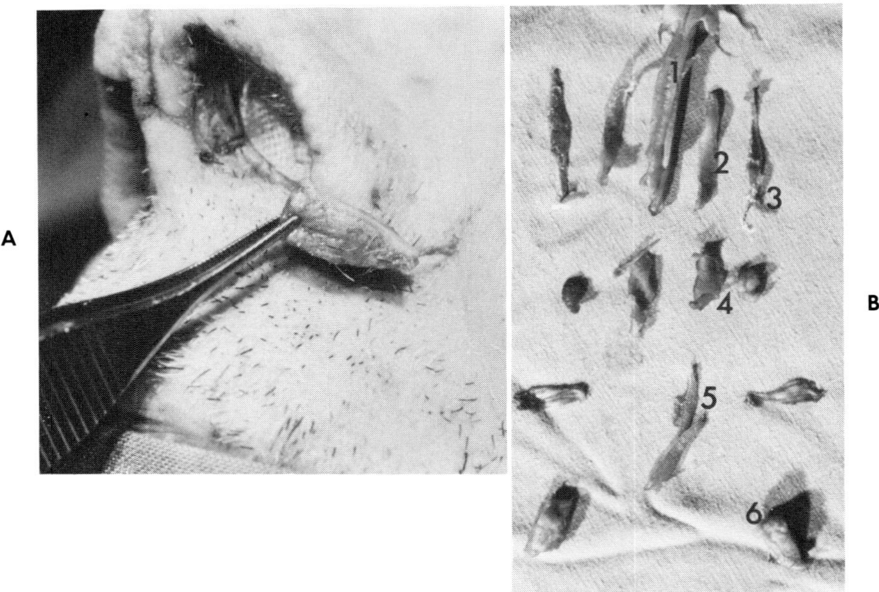

Fig. 5-8. A, One of many possible alar rim resections to shape nostril and alar wing. B, Fragments removed in rhinoplasty: *1*, bony nasal hump; *2*, upper lateral cartilage height; *3*, length; *4*, lower lateral cartilage and fat; *5*, distal septal length; and *6*, alar wing wedges.

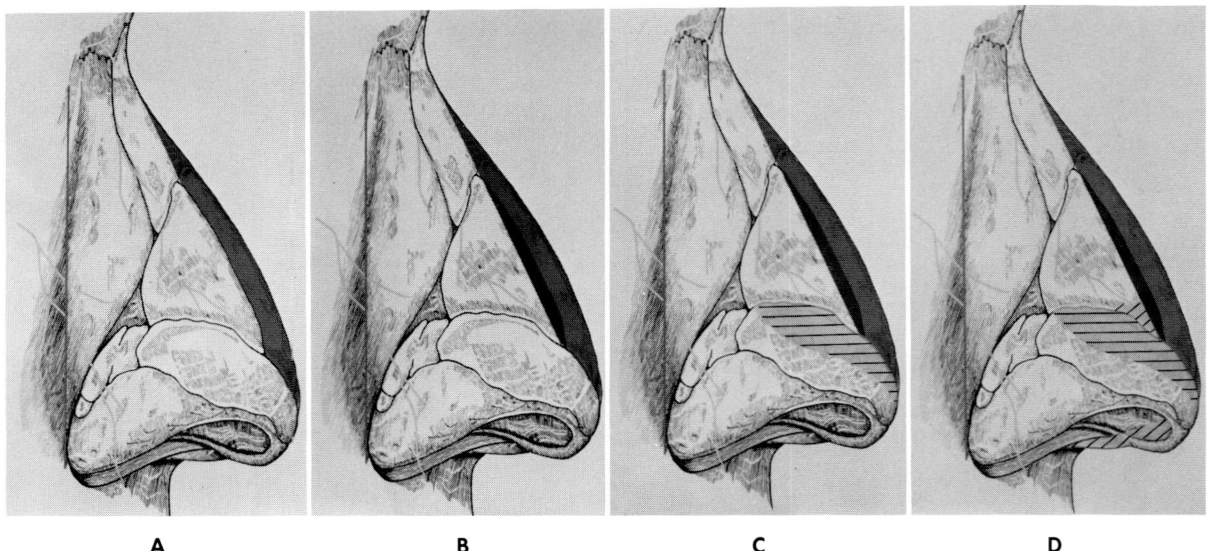

Fig. 5-9. Rhinoplasty requires management in these main areas: **A**, Hump removal and narrowing of nasal arch (note line of bony cut for lateral osteotomy). **B**, Upper lateral cartilage height. **C**, Narrowing and lowering of lower lateral cartilage (tip). **D**, Shortening and tilting of distal septum and shortening of tip of upper lateral cartilages.

ligament and the posterior arm of the canthal ligament deep to the tear sac (Fig. 5-2, *C*). The saw cut should be complete (Fig. 5-7, *B*) so that fracturing will not leave either side of the nasal arch springing out. An osteotome is driven between the nasal bone and septum, seated, and rocked out to fracture and produce a complete segment of nasal bone and upper lateral cartilage (Fig. 5-7, *C*). Digital infracturing can then easily be done and the nasal arch molded and narrowed (Fig. 5-7, *D*).

It is often desirable to reduce the fleshiness or the flare of the alar wing by removing a wedge of tissue from the rim as it approaches the nasal floor. Various configurations can be used for shaping the nostril and the alar wing (Fig. 5-8, *A*).

All incisions are carefully sutured, leaving no room for hematoma formation or extrusion of fatty tissue that may later necrose and result in nasal hemorrhage. Exact closure of the incision along the lower edge of the lower lateral cartilage and in the nostril dome is essential to maintain proper alignment and height; and nostril rim form.

Fragments removed during rhinoplasty are the bony-cartilaginous nasal hump, sections of height and length of the upper lateral cartilages, lower lateral cartilages and overlying fat, sections of nasal septal length, and excessive nasal mucosa and wedges from the alar wing (Fig. 5-8, *B*).

DRESSING

The elasticity of the mobilized soft tissue will facilitate draping, shrinkage, and contouring over the reconstructed nasal framework. The mobility of the tissues involved, however, following a complete rhinoplasty make it easy to appreciate the need for a form-fitting dressing and splint to hold the tissues until the nose is reasonably well set, usually 2 weeks. Packs are minimized and, if used, are removed 24 hours postoperatively. Excessive ointment on the pack should be avoided if one is to avoid paraffinoma from ointment squeezed into the incisions.

SUMMARY

Great latitude is given the concept of individual variations in part and in the entirety, but generally corrective rhinoplasty requires anatomic alteration in five areas: removal of a dorsal hump (Fig. 5-9, *A*), narrowing of the bony nasal arch (Fig. 5-9, *A*), lowering and/or shortening of the upper lateral cartilages (Fig. 5-9, *B*), narrowing and lowering of the nasal tip (lower lateral cartilage) (Fig. 5-9, *C*), and shortening and tilting of the cartilaginous septum (Fig. 5-9, *D*). Proper appreciation of the contribution of each of these anatomic parts to the entire unit of nasal form will aid the surgeon in more consistently reaching the desired functional and cosmetic goal in corrective rhinoplasty.

Chapter 6

Osteoplastic rhinoplasty—osteotome technique

Charles E. Horton, M.D., F.A.C.S.

Obvious advantages are inherent in the use of either the osteotome or saw in nasal surgery. Osteoplastic nasal surgery is an art, and the mechanical aids required to produce the finished aesthetic result varies with the individual surgeon, who may have great flair with the osteotome and little with the saw, or vice versa. There are, however, advantages to the osteotome technique that can be enumerated and evaluated.

Historically, Roe and Joseph, when first attempting nasal surgery, certainly did not have the well-known "Joseph nasal saw" or other prototypes. We can therefore assume that osteotomes or chisels were their original instruments for cutting bone. When the saw was developed, it received due recognition and wide usage by plastic surgeons for the specific purpose of removing the dorsal hump and sectioning the nasal bones at their bases. Just as new designs of saws were developed, new osteotomes and chisels have been made available in many shapes and sizes. In addition to width variation, the thickness of the tip may vary, the length of the handle can be changed, and the grip may be round, square, or of other shapes. The tip may have guards on one or both sides and may be straight or curved slightly. The metal may have a rough or shiny surface.

THE IDEAL INSTRUMENT

The ideal osteotome should be straight, sharp, and thin at the tip, so that minimal soft tissue displacement is produced. The strength of the distal end may be sacrificed to produce a thinner instrument tip. The width of the instrument may vary according to the local anatomic requirement. If the operator desires to remove the nasal hump by cutting both nasal bones at one time, a 10- to 12-mm. width is necessary. Some surgeons section each bone individually and use a 3- to 4-mm. width osteotome. For lateral osteotomies, a 4-mm. width seems ideal, so that the bone can be sectioned and the surrounding mucous membrane (internally) and the skin (externally) are preserved. Tip guides to protect the sharp edges of the osteotome are not necessary but may be desirable if the operator wishes to palpate against the blunt guide edge. An osteotome cuts directly in front of the tip, while a chisel, by virtue of the angled tip, tends to direct itself above or below the line of sectioning. Certain surgeons prefer one, some the other.

The tip and grip should both be elongated so that, as the osteotome enters deeply into the nose, the operator's hand does not cover the end of the grip and the instrument can be struck with the mallet without hitting the overriding soft tissue of the hypothenar tissue. The grip should not slip easily, for the osteotome should be grasped firmly and not allowed to twist into abnormal positions. In addition, the sides of the osteotome should not be sharp, since the osteotome frequently rests on the soft tissue of the upper lip and can cut the lip if pressure is applied.

Hundreds of variations of osteotomes and chisels are available over the world. Each surgeon should have a selection with which he is familiar and should use the appropriate "favorite" as the golfer does the favorite putter, expecting to become more proficient with use and practice.

OSTEOTOMY OF THE NASAL HUMP

By preoperative skin marking along the desired profile line, the level of sectioning of the septal cartilage and nasal hump bone is more easily identifiable. The cartilage is cut by scissors or knife back to the tip of the nasal bones. By placing the left thumb and forefinger on the dorsum of the nose, the surgeon can identify the proper level to place the osteotome underneath the nasal skin. The chisel or osteotome is grasped more effectively by using the thumb and forefinger toward the tip, and although introduction may be facilitated with the reverse grip, it is best to shift soon after sectioning begins for better control.

Using the left hand for position evaluation and control, the operator allows the assistant to strike the osteotome base regularly and rhythmically to provide better control of the mallet force. The entire hump can be removed with a wide osteotome. Using a narrower instrument, one nasal bone can be sectioned first, then the opposite bone sectioned at the same level. Finally the cartilaginous septum is cut with manual pressure. A rasp is required in most cases to smooth irregular surfaces. By using a sharp-tipped osteotome, the bone is cut, not crushed, and the proper level of bone division is maintained to the glabellar angle. This technique appears to have advantages over the saw because (1) it allows a deeper glabellar angle cut, and (2) in cases with a minimal bony dorsal hump, the sharp osteotome allows as much or as little bone to be removed as desired. Great accuracy can be obtained without anticipating loss of bony substance from the saw cut and dust. It is also difficult to obtain purchase on a small delicate hump with a saw (Fig. 6-1).

If the glabellar angle has still not been deepened adequately following hump removal, the osteotome or chisel can be used as a gouge, cutting from the nasal bone toward the frontal area with firm hand pressure, and many small cuts of bone can be easily removed to deepen the angle. These small bone fragments should be carefully identified and removed from the wound with a hemostat, or rasp, and suction (Fig. 6-2).

The osteotome can also be used to remove the nasal bone diploe at the glabella, to allow a higher

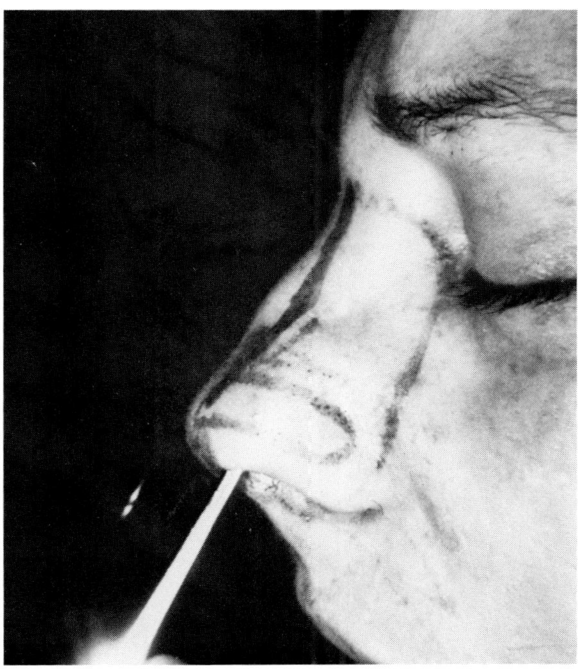

Fig. 6-1. Use of osteotome to revise dorsal hump.

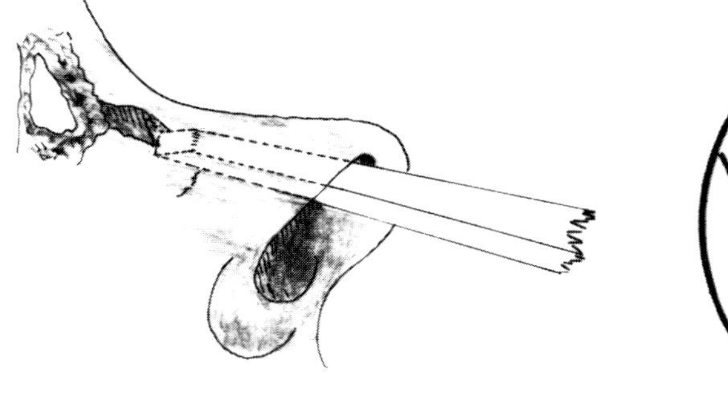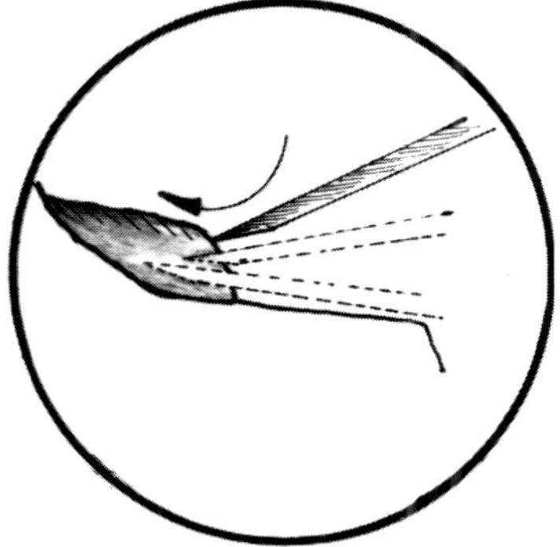

Fig. 6-2. Use of osteotome or chisel as a gauge to lower glabellar angle.

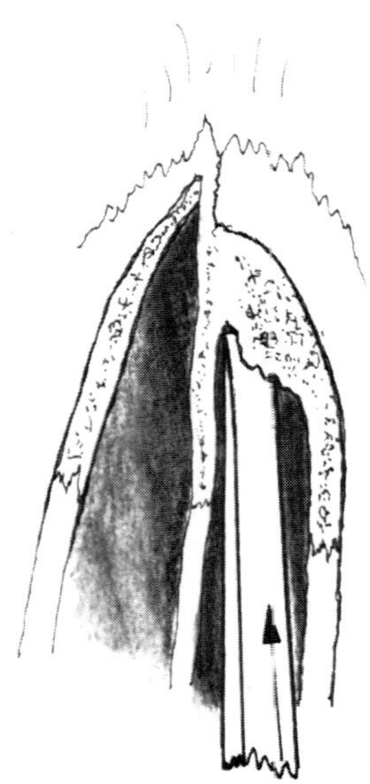

Fig. 6-3. Use of osteotome to remove triangle of bone at superior attachment of nasal bones. This allows a higher fracture and produces a thinner nose.

lateral fracture than otherwise might be accomplished.[1] If the hump is small, however, and the space narrow, a saw has advantages over the osteotome since it can be used to cut the narrow space to the desired length, and the cut itself will remove 1 mm. of tissue in the form of bone dust. When the nose is very wide and a large triangle of bone must be removed, the saw cut, taking only 1 mm. of bone dust, is not adequate (Fig. 6-3).

THE LATERAL OSTEOTOMY

It is essential to make the lateral bone cuts flush with the maxilla in order to obviate a step-like appearance of the lateral sides of the nose. Although this can be done adequately with either the osteotome or saw, the saw creates only a straight line cut and can extend superiorly only to the end of the soft tissue dissection. An osteotome, however, can curve the bone section along the base of the lateral nose, to be low at one area and higher to avoid the inner canthal area as necessary. This tends to produce a lower fracture at the cheek level and yet allows the operator to avoid injury to the lacrimal and canthal areas by simply changing the direction of the cut as these areas are approached (Fig. 6-4).

The osteotome can be introduced in many ways to section the base of the nasal pyramid.

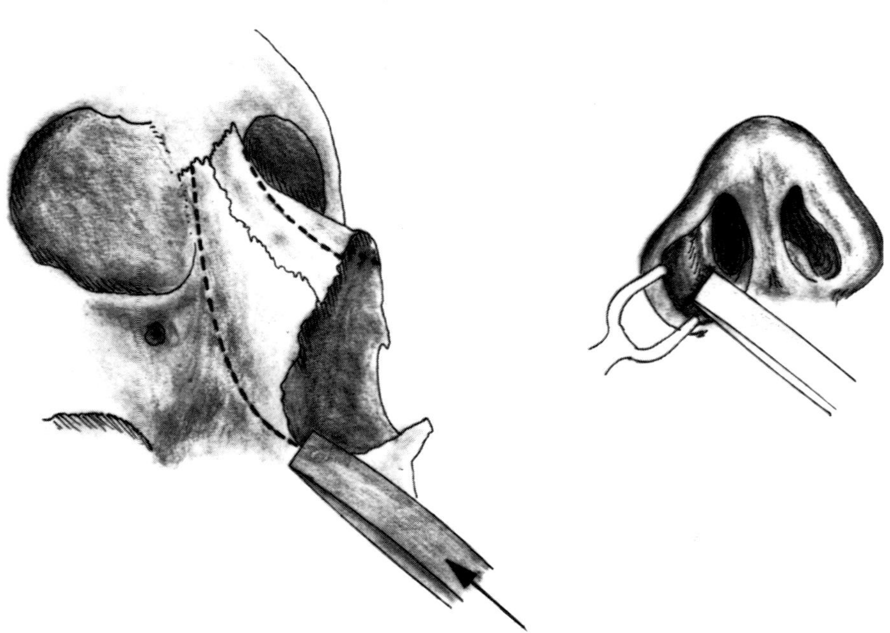

Fig. 6-4. Use of osteotome to separate nasal bones from maxilla. Approach is via piriform angle intranasally.

External incisions, lateral and above the alar rim, have been advocated. Since these areas heal well, there is little scarring (Fig. 6-5, *A*). An intraoral approach is recommended by some operators (Fig. 6-5, *B*), but the most frequently used technique is to place the osteotome at the base of the nose at the piriform angle intranasally at the bone level desired (Fig. 6-5, *C*). Using an assistant to tap the osteotome with regular firm strokes, the operator can evaluate progress with the index finger and thumb of one hand as the other guides the osteotome. If the level of cut proves undesirable, it may be restarted or repeated. Some surgeons advocate making multiple serrations at the base of the nasal bones rather than a single cut.

After sectioning, if the level of the lateral osteotomy is too high, a second cut can be made below the first, to lower the bone base appropriately. This does not cause asymmetry and, although not desirable, does leave an "escape mechanism" to prevent a step deformity. Shattering and comminution of nasal bones is not a common problem. Should this occur, the bones can be packed and splinted into position, and healing is usually uneventful.

Since the nasal bones are suspended as in a sandwich, between the mucous membrane internally and the skin externally, and bone cuts are made inferiorly and superiorly, it is important not to disturb the bone surface covering. Therefore, dissection of the soft tissue lateral to the nasal hump should be minimal, as should dissection at the base of the nose. If the osteotome is used, less soft tissue dissection is required.

Mobilizing the nasal bones is accomplished by seating the osteotome lateral to the septum and by exerting pressure laterally on the partially freed nasal bone. This will cause a fracture of the bones high in the glabellar area at the frontal bone attachment. Occasionally the nasal bones are thick at the glabellar angle and the fracture line may be poor, leaving an irregular fragment attached near the frontal bone area, with a lower fracture of the nasal bone than desired. The nasal septum, which is bony at this level, may also be thickened and/or deviated, so that one of the nasal bones, although adequately freed elsewhere, will not move medially to the proper position. An osteotome can be introduced through the skin at the glabellar angle to cut the septum and nasal bones allowing more freedom of movement. The tiny external skin incision heals without deformity (Fig. 6-6).

BONY AIRWAY OBSTRUCTION

In 1950 Beverly Douglas described the problem created when the lateral vestibular bony walls

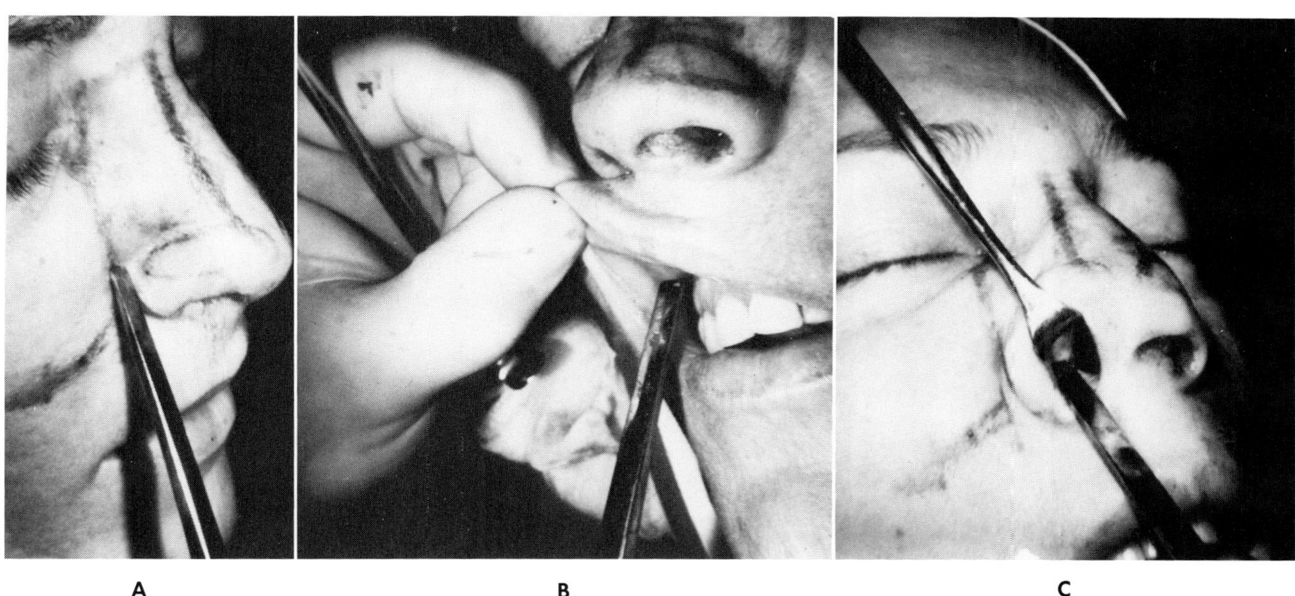

A   B   C

Fig. 6-5. An external, **A,** or intraoral, **B,** approach with the osteotome may be used, but an intranasal approach, **C,** is preferred.

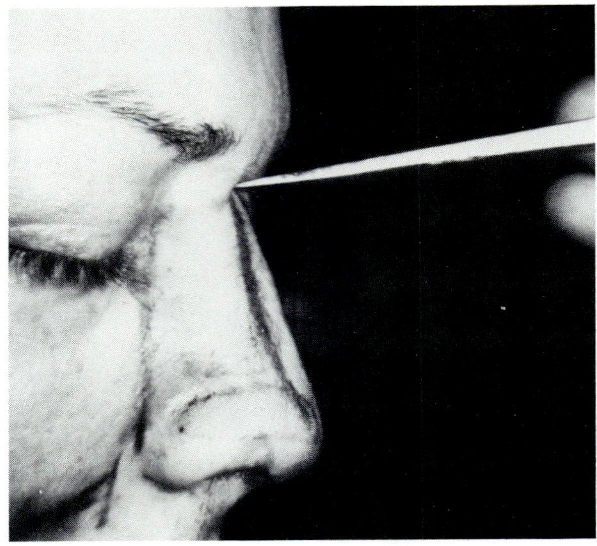

Fig. 6-6. Sectioning the base at the glabellar angle may be desirable.

impinge upon the nasal airway and recommended partial resection of the nasal process of the superior maxilla for correction of this problem.[2] Although this condition is rare, the osteotome and rongeur should be used for correction. Occasionally an excessively enlarged turbinate is present, and a submucous resection of part or the entire turbinate may be necessary. The osteotome and rongeur are essential if thick bone is encountered within the turbinate.

Finally, but most important, when the septum is deviated and distorted with bony spurs, it is essential to correct this problem during or before rhinoplasty. While cartilaginous deformities can usually be removed with the Ballinger knife or a punch, the more difficult inferior bone spurs can be removed with less trauma by elevating the mucoperiosteum and sharp sectioning of the bone with the osteotome or chisel.

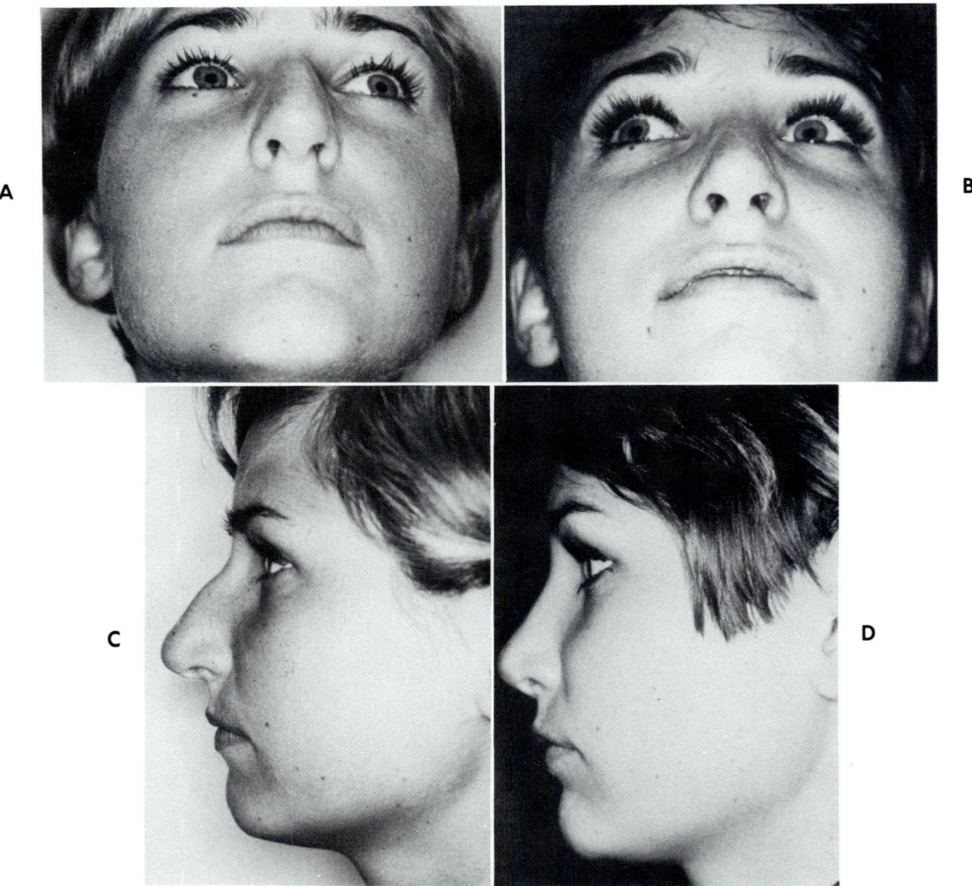

Fig. 6-7. A, Preoperative front view. **B,** Postoperative front view. **C,** Preoperative side view. **D,** Postoperative side view.

COMPLICATIONS

Few complications occur with the use of the osteotome. Bleeding is usually brisk after lateral osteotomy, and, therefore, this part of the operation is preferably done at the later part of the rhinoplasty procedure. If one bone cut is higher than the other, a new lower cut can be done, or on the dorsum, the rasp can be used to level the high areas.

Damage to the lacrimal sac may occur with a cut made too low in the canthal area. This should be suspected when fresh blood is seen coming from the punctum. Anderson[3] has reported frequent injury during rhinoplasty with no permanent deleterious results, and no treatment is needed in the majority of cases.

If the operator is careless, the sharp osteotome can puncture the skin, which is particularly thin at the canthal areas. Small sutures may be used if the skin wound should gap, otherwise the healing in this area is so good that no troublesome scar usually occurs.

When using a thin osteotome with great force, it is possible for the tip to break intranasally. Should this happen, the broken piece should be removed when feasible. If the fragment appears to be lost and is over 4 to 5 mm. in size, x-rays with stereotactic localization may be desirable.[4] It is important to tell the patient what has happened and to explain the course of action to be taken. If the fragment is small, it will usually cause no difficulty and will be retained asymptomatically within the nasal tissues.[5]

Osteotome technique in rhinoplasty may be difficult for certain operators, primarily because facility with the instrument can only be obtained with familiarity. Many rhinoplasties are performed without the use of the osteotome, and good results can be achieved. Once the operator is familiar with both the saw and the osteotome technique, however, most surgeons seem to favor the latter because of the ease of the technique and because of the controlled results that can be obtained. Since osteotomy is performed under a curtain of skin, it is difficult to achieve confidence in a sharp instrument that is forced toward structures that should not be damaged. However, careful control, coupled with judgment and familiarity, will allow faster, more efficient, and more delicate surgical contouring (Fig. 6-7).

REFERENCES

1. Aufright, G.: Surgery of the radix and the bony nose: preliminary report of a new type of nasal clamp, J. Plast. Reconstr. Surg. **22**:315, 1958.
2. Douglas, B.: Relief of vestibular nasal obstruction by partial resection of nasal process of superior maxilla, J. Plast. Reconstr. Surg. **9**:42, 1952.
3. Anderson, R.: Injury to the lacrimal apparatus during rhinoplasty (with Flowers), J. Plast. Reconstr. Surg. **42**:577, 1968.
4. Horton, C. E., and McFadden, J. T.: Stereotatic localization of a facial foreign body: case report, J. Plast. Reconstr. Surg. **47**:598, 1971.
5. McGregor, M. W., and others: Complications of rhinoplasty: I. Skin and subcutaneous tissue, J. Plast. Reconstr. Surg. **22**:574, 1958.

Chapter 7

Surgery of the nasal tip

George C. Peck, M.D.

The aim of the rhinoplastic surgeon is to produce consistently a nose that is aesthetically pleasing and anatomically normal. It is the purpose of this paper to demonstrate a technique for nasal tip surgery that will help achieve these goals.

In 1931, Jacques Joseph wrote his famous book on rhinoplasty that included a technique of nasal tip surgery. This has become the method of choice for most aesthetic surgeons, and to understand nasal tip surgery, there is no better place to begin than with Joseph's actual description.

In the Joseph technique, surgery of the tip begins with an intercartilaginous incision shown below and a rimming incision shown above (Plate 1). An intracartilaginous incision is made extending from the dome through the lateral crus, usually midway between incisions. A triangle or rectangle is taken from the remaining alar rim cartilage. Joseph wanted to narrow the tip. At this point we can see the appearance of the dome from which the triangle has been taken; the dotted lines represent the alar cartilage that will be removed superiorly. The external skin is undermined from the alar cartilage. The vestibulum skin is separated from the nasal side of the alar cartilage. The cartilage is dissected free and cut medially. The upper diagram (Plate 1, *I*) represents, by the shaded area, the cartilage that has been removed. We can visualize in the lower diagram that irregularities in the tip can result with this technique. More important is the fact that we have transected the alar cartilage at the dome, thereby destroying the normal anatomy of the tip. Once the cartilage is transected, we lose the symmetrical architectural relationship of the alar cartilage. The dotted area represents cartilage in the supratip area that can later produce supratip prominence, sometimes called supratip swelling. The final result utilizing Joseph's method is shown in Plate 2, *B*.

In the technique that is used, surgery begins with elevation and sculpturing of the nasal tip, followed by reduction of the nasal bridge. It was found that more consistent good profiles would result, since we can better control the reduction of the nasal bridge than that of the tip. We are, therefore, merely accommodating the bridge to the new nasal tip.

Initially the surgeon must not infiltrate local anesthesia in the nasal tip or bridge. The area that must not receive infiltration, so that there will be no distortion of the tip or bridge, is marked in red. The infiltration (approximately 8 to 12 ml.) is confined to the area of the inner canthus, the infraorbital nerve, the alar base, and the nasal spine. Next, infiltration is confined to the area of the intercartilaginous incision. The intercartilaginous line is found by finding the red-white line created by the junction of the white vestibulum skin and the red nasal mucosa.

Surgery begins with the intercartilaginous incision between the upper lateral and the alar cartilages, extending around the septal columella in one sweep of the scapel. The nasal tip is shortened when necessary. The columella is always shortened with a resulting obtuse angle, which produces an elevation of the tip without a subsequent elevation of the columella. This eliminates the retracted columella, or "pig-like" appearance. The shaded area represents the area of excision. Good nasal aesthetics should have the columella slightly lower in profile than the alar rims. The nasolabial angle in a woman should always be 100 to 110 degrees.

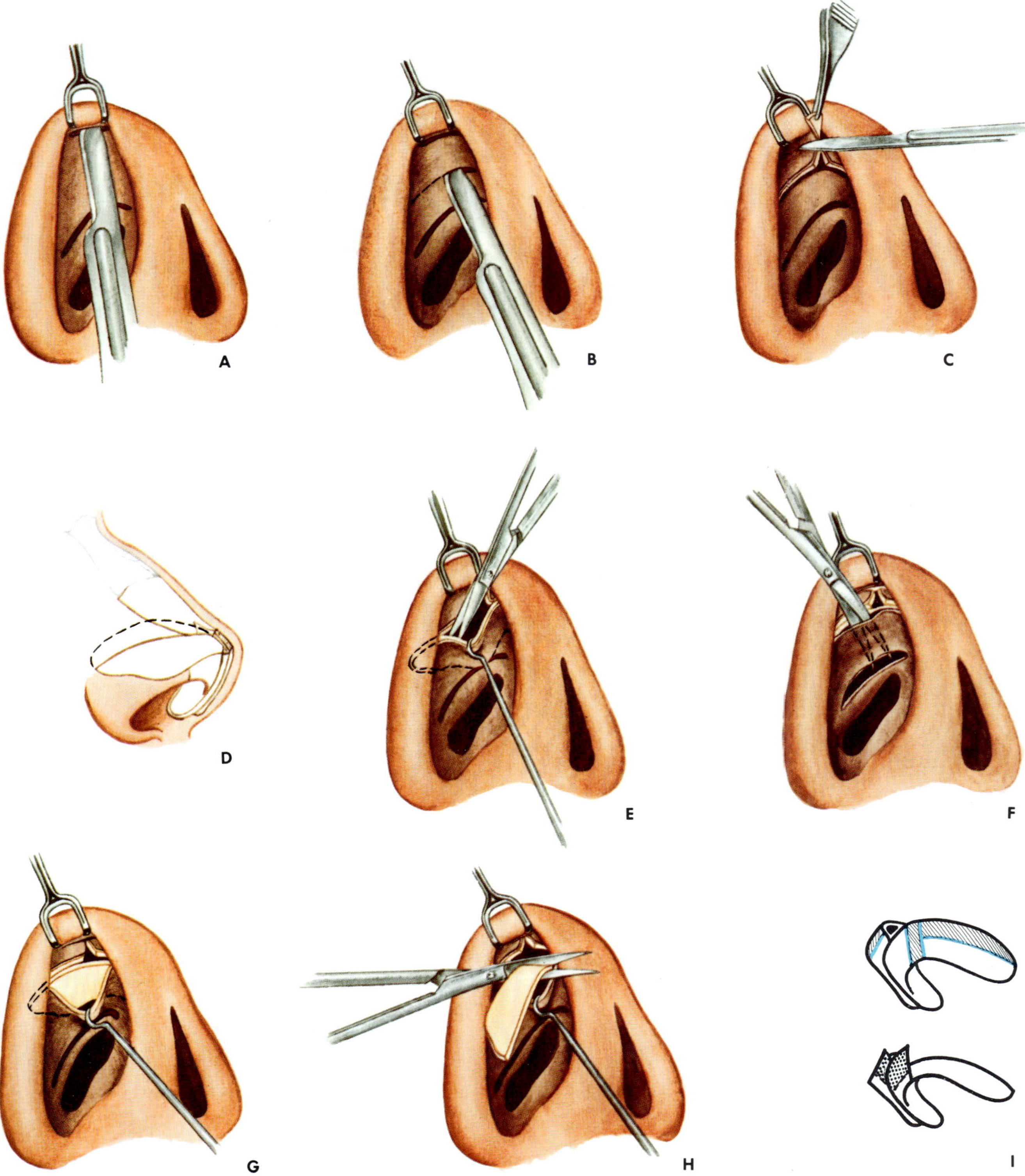

Plate 1. Joseph technique of nasal tip surgery. **A,** The technique begins with an intercartilaginous incision shown below and a rimming incision shown above. **B,** The next step is an intracartilaginous incision made from the dome through the lateral crus. **C,** In the third step a triangle or rectangle is taken from the dome of the remaining alar rim cartilage. **D,** The triangular defect can be seen in the dome area. The dotted lines represent the alar cartilage that will be removed superiorly. **E,** The external skin is undermined from the alar cartilage. **F,** The vestibulum skin is then separated from the nasal side of the alar cartilage. **G,** The cartilage is now dissected free. **H,** It is then cut medially. **I,** The shaded area in the upper sketch shows removed cartilage. In the lower sketch some irregularities resulting from this technique can be seen.

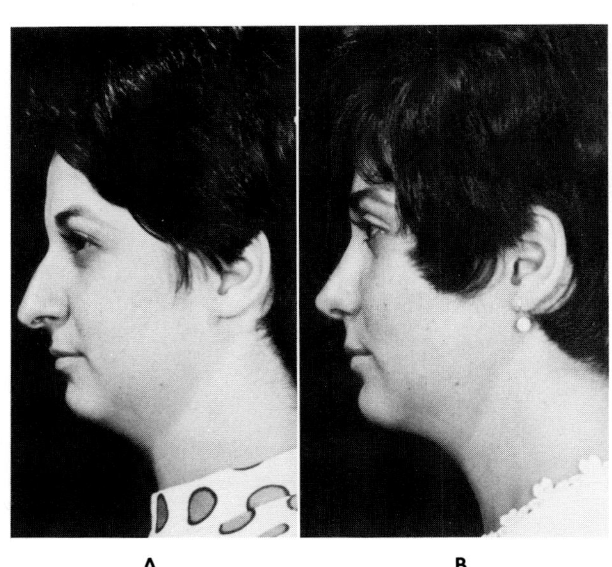

A   **B**

Fig. 7-1. A, Preoperative appearance of a 26-year-old woman; **B,** postoperative results.

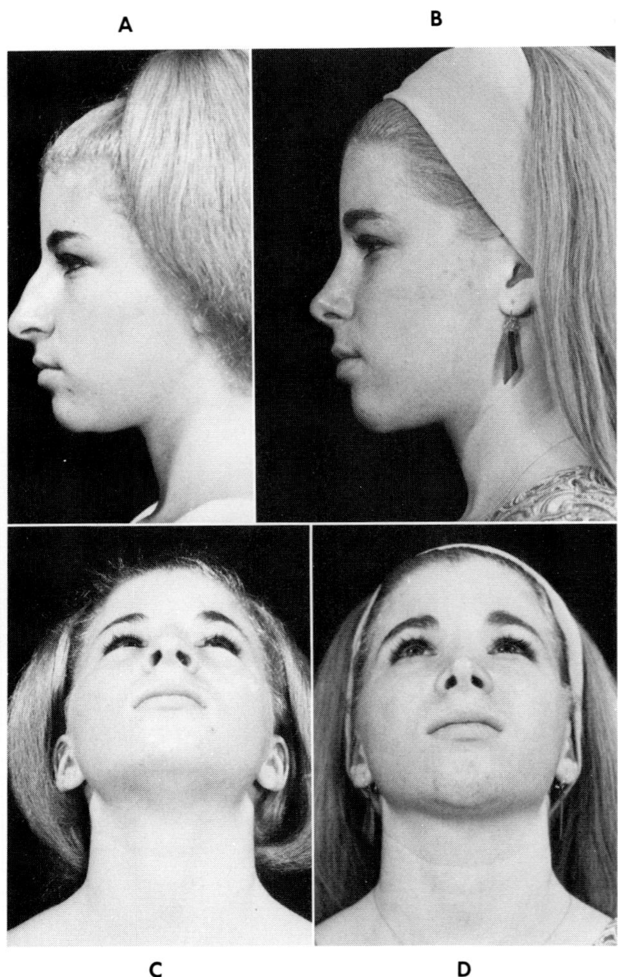

C   **D**

Fig. 7-2. A, Preoperative appearance of 16-year-old girl; **B,** her postoperative appearance 1 year later. **C,** Alar flaring in the preoperative photograph; **D,** correction by the modified Weirs' procedure used at the time of the rhinoplasty.

Plate 2. A, The dotted area shows cartilage in the supratip area that can later produce supratip prominences. **B,** This is the final result of the Joseph technique. Technique of Peck follows. **C,** The tip or bridge marked in red will not be injected, to prevent distortion. **D,** Infiltration is confined to the area of intercartilaginous incision. **E,** An intercartilaginous incision is made between the upper lateral and alar cartilages, extending around the septal columella. **F,** To produce an elevation of the tip but not of the columella, the columella is shortened with a resulting obtuse angle. **G,** The straight-line excision of the columella septum is done in the Joseph technique. **H,** The sculpturing of the alar cartilages is done with a curved incision made at the beginning of the dome and extending medially across the medial crus for about 5 mm. and laterally across the lateral crus. **I,** This shows the position of the incision together with the preserved cartilage below and the cartilage to be removed.

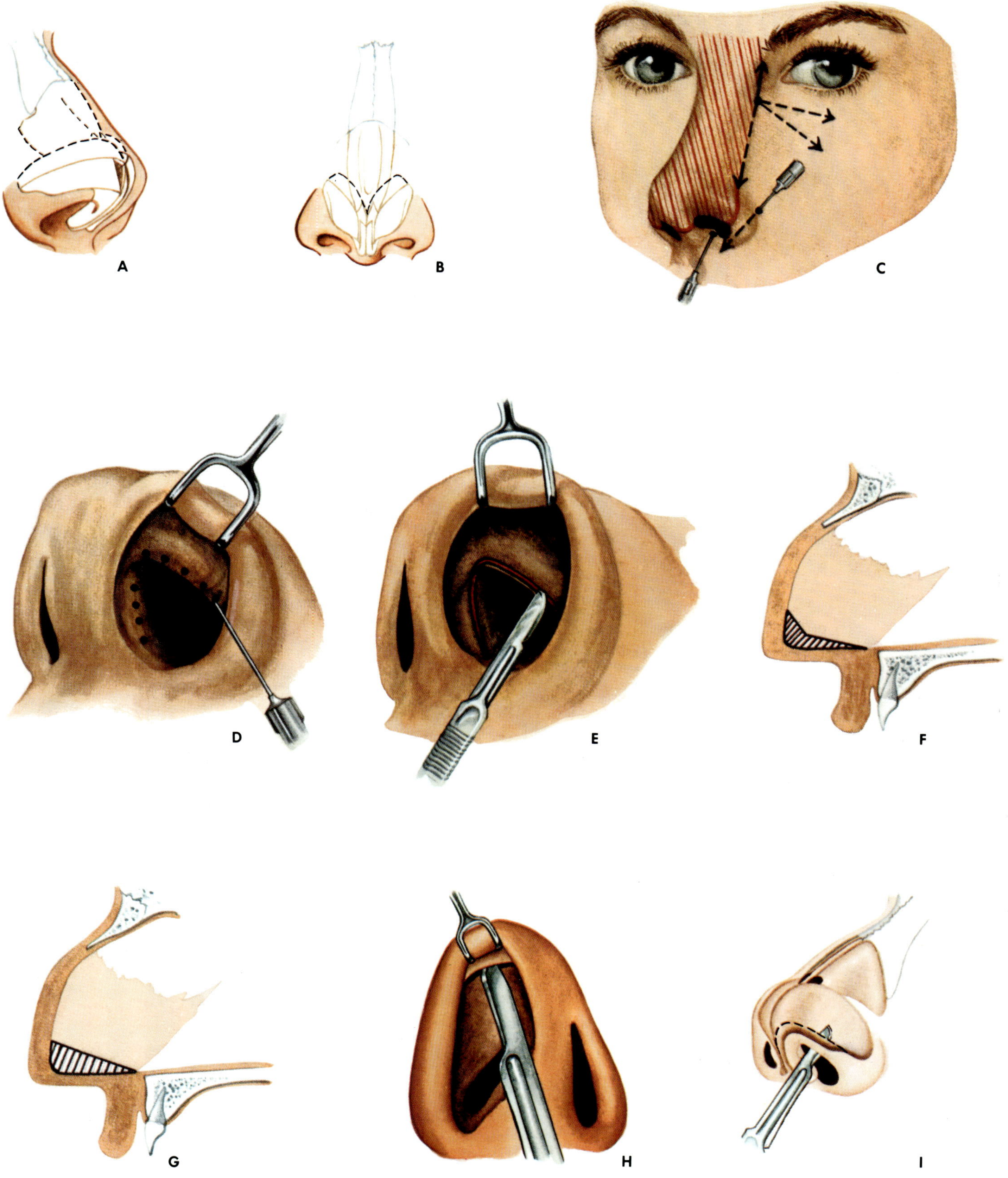

Plate 2

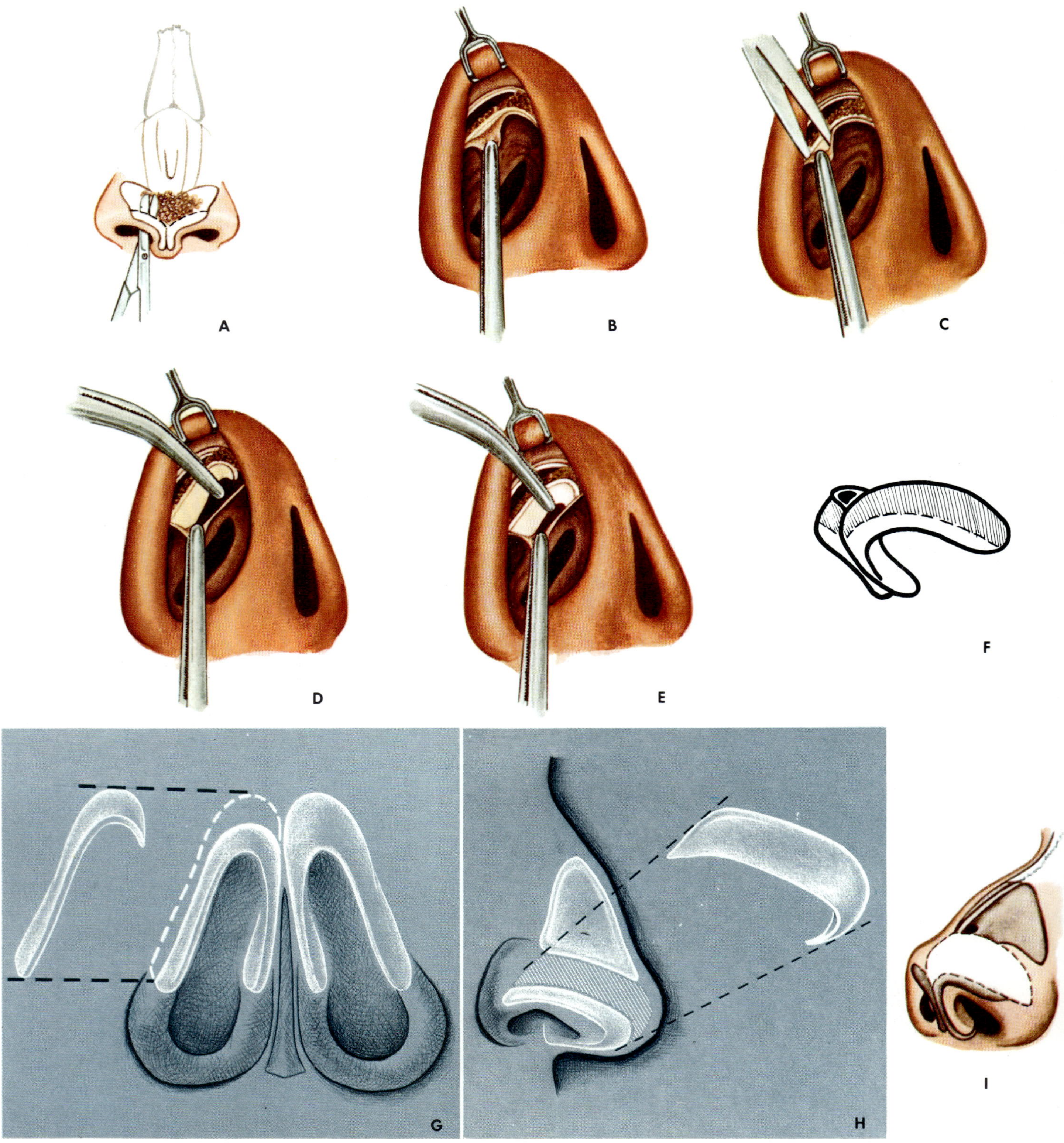

Plate 3. **A,** To remove the alar cartilage above the incision, first the skin from the underlying fat and alar cartilage is undermined with blunt scissors. **B,** The cartilage with the overlying fat is exposed when a hemostat is used to grasp the nasal lining and pull it inferiorly. **C,** The alar cartilage that is to be removed is separated from the nasal lining. **D,** The cartilage is then removed. **E,** The fat is removed in continuity with the cartilage. **F,** This is a typical specimen of the cartilage together with the medial crus, dome, and lateral crus. **G,** The resected cartilage is a three-dimensional structure. Following excision, there is a reduction of the nasal tip in all planes. **H,** Because of the loss in volume of resilient cartilage, the tip has reduced size in its entire pyramid. **I,** The final relationship of remaining cartilage in the nasal tip and how it stands out from the supratip area is the desired aesthetic characteristic.

Plate 2, *G,* shows the straight line excision of the columella septum as shown in Joseph's textbook, which tends to produce a retracted columella with the "pig-like look." The sculpturing of the alar cartilages must be performed after the tip has been elevated, because the position of these cartilages is changed by the elevation of the tip. Sculpturing must always be done with the tip in as close to a final position as possible. With the double hook, the nasal tip is everted and the outline of the alar cartilage is determined. The inferior margin of the cartilage protrudes through the overlying vestibulum skin and is easily seen. If it is not, slight pressure with the point of the No. 15 blade will locate the lower margin. A curved incision is made beginning at the dome and extends medially across the medial crus for about 5 mm. and laterally across the lateral crus for the full extent of the latter. The incision is made through and through the nasal lining and cartilage, leaving a 2- to 3-mm. width of cartilage inferiorly. The position of the incision together with the preserved cartilage below and the cartilage to be removed above is pictured in Plate 2, *I.*

The alar cartilage above the incision is now removed (Plate 3). This is accomplished by first undermining the skin from the underlying fat and alar cartilage with blunt scissors. It is important to remove all fat lying in the V between the two medial crura and the two domes. In Plate 3, *B,* a hemostat has grasped the nasal lining and, by pulling inferiorly, the cartilage with the overlying fat is exposed. By sharp dissection the alar cartilage that is to be removed is separated from the nasal lining. The lining is now a bipedicle flap attached medially and laterally. The cartilage is then removed. Notice that the fat is removed in continuity with the cartilage.

Plate 3, *F,* shows a typical specimen of the cartilage together with the medial crus, dome, and lateral crus. It is important to note that the domes have not been transected by removal of wedges. It is not necessary to do dome-wedge resections. The box-like nasal tip is eliminated, as are the sharp bumps of protruding edges of alar cartilage. The normal anatomy and lines of the nasal tip have been preserved. One must remember that it is not the size of the nose that is important but the aesthetic lines that are achieved. Certainly a slightly larger nose with excellent lines and anatomic symmetry is more to be desired if smooth lines result. It is important to preserve normal anatomic lines to prevent the possibility of bumps or irregularities.

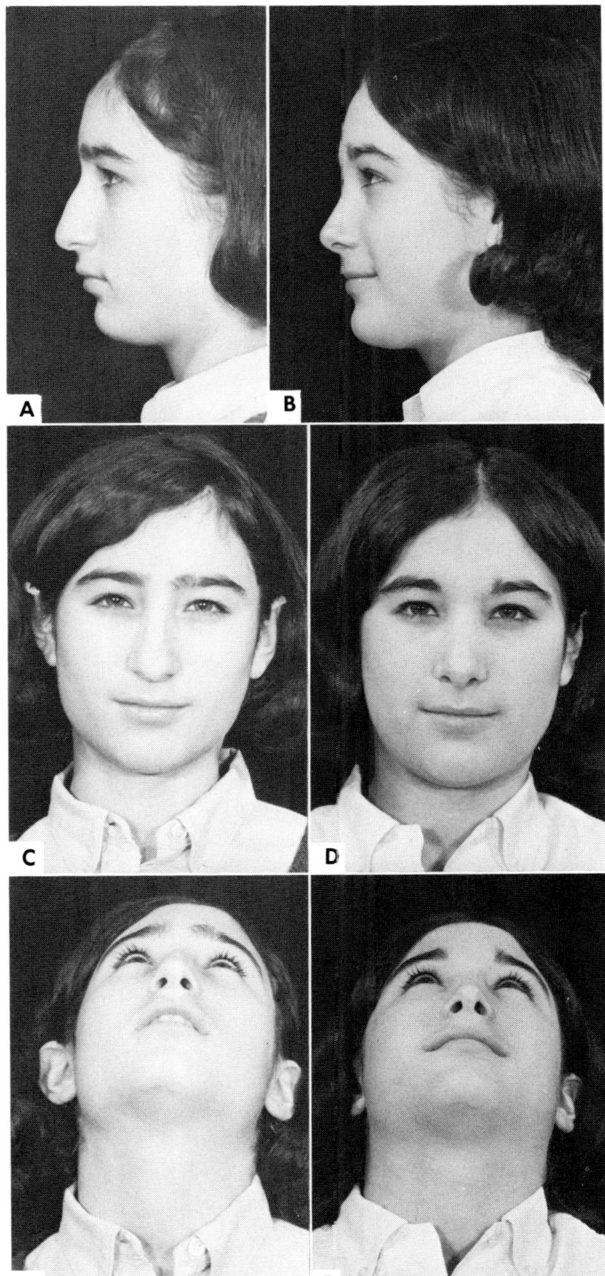

Fig. 7-3. A, Preoperative photograph of a 14-year-old girl; **B,** her postoperative appearance 2 years later. Front view: **C,** preoperative; **D,** postoperative. Worm's eye view: **E,** preoperative; **F,** postoperative.

An analytic evaluation of the geometry involved reveals that the alars are not flat structures shaped as an inverted U. Many anatomic diagrams would lead us to believe that this is true. In Plate 3 it can be seen, however, that the resected cartilage is a three-dimensional structure and, following ex-

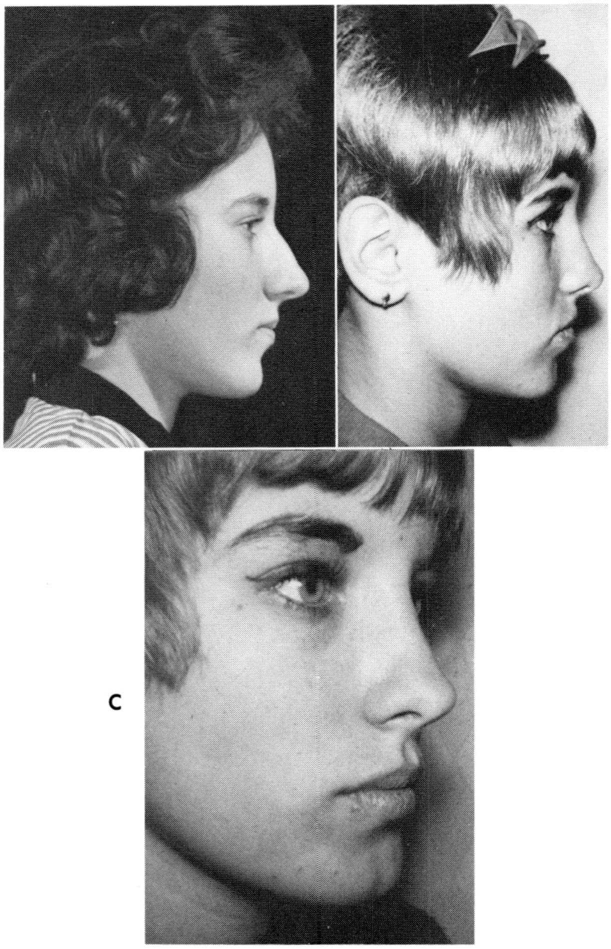

Fig. 7-4. A, Preoperative photograph of 18-year-old girl; B and C, postoperative views taken 2 years later.

cision, as shown, there is a reduction of the nasal tip in all planes. Upon removal of the cartilage, the tip has reduced size in its entire pyramid. This is directly related to a loss in volume of resilient cartilage. A reduction in entire nasal tip size has been achieved without sacrifice of anatomic symmetry, and there is an obvious reduction in tip height. Plate 3, *I*, shows the final relationship of remaining cartilage in the nasal tip. This new relationship of the nasal tip and how it stands out from the supratip area is an aesthetic characteristic that is most desirable—that is, the nasal tip stands out slightly and gracefully from the bridge above.

Fig. 7-1 shows preoperative and postoperative results in a 26-year-old woman. Note the definition and normal anatomic appearance of the alar cartilage. The patient had a receding chin that was corrected by a Silastic implant.

Fig. 7-2, *A* and *B*, shows a 16-year-old girl and her postoperative appearance after 1 year. Alar flaring is shown in the preoperative photographs. This was corrected by the modified Weir's procedure done at the time of the rhinoplasty.

Fig. 7-3 shows a 14-year-old girl and her postoperative appearance after 2 years. A slight chin implant with Silastic was used to augment the chin line.

Fig. 7-4 shows an 18-year-old girl 2 years postoperatively. Note the close-up, the nasal tip, and the normal anatomic lines and symmetry of the alar cartilage.

Chapter 8

Twenty-five helpful hints in corrective rhinoplasty

D. Ralph Millard, Jr., M.D., F.A.C.S.

1. Photographs (anteroposterior, profile, and nostril views) should be available during surgery as a base line for referral as the nose changes with the stages of reduction and the gradual increase in swelling.

2. The nose should be packed with petrolatum gauze to prevent blood from getting into the throat and being swallowed. If blood gets into the stomach, nausea and vomiting may rudely interrupt the surgery.

3. Local anesthetic should be injected around the periphery of the nose to avoid distortion of areas requiring careful shaping, such as the bridge and the tip.

4. Anterior vestibular incisions 0.75 cm. from the margin leave an intact strip of alar cartilage distal to the incision of about 2 to 3 mm. in width. All alar cartilage proximal to this is excised.

5. If the tip is wide, the tip cartilages are turned inside out with exposure through one nostril, the subcutaneous tissue is excised between the medial crus at the tip, and the cartilages are sutured together.

6. If the hump is small, it should be removed with a sharp chisel. If it is large and has a glabella notch, this should be excised with a saw and then shaped with the chisel and rasp. It is always better to take too little than to take too much, for along with the bridge goes the character of the nose and the face.

7. The excess bone in the glabellar area should be lowered with a chisel.

8. The cartilaginous septal bridge should be lowered with scissors and scalpel and overcorrected in the tip and supratip area to avoid the parrot's beak.

9. The septal mucosa should be trimmed along the bridge to avoid overlapping.

10. Lining sidewalls should be trimmed even with the bridge and thinned by excision of subcutaneous tissue.

11. The nose may be shortened by anterior septal resection. A man's nose should not be turned up too much. An older patient should not have his nose shortened too much. No nose should be shortened too much! The nasal spine should be reduced if the nasolabial angle is too wide.

12. If the feet of the medial crus flare, they may be amputated or sutured together.

13. In submucous resection, once the bridge has been lowered and the anterior septum resected, then the remaining septum is ready for resection and straightening.

Membranous septal incision gives access to mucosal dissection from the septal cartilage on one side. Anterior incision in cartilage leaving adequate anterior strut gives access to mucosal dissection on the opposite side. Swivel knife above and chisel along the vomerine groove remove the major portion of the cartilage, if indicated. To correct a deviation, the cartilage is scored on the convex side after freeing the mucoperichondrium. The septal prow is freed at the nasal spine, moved into central position, and fixed with catgut suture when indicated.

14. The anterior edge of the nasal lining flaps

should be reducd slightly less than the amount of anterior septal resection required to shorten the nose.

15. Bilateral osteotomy is made through lateral stab incisions within the vestibule. Periosteal dissection is followed by saw division of the bones flush with the maxilla, completed with a 4-mm. chisel through the same incision and along the saw cuts. Osteotomies should be completed with wide osteotomes along each side of the septum to outfracture and then to move bone components into narrowed position.

16. Lining should be sutured with 4-0 chromic catgut under no tension.

17. One suture on each side just proximal to the tip should stitch the sidewall to the septum to prevent overlap and to reduce the possible area of granulation with ultimate reduction in the chance of supratip swelling.

18. If the original nose had a hooked tip and if there is tendency for nasal tip to be depressed after surgery, then an orthopedic "saddle" suture is taken through the full thickness of the posterior columella and the full thickness of the anterior edge of the septum. The columella should slide and be fixed up along the septum to lift the tip in an overcorrected position. This is not permanent, but if the bridge has been shaped correctly, this slight lift will be sufficient to allow the nasal tip to settle into an ideal position.

19. Alar base flare should be noted, marked, excised, and sutured with 5-0 catgut and 6-0 silk.

20. Alar margin excision may be carried out in conjunction with alar base excision if the sidewalls overhang. It should be closed with continuous 6-0 silk.

21. Previous packs should be removed and the nasopharynx suctioned. Then each vestibule is packed with petrolatum gauze to ensure septal mucosal apposition with prevention of a submucosal hematoma.

22. Benzoin is painted over the nose and 3M tape is used to splint and shape the skin over the bridge and tip, with special taping in the supratip area to press out any dead space.

23. One layer of Telfa is covered with a nine-layer splint of quick-setting plaster of Paris to mold the nose and hold the bones snugly together.

24. Ears and eyes are padded with cotton and gentle pressure is applied with an Ace bandage to the entire face to reduce postoperative bleeding, ecchymosis, and swelling. The nostrils should be checked after the face bandage has been applied to see that there is no distortion such as telescoping of the nose with retraction of the columella caused by down-sliding of the dressing. This outer pressure bandage is removed in about 5 to 6 hours and ice compresses applied to the eyes.

25. Note the order of procedures: The tip is met first and is shaped first. The bridge and septum are shaped before the submucous resection to leave an adequate L-shaped support. The osteotomies are carried out as near the end of surgery as possible so that hemorrhage from the nasal fractures is postponed until splinting and pressure can prevent subsequent ecchymosis. Alar wedge excisions must come last, since only then do the nostrils show their new true size.

• • •

Each nose is different, each face is different, and each patient is different. With this many variables there is no way to have a routine and stick to it. The ultimate is to produce an attractive nose that fits its own face and, first and foremost, looks natural and happy in its habitat.

Chapter 9

Problem cases in rhinoplasty

D. Ralph Millard, Jr., M.D., F.A.C.S.

At this time in history in a portion of the world where movies and television have brainwashed the public, the most popular nose is a fine Caucasian type. The female ideal has a reasonably high, straight, and slender bridge and a slim, sculptured tip, is slightly up-tilted, and presents narrow, delicate nostrils. The male version is fine, straight, and sculptured but stronger in character with less up-tilt and less delicate qualities. For social, psychologic, and economic reasons most people living in our European-Anglo-American culture are desirous of having the popular "ideal" nose. This explains why minority groups as well as the deformed, in search of wider public acceptance, find their way to the plastic surgeon.

THE NEGROID NOSE

In general, it would seem neither feasible nor desirable (and might even be considered ridiculous) to transform a Negroid nose, especially when on a Negro, into an aquiline nose better befitting the classical British butler. The same seems true of an Oriental nose on an Oriental. There are, however, several loopholes in this dictum.

1. If the Negroid nose happens to be on a Caucasian, corrective surgery seems justified.
2. There are variations of the Negroid nose, and an exaggerated extreme certainly merits modification to within the norm.
3. Each patient has his or her own reasons and desires for change. These must be evaluated carefully by the surgeon and, if they are sane, sound, and within reason, surgery should not be withheld.

The typical Negroid nose has a flat, broad bridge and a flat, wide tip with thick, flaring nostrils. Of course, each Negroid nose varies. Certain surgical maneuvers have been found effective toward partial alteration of these characteristics.[1-6]

Bridge. Osteotomy narrows the broad, bony base and a long, shaped implant produces a straight, thinner, and higher profile line.

Tip. Radical reduction of the nasal alar cartilages with suturing of the remaining crus together at the tip only begins the shaping. Probably the most important action is to provide enough projection to lift the tip high enough to catch a highlight. Although the Negro nose is often flat and the amount of septal cartilage disappointing, a careful submucous resection should provide enough cartilage with which to construct an adequate strut. It will usually take two 2.5-cm. strips inserted through a stab incision at the columella base to lift this type of tip. When possible these struts should be inserted under the remaining alar cartilages at the tip to give insulation and protection against the projection.

Nostrils. The flare of the nostrils can be corrected by alar base wedge excisions and, in rare instances, full Weir wedges may be indicated. Thickened and overhanging sidewalls can be thinned and sculptured by parallel marginal wedge excisions in continuity with alar base excisions. There have been no reports of keloid formation occurring in scars of the alar margin or alar base areas (Fig. 9-1).

CONGENITAL RETRUSION OF THE NASAL TIP

Congenital retrusion is an odd deformity in which the lower half of the nose has not proceeded

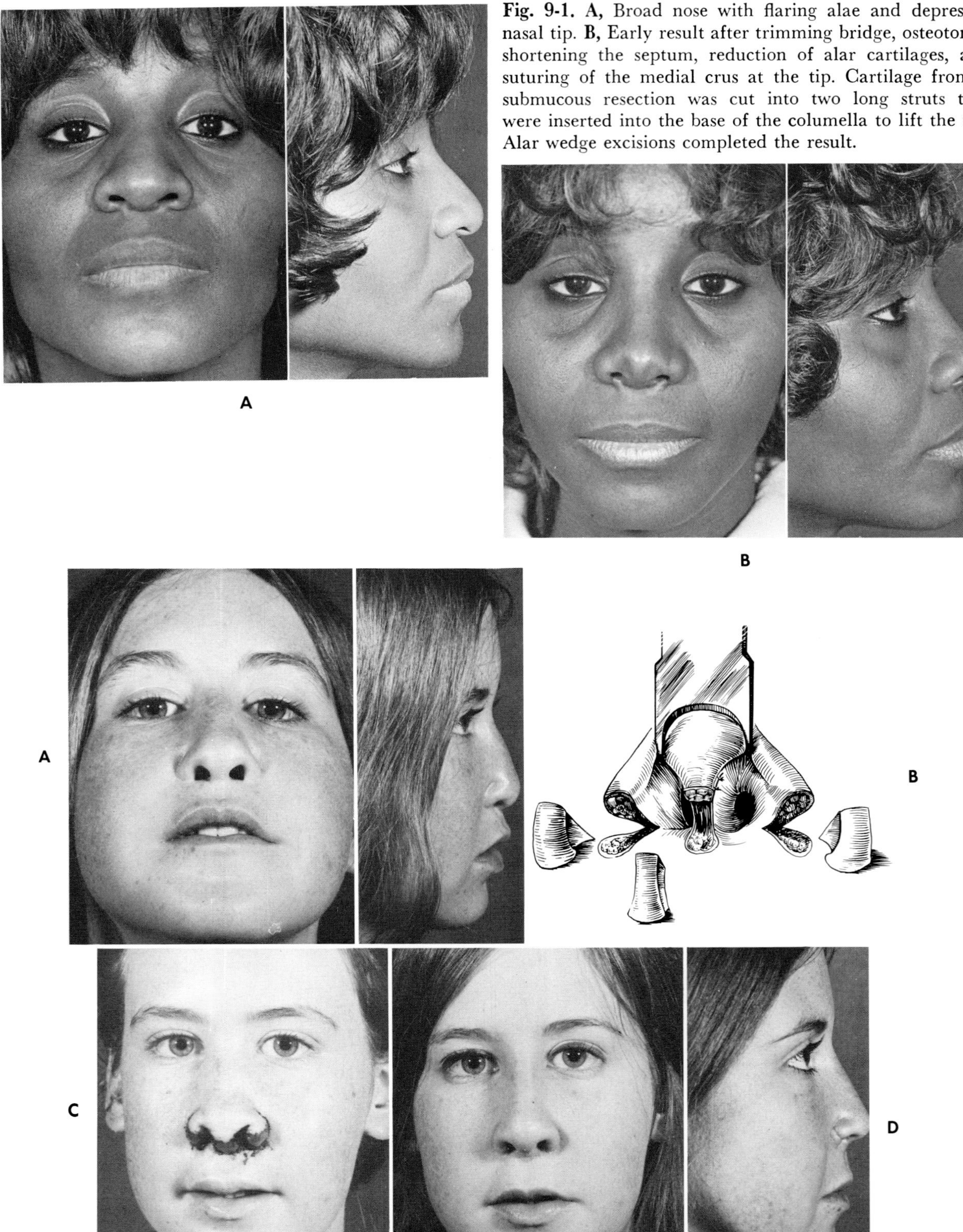

Fig. 9-1. A, Broad nose with flaring alae and depressed nasal tip. **B,** Early result after trimming bridge, osteotomy, shortening the septum, reduction of alar cartilages, and suturing of the medial crus at the tip. Cartilage from a submucous resection was cut into two long struts that were inserted into the base of the columella to lift the tip. Alar wedge excisions completed the result.

Fig. 9-2. A, Congenital retrusion of the nasal tip with short sidewalls and columella. **B,** Triple release filled with three composite ear grafts. **C,** Successful "take" of the grafts. **D,** Final trimming with marginal excisions and implants to the bridge and subalar lip area.

with normal projectile growth. The length of the columella and both sidewalls, as measured from the height of the alar arch to the lip, is about half that of the normal.

There are several possible methods of corrections, but the preferred approach was a transverse division from the lip of the three alar and columellar bases, allowing ample release of the tip. Into these gaping defects were fitted three composite auricular wedges that enjoyed a successful triple-header "take." The new nasal tip, elevated on ear stilts, presented a more normal and thus pleasing effect[7] (Fig. 9-2).

COLUMELLAR RETRACTION

Columellar retraction has many causes, most of which are manmade. Minor degrees can be treated by the insertion of a composite auricular graft into a releasing incision in the membranous septum.

For more severe retraction, the Dingman "banana-split" auricular graft has merit.[8]

There is also the bilateral chondromucocutaneous flap procedure that has been found extremely effective.[9, 10] Long, narrow flaps, taking skin and mucosa of the vestibule along with a thin strip of alar cartilage, are based at the tip behind the columella. The posterior incision of the parallel incisions that create the flap continues as the membranous septal releasing incision. Then, as the columella advances forward, these alar flaps ride with the advancing tip, swing down, and clasp each other through the membranous septal gap, maintaining the correction gained. These flaps are extremely viable in spite of previous scarring in the area and the disproportion of their length and width. The presence of cartilage undoubtedly protects the vascularity and prevents flap contraction with reduction of arterial and venous blood flow. These flaps have been used numerous times without loss and almost no circulatory embarrassment (Fig. 9-3).

The transposition of these flaps from the sidewall axis to the septal position has a double effect:

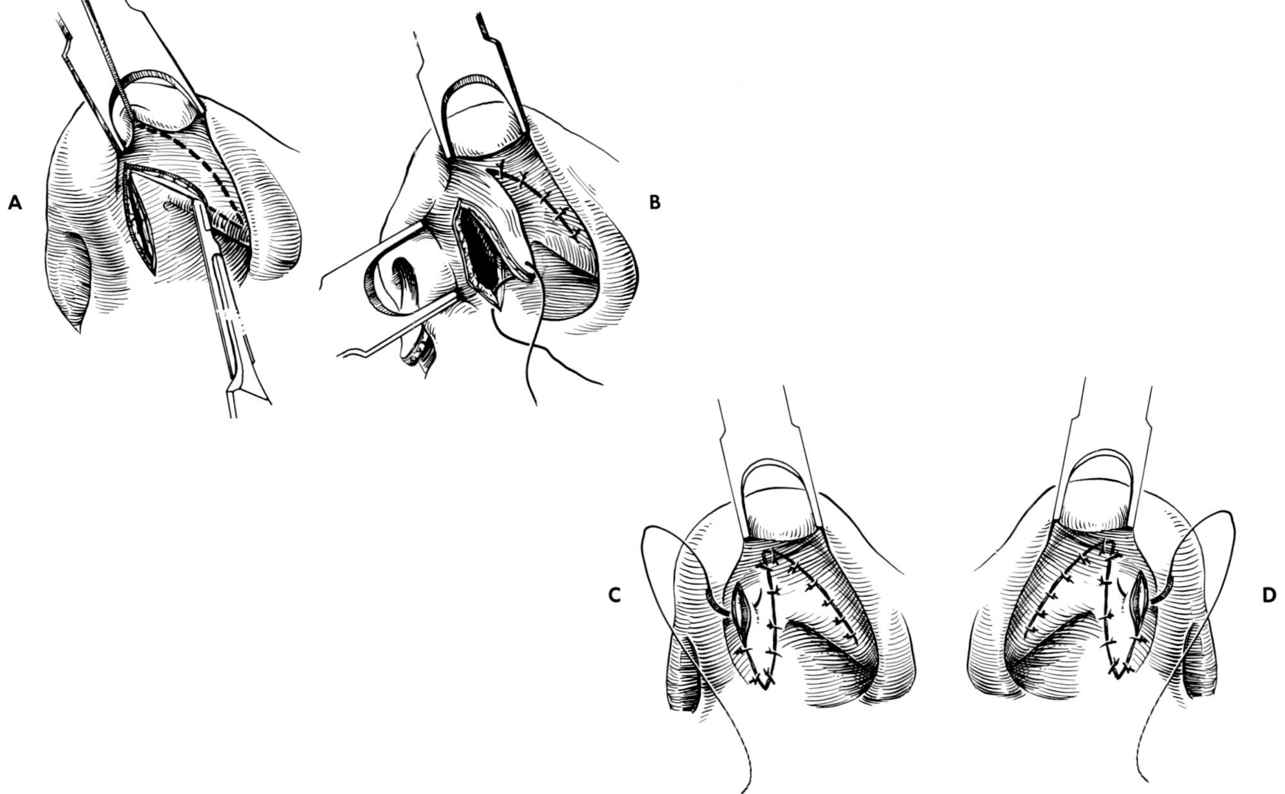

Fig. 9-3. **A,** Membranous septal releasing incision and chondromucocutaneous flap marked and cut in lateral alar area. **B,** Flap being transposed into septal gap. **C,** Flap being sutured and donor area closed. **D,** Similar procedure on the opposite side.

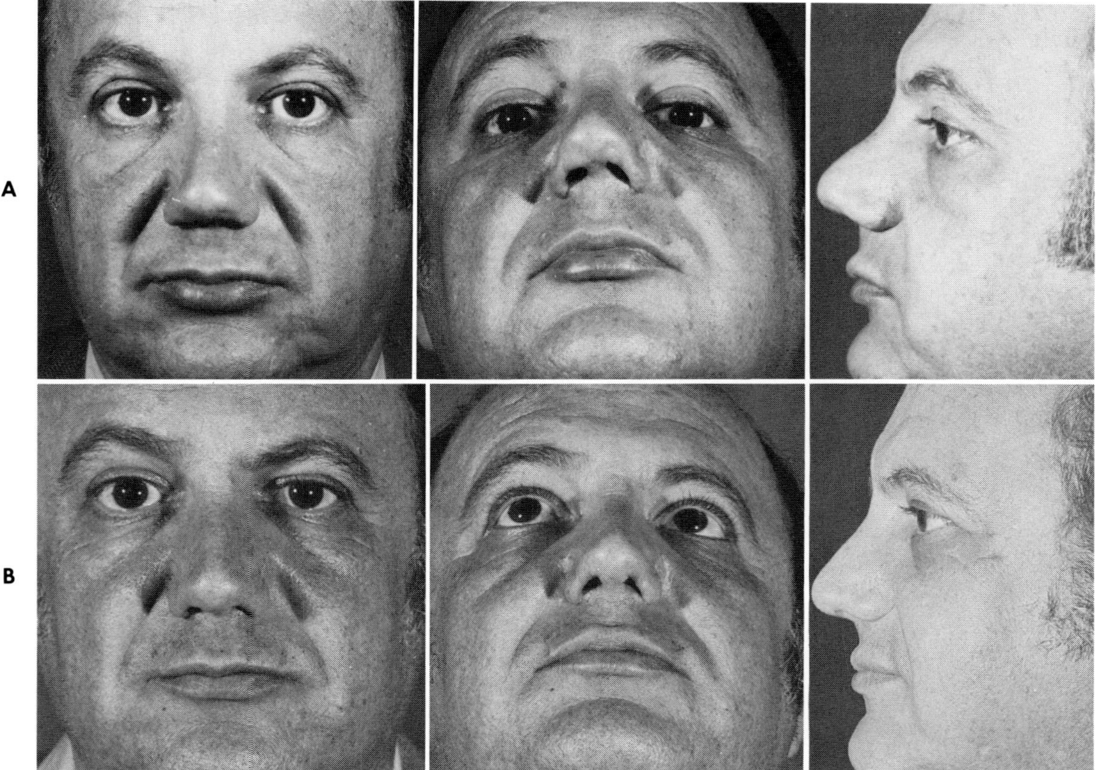

Fig. 9-4. A, Retracted columella, overhanging sidewalls, and "parrot's beak" result of reduction rhinoplasty elsewhere. **B,** Bilateral chondromucocutaneous flaps to the membranous septum and correction of "parrot's peak" improved the result.

reduction of the long sidewalls and release of the retracted columella. An overly short nose also is benefited as the snub tip is released in the process (Fig. 9-4).

COMMON POSTOPERATIVE NASAL DEFORMITIES

In the wake of the ever-increasing stream of corrective rhinoplasties, there is a secondary wave of postoperative deformities. Some of these have occurred often enough to have earned laymen nicknames such as "ski jump," "pig's snout," and "parrot's beak" or "ram's horn."

Parrot's beak. Probably the most notorious deformity is the "parrot's beak," where the bridge, instead of running gracefully straight until it takes a slight kick-up at the end of the tip, rather rises in a slow swell in the supratip, rounding off into a hook at the tip.

Numerous causes have been cited and surgeons have advocated various measures to combat them. Aufricht[11] scoops the distal septum, Safian[12] warns of overlapping sidewalls, Webster[13] lowers the septal tip, Rees[14] points to contracture of granulation tissue in this area, and Peck[15] maintains an intact alar cartilage arch.

The best treatment, of course, is prevention, and all of these suggestions should be used. Timing the secondary surgery is also important.[16]

Adequate lowering, even to the point of overcorrection, of the septal bridge and maintenance of at least a thin strip of intact alar cartilage arch are important. The sidewalls must be tailored, and reduction of granulating areas by careful lining closure is essential. Prevention of a hematoma in this area is vital.

In the last few years several additional measures have been added that seem to discourage the incidence of this problem.

1. Suturing of the sidewalls to the septum 1 cm. from the tip bilaterally with a stitch ensures the best apposition and the least area of granulation.
2. An orthopedic "saddle" suture between columella and septum lifts the tip in slightly overcorrected position when the nose was

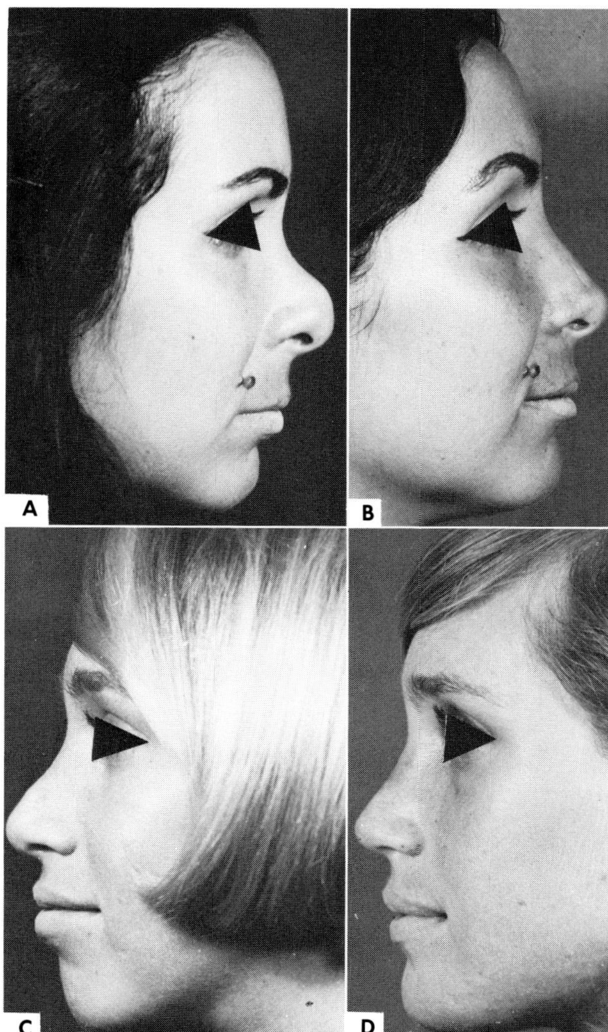

Fig. 9-5. A, Unbelievable "parrot's beak" postoperative deformity. **B,** Results after shaping the septum and alar cartilages. **C,** Another postoperative "parrot's beak" with a receding chin. **D,** Correction of the nose augmented by a chin implant.

severely hooked originally. The skin is thick, offering a greater threat toward an ultimate "parrot's beak." The full effect of this suture of course is temporary, but when the nose has healed there seems to be a slight residual gain that may be enough to win the game.

3. Careful postoperative taping of the skin of the nose forces the ideal profile and closes any dead spaces between skin, skeleton, and lining. The 3M taping in the area of the supratip is done with special enthusiasm. Then the entire nose is immobilized with a plaster of Paris splint.

Even after special precaution, the "parrot's beak" occasionally may creep surreptitiously into being. This requires surgery to free the skin, as in the original operation, and divides the sidewalls from the septum. Then the septum can be lowered in slightly overcorrected profile and the sidewalls trimmed to fit and thinned of scar. Finally, the same additional precautions of suturing the sidewalls to the septum, the orthopedic "saddle" suture, and careful skin taping should achieve the desired effect. Of course, time is necessary for the final result, since the local area of surgery will suffer reaction and retain swelling (Fig. 9-5).

REFERENCES

1. Millard, D. R., Jr.: External excisions in rhinoplasty, Brit. J. Plast. Surg. **12:**340, 1960.
2. Millard, D. R., Jr.: Adjuncts in augmentation mentoplasty and corrective rhinoplasty, Plast. Reconstr. Surg. **36:**48-61, 1965.
3. Millard, D. R., Jr.: Alar margin sculpturing, Plast. Reconstr. Surg. **40:**337, 1967.
4. Rees, T. D.: Nasal plastic surgery in the Negro, Plast. Reconstr. Surg. **43:**13, 1969.
5. Falces, E., Wesser, D., and Gorney, M.: Cosmetic surgery of the non-caucasian nose, Plast. Reconstr. Surg. **45:**317, 1970.
6. Snyder, G. B.: Rhinoplasty in the Negro, Plast. Reconstr. Surg. **47:**572, 1971.
7. Millard, D. R., Jr.: Congenital nasal tip retrusion and the three little composite ear grafts, presented at the Annual Meeting of the American Society for Aesthetic Plastic Surgery, May, 1971, Boston, Mass. In press.
8. Dingman, R. E., and Claus, W.: Use of composite ear grafts in correction of the short nose, Plast. Reconstr. Surg. **43:**117-124, 1969.
9. Millard, D. R., Jr.: The triad of columella deformities, Plast. Reconstr. Surg. **31:**370, 1963.
10. Millard, D. R., Jr.: Secondary corrective rhinoplasty, Plast. Reconstr. Surg. **44:**545-557, 1969.
11. Aufricht, G.: Personal communication.
12. Safian, J.: Fact and fallacy in rhinoplastic surgery, Brit. J. Plast. Surg. **11:**45, 1958.
13. Webster, G. V.: Random reflections in rhinoplasty, Plast. Reconstr. Surg. **39:**147-152, 1967.
14. Rees, T. D., Krupp, S., and Wood-Smith, D.: Secondary rhinoplasty, Plast. Reconstr. Surg. **46:**332, 1970.
15. Peck, G.: Study session in corrective rhinoplasty presented at the 39th annual meeting of the American Society of Plastic and Reconstructive Surgeons on October 6, 1970, Los Angeles, California.
16. Rogers, B. O.: The importance of "delay" in timing secondary and tertiary correction of post-rhinoplastic deformities, Excerpta Medica International Congress Series No. 174, Transactions of the 4th International Congress of Plastic and Reconstructive Surgery, Rome, October, 1967.

Chapter 10

Unsatisfactory cosmetic rhinoplasty

Mar W. McGregor, M.D.
Lars M. Vistnes, M.D., F.R.C.S.(C), F.A.C.S.

In order to ascertain how many people were dissatisfied with their operation and how this related to complications, in 1964 Klabunde and Falces[1] reviewed 300 rhinoplasties performed by various plastic surgeons. Of these 91% had been done under local anesthesia and 9% under general anesthesia, and one third of the cases had a simultaneous septoplasty. They found that the complication rate was 19% in those patients whose operation was performed under local anesthesia, 12% for those done under general anesthesia, and 16% where septoplasty was carried out at the same time. The dissatisfaction rates, however, were 32%, 44%, and 33% in the same categories. It is significant to note that in all categories there is a difference between complication and dissatisfaction, which emphasized the fact that the ultimate result of a cosmetic operation is patient satisfaction. A dissatisfied patient is thus a very serious complication of rhinoplasty and, therefore, is included in this discussion.

There are two main aspects involved in patient dissatisfaction.

1. Psychologic
   a. Unprepared patient
   b. Poor patient selection
2. Unsatisfactory result, that is, anatomic and/or aesthetic imperfection
   a. Surgeon factor (iatrogenic)
   b. Patient factor (inherent)

The remainder of the discussion will deal with the second main cause of dissatisfaction, the unsatisfactory result.

Following the broad outline given here, the main reasons for an unsatisfactory result will be delineated and by implication the methods of avoiding such complications given.

A. Skin and subcutaneous tissue
   1. Endogenous factors
      a. Thickness and elasticity
      b. Pigmentation
   2. Exogenous factors
      a. Contact (splint, tape)
      b. Infection
      c. Hematoma and bleeding
      d. Adhesions and webs
      e. Scars
B. Bony pyramid
   1. Width
   2. Angle
   3. Height
C. Septum
   1. Length
   2. Height
   3. "Straightness"
D. Supporting cartilages
   1. Upper lateral
   2. Lower lateral
E. Foreign materials
   1. Implants
   2. Packing

SKIN AND SUBCUTANEOUS TISSUE
Endogenous factors

Thickness and elasticity. The nasal skin of most patients has sufficient thickness and elasticity to allow good draping after reduction of the nasal framework. Some people, however, have nasal skin that is so thin that the slightest irregularity of the underlying supporting structures will be apparent. In such cases, it is imperative to rasp off all irregularities and to remove all loose fragments of bone and cartilage. It is also wise to stay

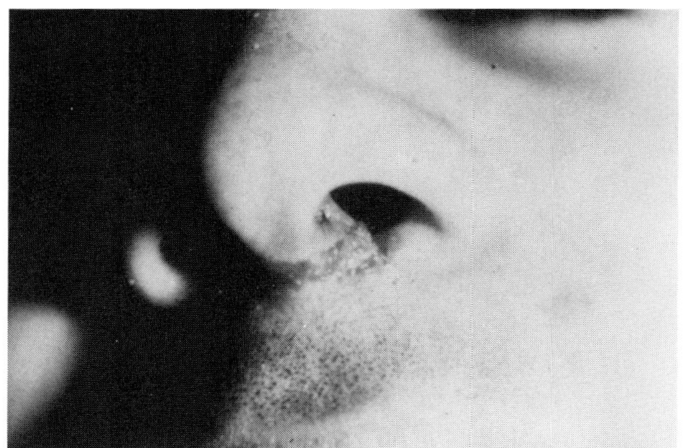

Fig. 10-1. Frank ulceration of columella produced by tape irritation, swelling, and pressure.

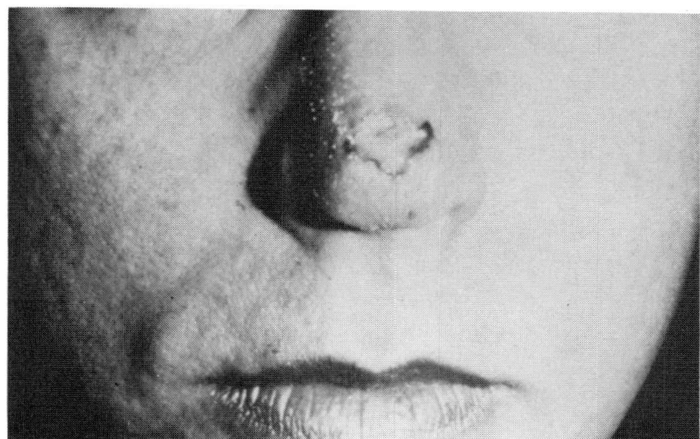

Fig. 10-2. Full-thickness loss of nasal skin revealed on removal of splint 7 days after rhinoplasty.

beneath the periosteum if at all possible. In other instances the skin is so thick and inelastic that it will not shrink to conform to the new framework that surgery created. It is important to recognize this preoperatively in order that the reduction may be modified to suit the skin.

Pigmentation. All plastic surgeons are familiar with the ecchymosis that may appear in the eyelids during and after nasal surgery. This usually subsides and disappears within 2 weeks after the operation. This dark pigmentation, however, may remain for many months and even permanently. This is particularly true in patients with brown eyes and dark brunette coloring. Since these patients often have some shadowing in their eyelids before surgery, such individuals must be forewarned that this pigmentation may not disappear following surgery but may in fact become more pronounced.

Exogenous factors

Contact. In many patients, the nasal skin under the splint has a tendency to become irritated. The cause of this has not been definitely determined, but it is felt that factors such as pressure from the splint, sensitivity to the adhesive tape, and dormant organisms within the glands beneath the nasal integument or any combination of these may be responsible. The irritation may be generalized or localized to a specific area. When localized, it is usually in the area of thickest and most sebaceous appearing skin.

In some cases (Fig. 10-1) it is almost certainly caused by tape irritation, combined with pressure produced by swelling. It is essential to check these points of potential trouble in the early postoperative

period to prevent problems. If ignored, these small problems can go on to become major complications, such as in frank necrosis of nasal skin (Fig. 10-2). When such irritation and even ulceration or necrosis occurs, it is best treated conservatively. Daily cleansing followed by the application of a topical antibiotic allows the area to heal spontaneously. On occasion the resulting scar may be improved upon by selective dermabrasion. Attempts to excise these ulcers in the hope of leaving a fine, thin scar have yielded disappointing results, and this method is not recommended.

Infection. Apart from the localized skin infections associated with pressure and irritations, a generalized nasal cellulitis may occur. The majority of the frank infections of the nose following rhinoplasty begin with redness and swelling in the area of the nasal bones, either dorsally or along the site of one or the other lateral osteotomy. If allowed to progress, a localized area of pus collection with fluctuation and compromised overlying skin may develop. The pus must be drained, but fortunately most infections are seen earlier and respond quickly to antibiotics and warm moist packs.

The small pustules sometimes seen on removal of nasal splint and tape are usually superficial. They are best treated by unroofing with a hypodermic needle, cleansing with peroxide, followed by application of a topical antibiotic. This occurs most frequently in sebaceous skin. The problem does not seem to occur as frequently when the splint is plaster of Paris applied directly to the skin, which seems to have a drying effect on the skin.

Hematoma and bleeding. Occasionally, a patient will develop a large hematoma between the

Fig. 10-3. **A,** Webbing in floor of nose producing a fluid trap. **B,** Webbing along the dorsal aspect of the nose in area of intercartilaginous incision.

skin and the underlying bony framework in spite of the careful application of a firm splint. This elevates the splint away from the nose, and the entire undermined area of soft tissue becomes thick and broad. Conservative management is usually successful. Some may give oral enzymes or inject these directly into the hematoma along with hyaluronidase. In spite of these measures, however, in the occasional patient the thickness does not completely disappear but remains either as thick, spongy tissue or as firm, irregular lumps beneath the skin. Blood dyscrasias, unless detected preoperatively, may give rise to severe bleeding during and after surgery. Since the specifics of management for each dyscrasia is beyond the scope of this chapter, suffice it to say that prevention is of more value than cure.

Although the occasional nasal hemorrhage during and after surgery can be managed by pressure and packing, minor postsurgical bleeding is often from a specific area either on the septum or in the suture line. Major bleeding is frequently both undetectable as to source and inaccessible. Both are frightening to the patient who needs a great deal of reassurance and sedation. In either case, control depends on the availability of adequate light and suction, as well as shrinking of the congested nasal mucosa by epinephrine and anesthesia by application of xylocaine or cocaine topically. The spot bleeder may then be controlled by chemical or electrical cauterization. Major bleeders may require both anterior and posterior packs for control. Temporary control of such a severe hemorrhage may be obtained posteriorly with a Foley catheter in the nostril and inflation of the bag posteriorly. One must be prepared to ruin a good surgical result since an effective packing usually separates the nasal bones.

Adhesions and webs. If opposing incisions are made on surfaces that come in contact, adhesions may form between those opposing surfaces.

The intercartilaginous incision that continues across the vault of the nose and onto the septum can produce this situation, resulting in irregular webs that trap fluid, giving a persistent running nose (Fig. 10-3, *B*). These webs may also occur in the floor of the nose in which fluid trapping is even more pronounced (Fig. 10-3, *A*).

The internal nasal problem, however, leads to a marked nasal deformity with binding down of the ala on the affected side. This bothersome problem can in large measure be prevented by firm pressure applied along the area of potential trouble with a cotton-tipped applicator at regular intervals for the first few weeks postoperatively. This will break up any adhesions and allow healing in a smooth even line.

Scars. Unless extreme care is exercised, the skin of the nose may be perforated with a knife, scissors, or saw while attention is focused where the instrument is doing its intended work. This in itself can lead to an unsightly scar, since the cut is usually in a completely unplanned area. Even planned incisions, such as the Weir excisions, can lead to unsightly scars unless properly placed and closed (Fig. 10-4).

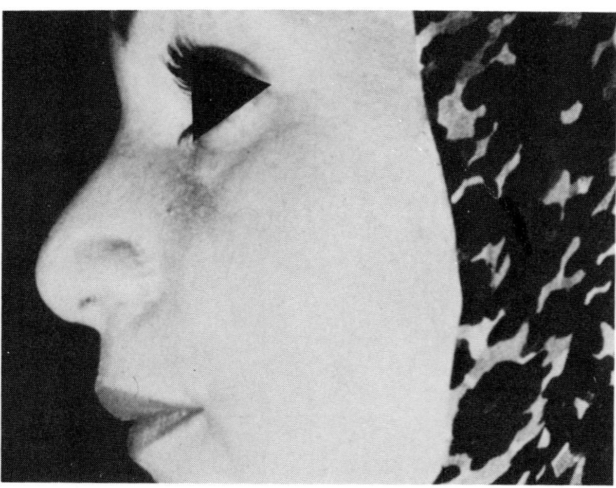

Fig. 10-4. Unsightly alar base scars after Weir type excision.

The alar rim retraction occasionally seen after rhinoplasty is usually caused either by excision of too much mucous membrane or by scar formation in the intranasal incision that pulls the alar rim directly upward, forming a large or small notch.

BONY PYRAMID

Width. When the dorsum of the nose has been lowered, the resulting flatness and apparent excessive wideness of the nose are corrected by lateral osteotomy, fracture of the nasal bones, and movement of the fractured components to the midline.

A nose that is too wide after rhinoplasty may be the result of insufficient mobilization of the bones at the time of infracture or loss of control of these components with secondary widening. Control of width of the nose is aided by lateral pressure applied for approximately 10 days. This can be achieved either by leaving the splint on for this length of time or by using a well-padded screw or spring-clip splint.

A nose may be suitably narrow near the dorsum but wide closer to the cheek. This is a relative widening and is produced by placing the lateral

osteotomy too high. It is unsightly, distressing, and difficult to correct.

Angle. When dealing with the bony pyramid, by far the most frequently occurring complication is the lack of symmetrical placement of the two sides of the pyramid. Excessive infiltration of local anesthesia, hematoma, or edema all make it difficult to determine the exact location of the bones.

If the skin is completely undermined without leaving a small portion attached to the fractured nasal bone, the fracture fragment is easy to dislocate or even lose (Fig. 10-5).

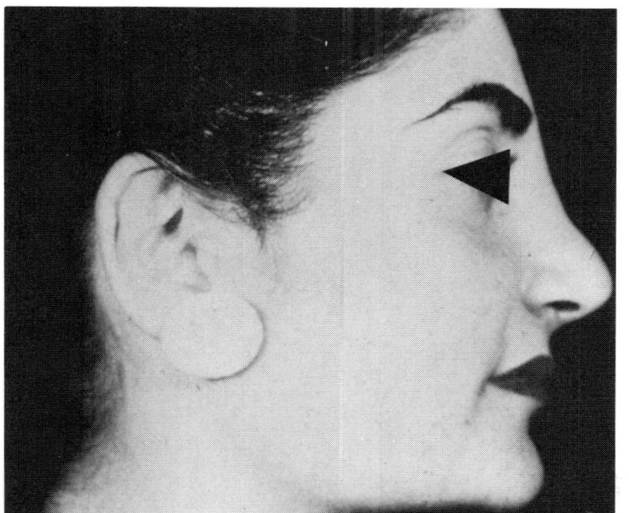

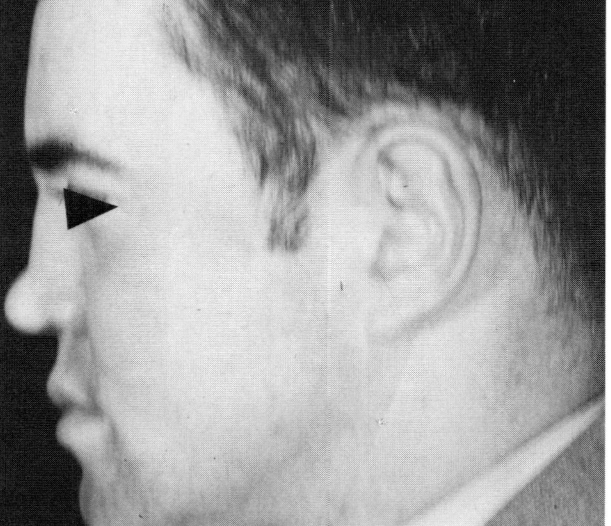

Fig. 10-6. A, Unsatisfactory nasofrontal angle caused by failure to remove sufficient bone in this area produces poor profile. **B,** Excess removal of skeletal and cartilaginous support results in a nose that is too low.

Fig. 10-5. Dislocation of nasal bone fragment resulting in uneven nasal contour and light reflex distribution.

Occasionally the situation will be encountered when the nose appears satisfactory following removal of the nasal splint. It is symmetrical, has good shape and proportion, but does not appear to fit the face. At that point the asymmetry of the entire face is discovered. A line drawn through the midline of the nose and forehead ends up on the side of the chin or a line drawn through the pupil of the eye is not at right angles to a line drawn through the middle of the face. It is essential to note this asymmetry preoperatively, so that placement of the nasal bones at the time of surgery can be the compromise that will achieve the best contour for that particular face.

Height. In order to produce satisfactory aesthetic proportion between the nose and the rest of the face, considerable care must be taken to produce a proper nasofrontal angle. If insufficient bone is removed from this area, the nose appears and is too high (Fig. 10-6, *A*). On the other hand, removal of excess tissue from this area produces a nose that is too low (Fig. 10-6, *B*). The easiest time to correct this latter deformity is at the time of surgery, by replacement of a portion of the removed hump. Failure to do so allows a postoperative deformity that must be corrected by some type of implant, such as a cancellous bone graft.

Foreign materials used for such implants have shown a high percentage of eventual extrusion (Fig. 10-7). The smaller the implant and the less it is under any pressure, the more likely it is to remain in the nose without infection or extrusion. Homologous rib cartilage has been discontinued as an implant because of its tendency to curve.

SEPTUM

Length. The length of the septum affects the entire appearance of the nose, as well as the columella specifically. Many noses demand the excision of a triangular wedge of cartilaginous septum with its apex at the vomer to raise the tip. If this resection is performed too enthusiastically and the base of the triangle is too wide, the tip is raised too much. The nasolabial angle becomes incorrect and the frontal view gives one the unpleasant experience of looking directly into the airway.

If instead of removing a triangle of cartilage from the lower portion of the septum one removes a generous rectangle, the result is a retracted columella. If the resection takes the form of a triangle added to the rectangle, a combination of the two aforementioned deformities results with a tip that is too high plus a retracted columella (Fig. 10-8).

Failure to remove sufficient cartilage at the inferior border of the septum, on the other hand, results in a nose that is entirely too long and a tip that droops.

Height. Just as the length of the septum inferiorly affects both the shape of the nose and columella, so does the height of the septum affect the dorsum of the nose. After removal of the hump and infracture, a corresponding amount of dorsal septum must be removed. It is not enough to provide a smooth transition between the bony pyramid and the septum, but the entire angle of resection along the dorsum must be correct (Fig. 10-9).

An unequal level between bony pyramid and septum usually results when the septum has been left too high (Fig. 10-10).

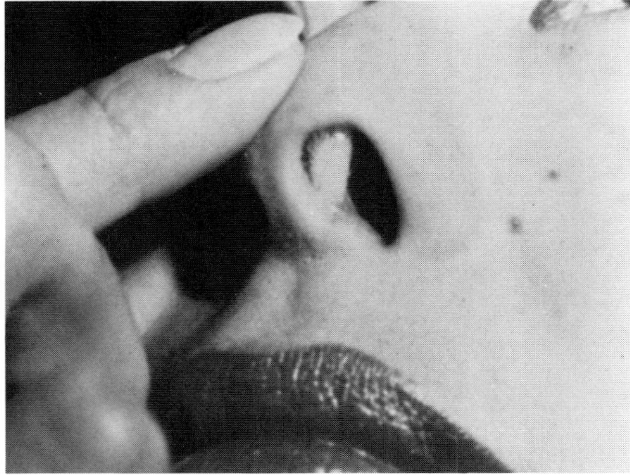

Fig. 10-7. Extrusion of Marlex implant through nasal mucosa.

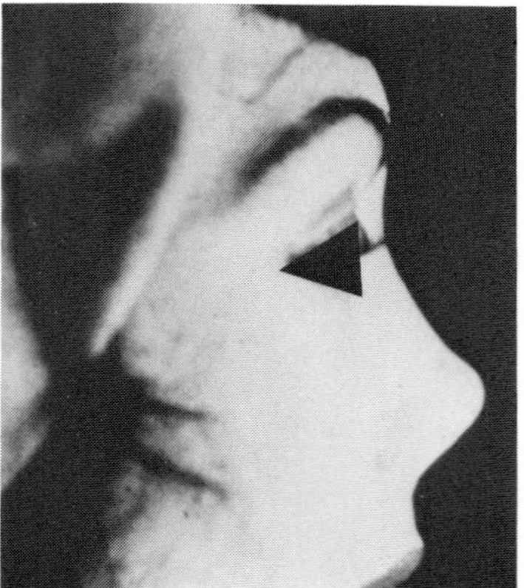

Fig. 10-8. Excessive resection of septum inferiorly producing retracted columella and excessively elevated nasal tip.

When the level of resection has been correct superiorly at the junction with the bony pyramid but resection along the dorsum has been insufficient further inferiorly, a tip that is too high and an elongated columella are produced. Similarly, when the entire angle of resection of both bony hump and septum has been incorrect, a nose that is too high may result on the one hand, or at the other extreme, a nose that is too low with a foreshortened columella (Fig. 10-11).

Finally, if the septum is not resected sufficiently along the dorsum, the nasal bones will not meet in the midline after infracture, and a noticeable dorsal ridge results. The problems associated with injudicious septal resection comprise the most common source of dissatisfaction with rhinoplasty. A suitable resection follows a line from the top of the existing glabellar notch to the top of the existing nostril. The advice of Dr. George Webster, who suggests removal of the dorsum and inferior edge of the septum to the desired result, then removing an additional 3 to 4 mm. from each edge, is valuable.

Straightness. Nasal deformity produced by a deviated septum is not straightened by rhinoplasty alone. To paraphrase a political expression: "As goes the septum, so goes the external nose." It is thus imperative to do an adequate septal reconstruction with straightening of both nose and septum at the same time.

SUPPORTING CARTILAGES

Upper lateral. The upper lateral cartilages fill out and thus give the appearance of thickness to the dorsum of the nose at the junction between the middle and lower thirds of the nose. Insufficient removal of upper lateral cartilage along with the dorsal hump results in a nose that is too wide and thick in this area.

A study of the normal arrangement of upper and lower lateral cartilages show a continuous cartilaginous support of the soft tissues of the tip, with

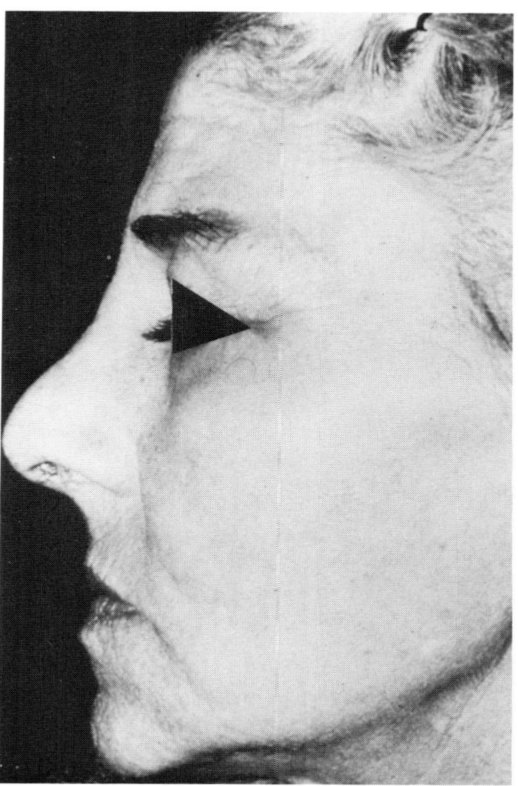

Fig. 10-9. The entire angle of resection of nasal supporting skeleton must be correct lest the nose become too short or too long.

Fig. 10-10. Insufficient resection of dorsal cartilaginous septum produces this result.

the upper edge of the lower cartilage overlapping the lower edge of the upper cartilage. It is obvious from this that any operation that removes too much cartilage from this junctional area results in a gap of support of the soft tissue (Fig. 10-12). It is immaterial whether there is excessive resection of either the upper or lower lateral cartilages, or a combination of both. The result is the same: alar collapse (Fig. 10-13). In addition the aesthetic deformity of obstruction on inspiration may occur. Consideration of the anatomy of the area helps to prevent this complication.

Lower lateral. The lower lateral cartilages provide support for the inferior portion of the tip and alae. A tip that is too wide with a flaring nostril can be corrected, either by the resection of a wedge of cartilage near the tip or by a resection of soft tissue near the base. The former method may produce a nose that is too pointed, whereas the latter method may not narrow the tip enough.

It is generally recognized that whichever incision is used to expose the lower lateral cartilages, a strip of the cartilage along the alar rim should be left for support. It is, however, important to dissect this strip free and extend this dissection around the genu and dome of the remaining cartilage in order that it may seek its new position relative to the soft tissues of the tip.

FOREIGN MATERIALS

Implants. For information about implants see the section on the bony pyramid.

Packing. Packing of the nose must be performed carefully so that none is placed in the dissected space between skin and bone. Similarly, it

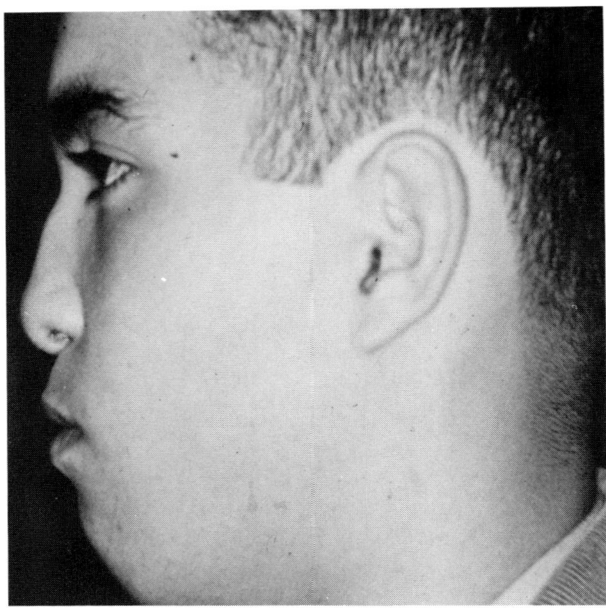

Fig. 10-11. Wrong angle of resection, removing too much bone *and* cartilage.

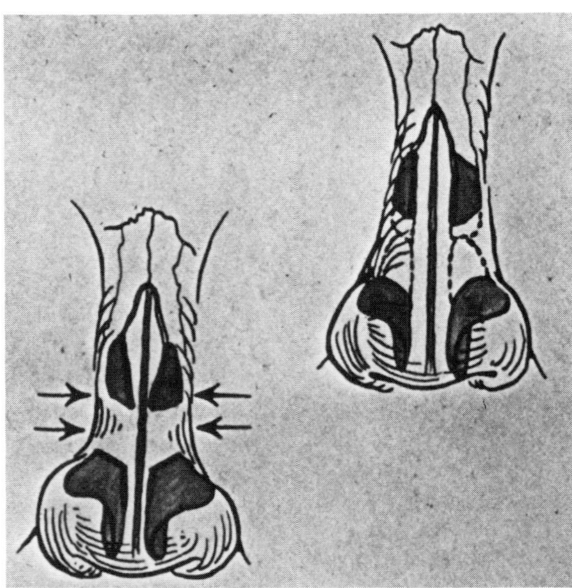

Fig. 10-12. Creation of an excessive gap in cartilaginous support between the upper and lower lateral cartilage produces alar collapse in the area.

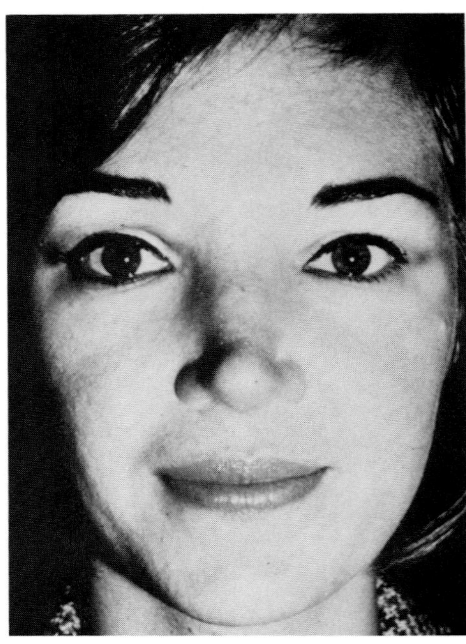

Fig. 10-13. Alar collapse caused by removal of underlying cartilaginous support.

is important on removal of the pack that no remnants be left behind.

The removal of retained packing from a swollen nose with purulent discharge a week after removal of the main pack causes much unhappiness and is a complication that can be avoided easily.

REFERENCES

1. Klabunde, E. H., and Falces, E.: Incidence of complications in cosmetic rhinoplasties, Plast. Reconstr. Surg. **34:**192-196, 1964.
2. Campion, R.: Anosmia associated with corrective rhinoplasty, Brit. J. Plast. Surg. **19:**182-185, 1966.
3. Cottle, M. H.: Clinical benefits and disorders following nasal surgery, Southern Med. J. **61:**1281-1286, 1968.
4. Flowers, R. S.: Injury to the lacrimal apparatus during rhinoplasty, Plast. Reconstr. Surg. **42:**577-581, 1968.
5. Fry, H. J.: The distorted residual cartilage strut after submucous resection of the nasal septum, Brit. J. Plast. Surg. **21:**170-172, 1968.
6. Goldman, I. B.: Rhinoplastic sequelae causing nasal obstruction, Arch. Otolaryng. **83:**151-155, 1966.
7. Goldwyn, R. M., and others: The effects of submucous resection and rhinoplasty on the sense of smell, Plast. Reconstr. Surg. **41:**427-432, 1968.
8. Hage, J.: Collapsed alae strengthened by conchal cartilage, Brit. J. Plast. Surg. **18:**92-96, 1965.
9. O'Connor, G. B., and others: Alar collapse, Plast. Reconstr. Surg. **40:**49-52, 1967.
10. Rees, T. D., and others: Secondary rhinoplasty, Plast. Reconstr. Surg. **46:**332-340, 1970.

Chapter 11

Secondary rhinoplasty

Thomas D. Rees, M.D., F.A.C.S.

Relatively little has been published about the correction of secondary deformities following rhinoplasty, despite the fact that reconstruction of the nose is a common procedure. Few reports dealing with the incidence of secondary rhinoplasties have appeared in the medical literature. In one of the more comprehensive studies, T. W. Smith[1] reported that 12% of a series of 221 patients required secondary surgery, a figure that seems somewhat high. An experienced surgeon should not exceed a 5% secondary operation rate.

Many surgeons hesitate to perform secondary rhinoplasties on other than their own patients, since such patients may be dissatisfied with the results of the second operation as well. Of course, every surgeon has an obligation to attempt to correct residual problems in his own patients. Nevertheless, if the secondary deformity is severe and if extensive soft tissue scarring exists, there may be little or no benefit derived from further surgery.

Since it is not possible to cover all aspects of secondary rhinoplasty, this chapter will zero in on some of the more common causes of secondary deformities and will emphasize useful techniques for their correction.

Complications involving the skin of the nose—hyperpigmentation, hypopigmentation, erythema, cyanosis, telangiectasis, and scarring—have been excluded from this discussion. Only surgical failures will be considered.

It is important to note here that the severely "butchered nose" is extremely difficult to improve and presents a technical and psychologic problem of great complexity. The novice surgeon would be wise to refer these patients elsewhere. Even the experienced surgeon should be wary and should give careful study and consideration to each one before he attempts any surgical procedure.

Secondary deformities can be classified into four groups: (1) those involving the bony framework, (2) those involving the cartilaginous framework, excluding the tip, (3) those involving the tip, and (4) those involving the nostrils, vestibulum, and columella.

SECONDARY DEFORMITIES INVOLVING THE BONY FRAMEWORK

The form of the upper part of the nose is provided by the nasal bones, the bony septum, and their junctures with the maxillae and frontal bones. Postrhinoplasty problems involving this framework can be classified as follows.

Hump deformities. Secondary hump deformities are basically the result of too much or too little hump removal. Too little hump removal during rhinoplasty leaves a residual hump requiring further, though judicious, hump resection. Excessive hump removal can cause a bony saddle deformity that can be corrected only by an implant or by a bone or cartilage graft. Removal of too much hump occurs most often in those patients who originally had only a small defect. During primary or secondary surgery, small humps are best removed with a sharp osteotome, rather than a saw, because there is a greater chance of inadvertently removing too much tissue when the saw is utilized.

Since any hump deformity—primary or secondary—is so closely related to the remainder of the nasal framework, it can be produced by either a real or "imagined" hump. For example, a depression in the lower cartilaginous part of the nose can appear as if there were a hump above.

Because of this close anatomic relationship, it is often advisable to correct any other nasal deformities before removal of the hump itself, particularly if the hump is small. Since the size of the hump is also relative to the length of the nose, Lipsett[2] feels that a lesser amount of hump can be removed if the septum is shortened *before* the hump is removed. Similarly, the nose with a projecting tip requires less extensive hump removal than one with a recessed tip.

Another important anatomic area to be kept in mind during hump removal is the nasofrontal angle at the radix of the nose. Aufricht[3] pointed out the importance of the radix, especially in relationship

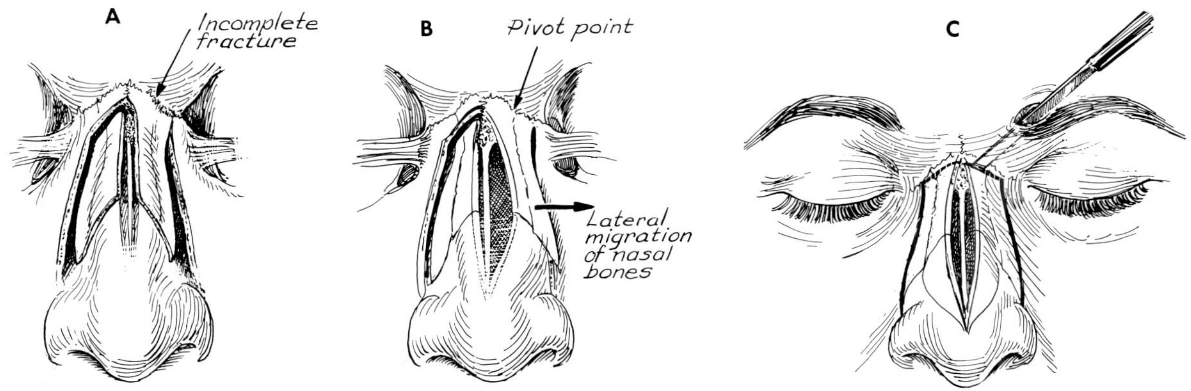

Fig. 11-1. A, Incomplete infracture is a common reason for secondary rhinoplasty. **B,** Usually the superior fracture line is incomplete, allowing the bones to drift laterally. **C,** The surgeon should create a superior osteotomy when necessary with a small osteotomy to ensure mobility of the nasal bones. This step may be all that is required at secondary operation. (From Rees, T. D., Krupp, S., and Wood-Smith, D.: Secondary rhinoplasty, Plast. Reconstr. Surg. **46:**332, 1970.)

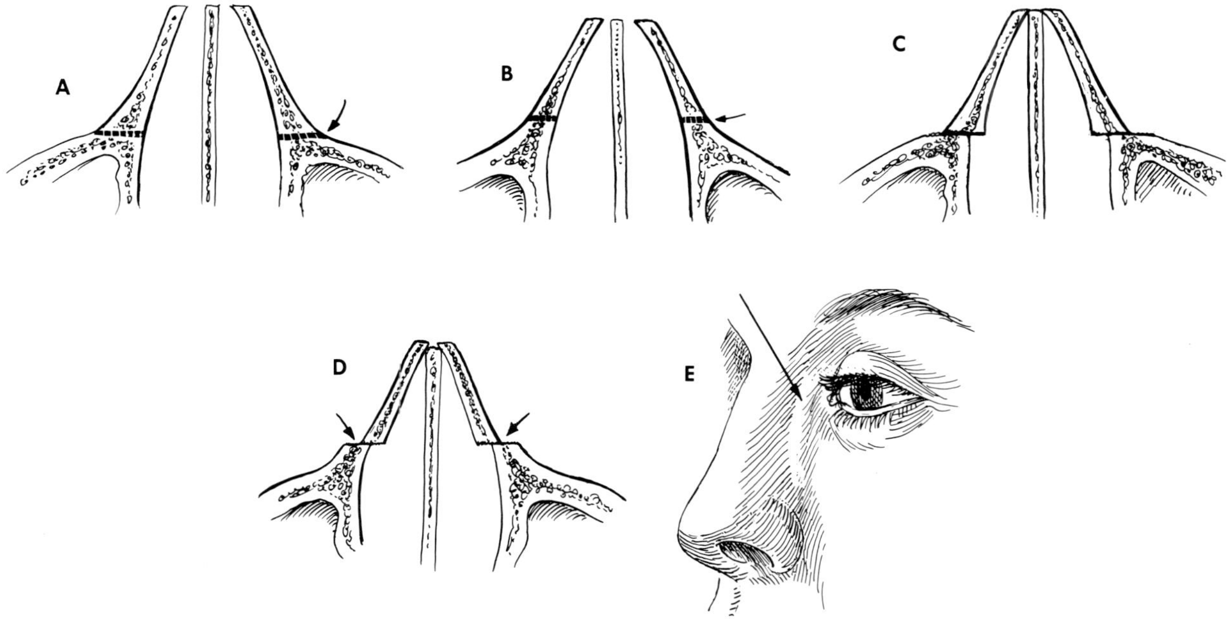

Fig. 11-2. The lateral osteotomy must be placed flush with the maxilla, **A, C,** so that the infracture will not result in a "stair-step" deformity (arrow). If the osteotomy is made at too high a level, **B,** such a stair step will result, **D** and **E.** (From Rees, T. D., Krupp, S., and Wood-Smith, D.: Secondary rhinoplasty, Plast. Reconstr. Surg. **46:**332, 1970.)

to the profile of the forehead and the nose. If the radix nasi is tangential to the frontosubnasal line, the angle should be left undisturbed. If, however, it lies in front of the frontosubnasal line, a deepening of the radix may be desirable. In some cases, a reconstruction of the forehead "bossing," as suggested by Gonzalez-Ulloa,[4] might be a more logical method of solving the problem, although alloplastic implants in this region are not favored by most plastic surgeons. Excessive hump removal and lowering of the radix may result in a distressing "bird beak" profile. Such a deformity is extremely difficult to correct and may require a bone graft.

Asymmetry or external deviation of the bony framework. Asymmetry is generally caused by a deviation of the bony septum, although it may be the result of an incomplete infracture or a failure to perform an adequate medial osteotomy on one side. Surgical comminution of the nasal bones, fracturing and moving the ethmoid plate and the nasal bones at the nasofrontal suture, as well as straightening of the septum may be required for correction. Percutaneous puncture with a 2-mm. osteotome is useful in accomplishing these fractures.

Excessive width of the bony framework. Postoperative widening of the bony framework is usually caused by a superior bend of the nasal septum that went unrecognized at the primary operation. Although this deformity can be hard to correct, fracturing and repositioning the septum in two or more parts may solve this problem. It may also be possible to produce incomplete narrowing of the nasal bones at the radix by performing an incomplete osteotomy at the frontal process of the maxillae, thus producing a "greenstick fracture" (Fig. 11-1).

Insufficient narrowing at the radix may occur from excessive thickness of the nasal bones. Selective resection of a medial triangle of bone (with osteotomes or bone-biting forceps) may be needed to allow room for enough medial displacement during the infracturing.[5, 6]

"Stair-step" deformity. "Stair-step" deformity is the result of placing the lateral osteotomy too high on the prefrontal process, rather than deep in the nasomaxillary groove through the thick portion of the nasofrontal process of the maxilla. "Stairstepping" may be corrected by comminuting the prefrontal process with a 2- or 3-mm. osteotome or by rasping off the bony ridge (Fig. 11-2).

Saddle nose. If infractured nasal bones drop into the piriform aperture, a saddle nose may re-

sult. To prevent this most serious complication, the periosteum should never be completely stripped off the nasal bones. Thus some soft tissue attachment will always remain to support the fractured bones.[7, 8] Repair of a saddle nose deformity is best accomplished with autogenous iliac bone grafts, since

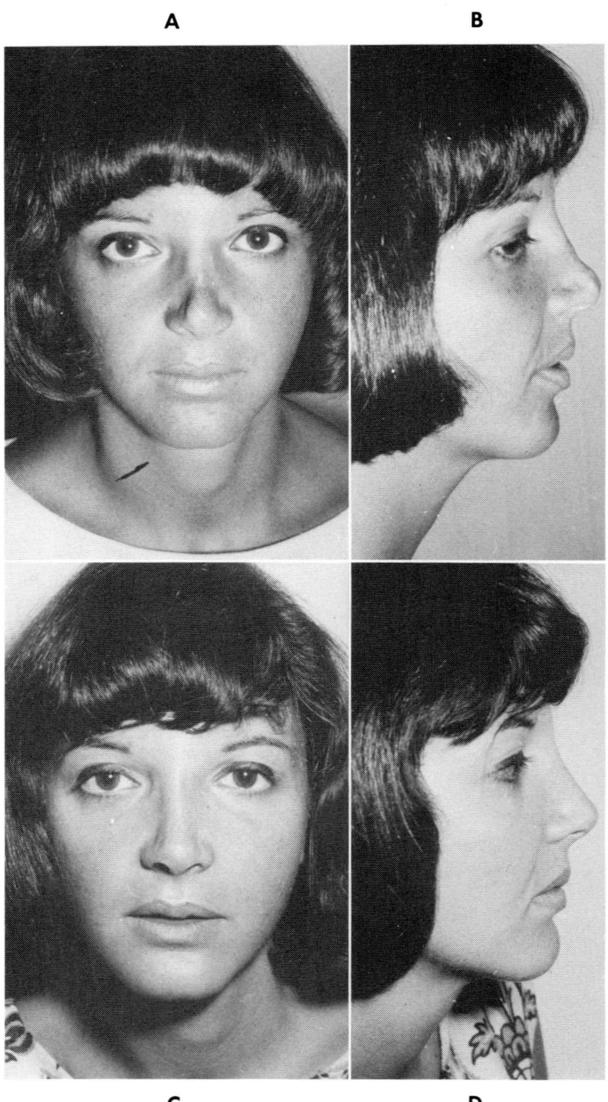

Fig. 11-3. A and **B,** This attractive young woman had several typical deformities following rhinoplasty performed elsewhere. The osteotomies were incomplete and apparently no infracturing was done. The septal angle was left at a high level and the alar cartilages asymmetrically trimmed. Excessive cartilage and dermis were removed from the right lateral crus with a pit depression. **C** and **D,** The postoperative results were obtained by lowering the septal angle, equalizing the alar cartilages, and obtaining adequate osteotomies with infracturing as shown in Fig. 11-1.

alloplastic materials are often extruded after insertion into the scarred soft tissue of the nasal dorsum.

Incomplete lateral osteotomy, incomplete fracture at the juncture of the nasal and frontal bones, and incomplete infracture. These are among the more common postrhinoplasty problems. Refracture with comminution of the bony complex will help in their repair (Fig. 11-3).

Palpable irregularities resulting from comminution of the nasal bones. If palpable irregularities are visible, they can be rasped or chiseled away. If they are not visible, they should be accepted, and no further correction should be attempted.

SECONDARY DEFORMITIES INVOLVING THE CARTILAGINOUS FRAMEWORK

The cartilaginous framework extends caudally from under the lower edge of the nasal bones. It supports approximately the lower half of the nose. Cartilage deformities are often more difficult to correct than those of the bone. Excluding deformities of the tip, which will be discussed in a separate section, deformities of the cartilaginous framework can be classified as follows.

External deviation of the nose. External deviation produced by deviations of the dorsal border of the septal cartilage forming C- or S-shaped curves that are very difficult to correct. Straightening these curvatures may involve complete freeing of the septum along the vomerine groove and multiple incisions in the deviated cartilage. It is preferable not to remove any cartilage unless it is necessary to obtain an adequate airway.

Radical submucous resection of the ethmoid plate and vomer, as well as the quadrangular cartilage, is not recommended during any rhinoplasty

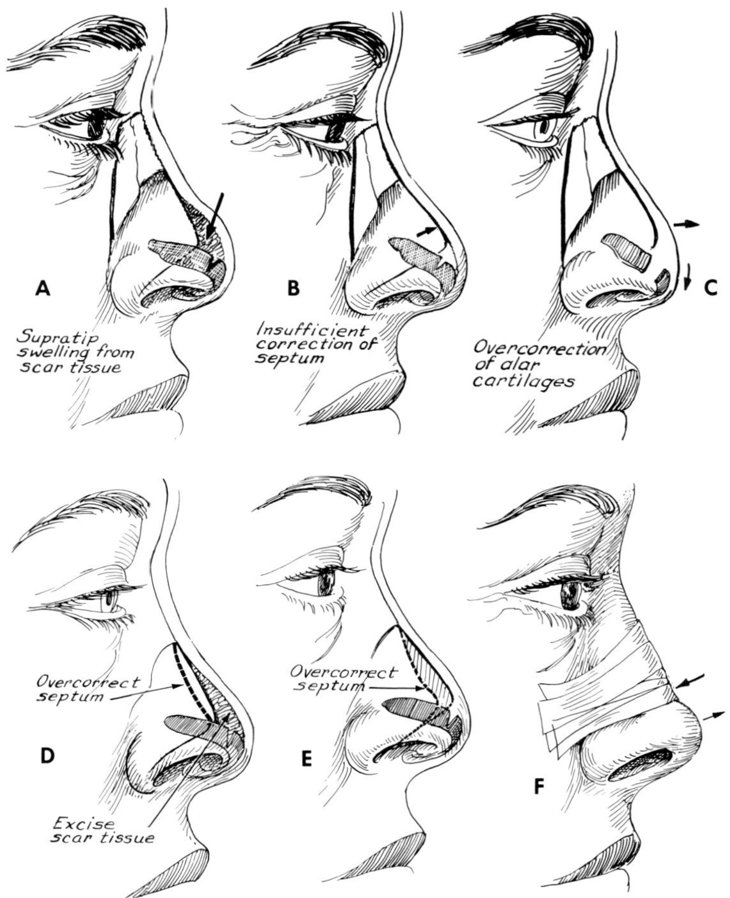

Fig. 11-4. A to **C,** The common causes of supratip swelling or "ram's tip." **D** to **F,** The remedy of this situation. Postoperative tape strapping is helpful in such cases. When the entire alar dome has been removed, correction is then difficult and sometimes impossible. (From Rees, T. D., Krupp, S., and Wood-Smith, D.: Secondary rhinoplasty, Plast. Reconstr. Surg. **46:**332, 1970.)

—primary or secondary—because of the danger of collapse of the nasal pyramid into the piriform aperture. If radical resection is indicated, it should be performed as a separate procedure at a later date. Usually, however, a crooked nose can be sufficiently straightened by repositioning the septal fragments, thus avoiding the more radical procedure.

Postoperative convexity of the supratip area. This convexity, frequently referred to as a "polly tip" or "parrot beak," may result if the septal dorsum was not sufficiently lowered during the primary procedure. It is probably the most common postrhinoplasty problem requiring secondary surgery. Although it is usually caused by a high septal border, it may also result from (1) insufficient trimming of the dorsal borders of the upper lateral cartilages, (2) excessive resection of the alar cartilage domes, (3) insufficient trimming of the septal mucosa, (4) inherent thickness of the skin and subcutaneous tissue, or (5) a short columella (Fig. 11-4). For example, if the septal mucoperichondrium is not trimmed lower than the dorsal border of the septum, it can grow over the dorsum to unite with the opposing mucosa and become interposed between the covering skin and the septal cartilages. Supratip swelling may also be accentuated by the formation of granulation tissue in the dead space between the dorsal skin flap and the septal border, with the granulation tissue resulting in scar formation. The thicker the skin of this area, the more obvious the deformity will be.

Since correction of the "polly tip" is difficult, many ways to avoid the deformity have been discussed. To avoid the "parrot beak" in the patient with thick skin, Wright[9] advised shaping the nasal tip before reduction of the septal height and removal of the bony hump. Pitanguy[10] believes the bulbous nose contains a fibrous ligament between the dermis and the alar cartilages that must be resected or divided to prevent supratip swelling after rhinoplasty.

Correcting a postoperative supratip elevation may be quite complicated and frequently involves trimming of the high dorsum, excision of scar tissue, and lowering of the upper lateral cartilages (Fig. 11-5). Rearranging the alar cartilages and lengthening the columella to increase tip projection may be helpful. It is not advisable to implant alloplastic materials or autogenous grafts to increase tip projection, since extrusion or resorption is the usual end result (Fig. 11-6).

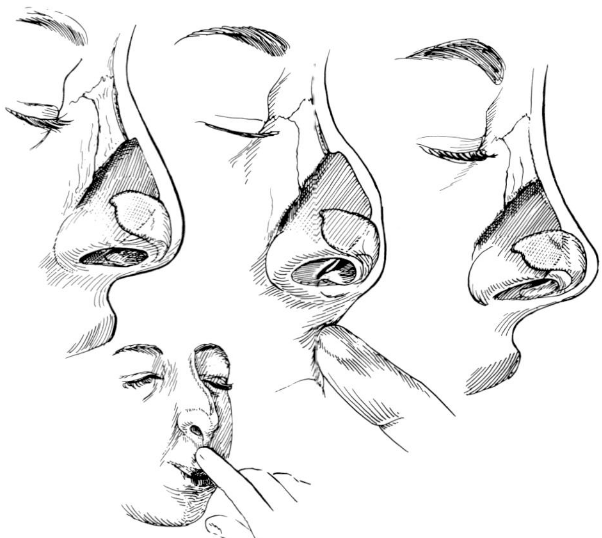

Fig. 11-5. A helpful test to determine whether the dorsal border of the cartilaginous septum has been sufficiently trimmed. If this border is redundant, downward traction on the nasal tip by the operator's finger readily demonstrates the excess, which should then be further resected.

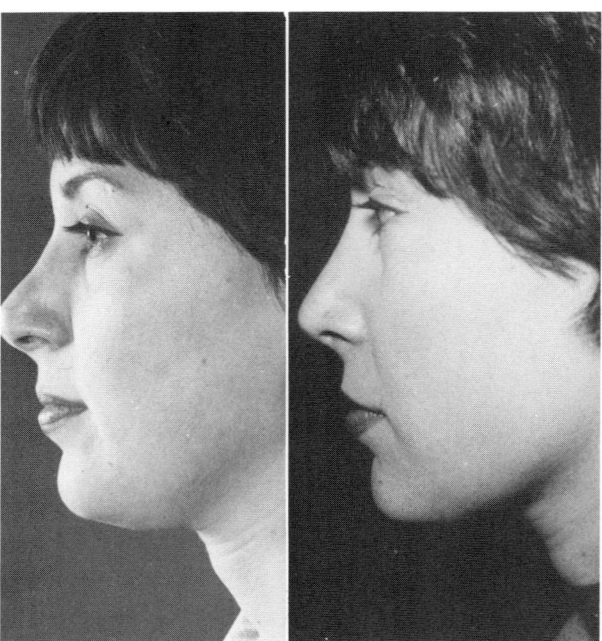

Fig. 11-6. This patient was left with a "ram's tip" after surgery resulting from insufficient resection of the dorsal septal edge. Partial correction was achieved by lowering the dorsal border; however, the results were limited because the alar domes had also been resected. Definition between the tip and the nasal dorsum could not be obtained.

Unsightly oblique grooves on either side of the nasal dorsum. Such grooving, usually caused by excessive removal of the upper lateral cartilages, is almost impossible to correct.

Palpable or visible irregularities caused by uneven trimming of the septum or upper lateral cartilages along their dorsal borders. These visible irregularities can be corrected by secondary trimming.

SECONDARY DEFORMITIES INVOLVING THE NASAL TIP

Asymmetry or pinching. Asymmetry or pinching are easily recognizable stigmata associated with plastic surgery of the nose. They are the result of injudicious carving of the alar cartilages along with, in the case of pinching, excision of vestibular lining.

Sharp points, asymmetry of the domes, and the "boxed" or square tip. Correction of these irregularities of the alar cartilage remnants requires secondary trimming of the offending tissue. The "boxed" tip may call for a total resection of the alar domes, a technique safer in the secondary than in the primary rhinoplasty since the added support of the cicatrix makes collapse or pinching less likely to occur. It is a known fact that patients with thick skin or subcutaneous tissue can afford to lose most or even all of the lateral crura of their alar cartilages without collapse of the nostrils. However, in patients with very thin skin, trimming of the alar cartilage must be careful, conservative, and accurate, with emphasis on shaping.

If lining has been sacrificed, correction is more difficult. Compensation for soft-tissue loss may be accomplished by Z-plasty, but local flaps, cartilage grafts, composite grafts, or full-thickness grafts may be needed to cover the raw surfaces and resection of the scar and may be required to provide support to the vault. These procedures often work better in the literature than in the living patient. Repairing the "crucified tip" is undoubtedly the most difficult, and usually the most disappointing task in secondary rhinoplasty. When complete vestibular collapse has occurred, support can only be provided by a prosthetic insert that provides an airway at the critical junction of the nasal cartilages and the septum.

The "drooped" tip. The drooped tip occurs if the upper lateral and alar cartilages have not been trimmed or sutured in suitable relationship to the shortened caudal end of the septum. The nose may also appear elongated if too much hump has been resected or if the nose was shortened without tilting the tip.[5] Overshortening of the nose in older patients in whom skin is less contractile may also cause a drooping tip.

Correcting the drooped tip may be accomplished by shortening the upper lateral cartilages or approximating them to the alar cartilages. Frequently, fixation of the medial crura of the alar cartilages to the lower septal border by an "orthopedic suture" through the medial crura is required. If the alar cartilages plunge downward naturally, it may be necessary to resect the domes and reposition the medial and lateral crura in a new relationship.

Recessed and elevated tip with undue nostril exposure. This secondary deformity may be caused by excessive shortening of the septum and/or the upper lateral and alar cartilages. In such patients, the entire membranous septum has often been resected. The difficulty in correcting this defect lies

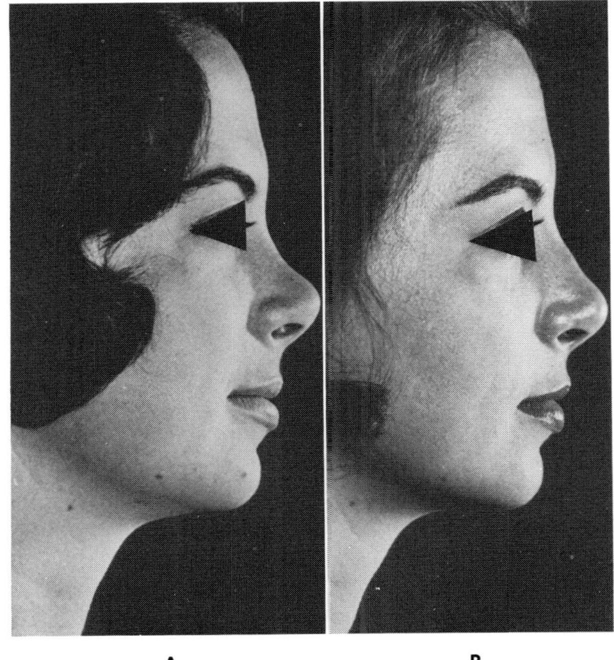

A   B

Fig. 11-7. A, Overshortening of the nose is one of the most difficult problems to correct at a secondary operation. Usually too much of the caudal edge of the septum was sacrificed. **B,** Some improvement was obtained in this patient by freeing the soft tissues from the nasal bony and cartilaginous skeleton, lowering the septal angle, and suturing the tip at a lower level. Sometimes composite grafts gain length when inserted into the membranous septum or turn-over flaps of lateral lining (Millard).

in the fact that usually too much tissue has been sacrificed in the primary procedure. If the nasal spine is prominent, trimming it may give the illusion of lengthening the nose, although this may often be accomplished only at the expense of an unwanted lengthening of the upper lip.

Undermining all of the soft tissue after a complete transfixion incision sometimes permits the degloved nose to be "stretched" downward and fixed in a lowered position with sutures (Fig. 11-7). Severe cases associated with saddling may require bone grafts to the dorsum or composite grafts[11] to the membranous septum to maintain additional length. Sometimes, if the septum has not been made too short, a septal flap can be fashioned for transfer to the columella.

Bifidity of the tip. Bifidity requires suturing the alar domes, or the remnants of the domes, together, a procedure more easily described than done.[5]

Many secondary deformities of the nasal tip

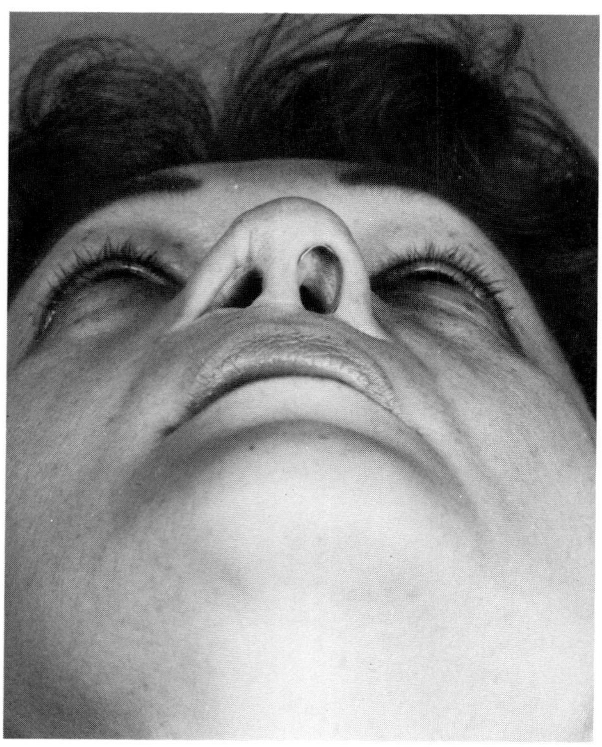

Fig. 11-8. The "mutilated tip" in which excessive tissue has been resected, usually including lining and alar cartilage, is perhaps the most difficult problem in secondary rhinoplasty. It is often insolvable. When the airway has been destroyed it is sometimes necessary to have the patient wear an acrylic hollow stent to ensure respiration.

cannot be surgically corrected. In fact, surgery can often compound the problem by adding to the already existing damage. Unfortunately, many patients with thickened, asymmetrical, or scarred nasal tips (particularly those in whom lining has been lost) must be advised against further surgery.

SECONDARY DEFORMITIES INVOLVING THE NOSTRILS, VESTIBULUM, AND COLUMELLA

Cicatricial stenosis, adhesions, synechiae, and retraction of the nostrils. These defects are usually caused by secondary healing of raw surfaces left when lining was sacrificed.[12-14] Correcting such scar defects is exceedingly difficult. Rearranging or replacing soft tissue lining by Z-plasty, local flaps, or skin grafts may be required, along with various procedures designed to reproduce the cartilaginous support. Most such techniques utilize cartilage or alloplastic implants, and the results are frequently disappointing (Fig. 11-8).

Excessive flaring of the nostrils. Flaring of the nostrils may be accentuated by nasal plastic surgery if the tip is recessed. Alar base resection can correct this defect.

"Hanging" columella. This convexity or roundness of the caudal margin of the medial crura of the alar cartilages often becomes more apparent after rhinoplasty. It can be remedied by marginal excision of the columella.

Flaring of the feet of the medial crura. This flaring, along with an abundance of soft tissue at the base of the columella, can obstruct inspiration. Resection of the soft tissue between the medial crura and the cartilaginous feet will correct the problem.

Retraction of the columella. Excessive removal of the membranous septum and columella may cause this deformity, which is very difficult, and often impossible, to repair. If the retraction is minor, a cartilage septal implant, as suggested by Millard,[15] may be effective. Replacing soft tissue and cartilage with a composite graft in the region of the missing membranous septum—the "banana split" operation[11]—is often the only effective method of repairing the retracted columella.

Aufricht[16] has recommended cartilage implants in the region of the nasal spine to correct an acute angle. Such a technique may also be of value in improving retraction of the columella resulting from removal of the nasal spine or correction of flatness of the upper lip that resulted from overambitious surgery of the nasal spine and inferior border of the septum.

Irregularities or thickening of the nostrils. These deformities, which may be accentuated after rhinoplasty, can be improved by selective resection of the nostril rims using techniques advocated by Millard[15] and by Pollet and Baudelot.[17]

SUMMARY

Secondary deformities following rhinoplasty present the plastic surgeon with many technical and psychologic problems, often of great complexity. Some of the more common defects that may follow nasal plastic surgery have been discussed, along with some corrective procedures.

REFERENCES

1. Smith, T. W.: As clay in the potter's hand, Ohio Med. J. **63**:1055, 1967.
2. Lipsett, E. M.: A new approach to surgery of the lower cartilaginous vault, Arch. Otolaryng. **70**:42, 1959.
3. Aufricht, G.: Surgery of the radix and bony nose: preliminary report of a new type of nasal clamp, Plast. Reconstr. Surg. **22**:315, 1958.
4. Gonzalez-Ulloa, M.: Quantitative principles in cosmetic surgery of the face (profileplasty), Plast. Reconstr. Surg. **29**:186, 1962.
5. Converse, J. M., and others: Deformities of the nose. In Converse, J. M., editor: Reconstructive plastic surgery, vol. II, Philadelphia, 1964, W. B. Saunders Co., pp. 694-828.
6. Levignac, J.: Des petits et des gros ennuis dans la rhinoplastie, Ann. Otolaryng. **75**:560, 1958.
7. O'Connor, G. B., and McGregor, M. W.: Secondary rhinoplasties: their cause and prevention, Plast. Reconstr. Surg. **15**:404, 1955.
8. Lewis, M. L.: Prevention and correction of cicatricial intranasal adhesions in rhinoplastic surgery, Arch. Otolaryng. **60**:215, 1954.
9. Wright, W. K.: Study on hump removal in rhinoplasty, Laryngoscope **77**:508, 1967.
10. Pitanguy, I.: Surgical importance of a cartilaginous ligament in bulbous noses, Plast. Reconstr. Surg. **26**:247, 1965.
11. Dingman, R., and Walter, C.: Use of composite ear grafts in correction of the short nose, Plast. Reconstr. Surg. **43**:117, 1969.
12. Wexler, M.: Post-rhinoplastic complications, Eye, Ear, Nose, Throat Monthly **31**:553, 1952.
13. Maliniac, J. W.: Prevention and treatment of sequelae in corrective rhinoplasty, Plast. Reconstr. Surg. **18**:213, 1956.
14. Cohen, S.: Complications following rhinoplasty, Plast. Reconstr. Surg. **18**:213, 1956.
15. Millard, D. R.: Adjuncts in augmentation mentoplasty and corrective rhinoplasty, Plast. Reconstr. Surg. **36**:48, 1965.
16. Aufricht, G.: Rhinoplasty and the face, Plast. Reconstr. Surg. **43**:219, 1969.
17. Pollet, J., and Baudelot, S.: Sequelles de la chirurgie ésthetique de la base du nez, Ann. Chir. Plast. **12**:185, 1967.

Chapter 12

Management of acute nasal trauma

William C. Grabb, M.D.

Since the nose is the most prominent part of the face and since the nasal bones require the least amount of force of all the facial bones to produce a fracture, it is not surprising that nasal fractures are the most common facial bone fracture. Schultz[1] reported that nasal bone fractures constituted 37% of over 1,000 consecutively diagnosed facial fractures. While the nasal bones are commonly fractured, the same amount of force allows the more flexible nasal cartilages to absorb the impact force.

Lacerations of the soft tissue about the nose are also common. They will be dealt with only briefly, since the principles of repair of these lacerations are common to all facial soft tissue injuries. It should be emphasized that all wounds should be closed by suture or, in the case of avulsion, by a split-thickness skin graft. This includes the oft-neglected intranasal laceration, which should be sutured with a few fine catgut sutures and the remaining mucosal edges approximated by intranasal petrolatum or Carbozine gauze packs left in place for 3 to 5 days. One often unrecognized intranasal laceration occurs at the caudal edge of the nasal bone, where the thin sharp bony edge is quite capable of cutting through mucosa when the lower half of the nose is forced up and backward.

TYPES OF NASAL FRACTURE

Before proceeding to the diagnosis and treatment of acute nasal bone fractures, it would be well to describe the various types of nasal fractures.

Laterally displaced nasal fracture. The common laterally displaced nasal fracture is usually caused by being struck with a fist or elbow on the side of the nose. One or both nasal bones are displaced laterally. Kazanjian and Converse[2] point out that the fracture of only one nasal bone usually occurs in children, since the nasal bones do not become fused in the midline until adolescence. Though most authorities state that only the nasal bones are fractured in these patients, it appears that in most instances the nasal bones as well as the frontal processes of the maxilla are fractured.

In the laterally displaced fracture, there is a spectrum from a minimal deviation with stability of the fragments to a marked lateral deviation with easily movable fragments (Fig. 12-1). Generally these fractures are stable, so that on palpation they are quite "sticky" and move only minimally between the examiner's thumb and index finger. The bone and cartilage of the nasal septum are also fractured in the more severely displaced fractures.

Depressed and comminuted nasal fractures. This smash fracture is most commonly caused by the trauma of an automobile accident. The violent force of the nose impacting with the interior of the vehicle is often sufficient to comminute the nasal bones, frontal process of the maxilla, nasal septum, lacrimal bones, and ethmoid sinuses. Lateral splaying of the eggshell fragments of the nasal bones and frontal process of the maxilla may also occur. It is convenient to divide depressed and comminuted nasal fractures under the following three headings: nasal fracture associated with maxillary and zygomatic fractures, nasoorbital fracture, and nasal fracture alone.

Nasal fracture associated with maxillary and zygomatic fractures. In a vehicular accident, when the blunt trauma of the face striking the vehicle's interior is severe enough to cause a maxillary fracture, it is common to also have a nasal fracture and

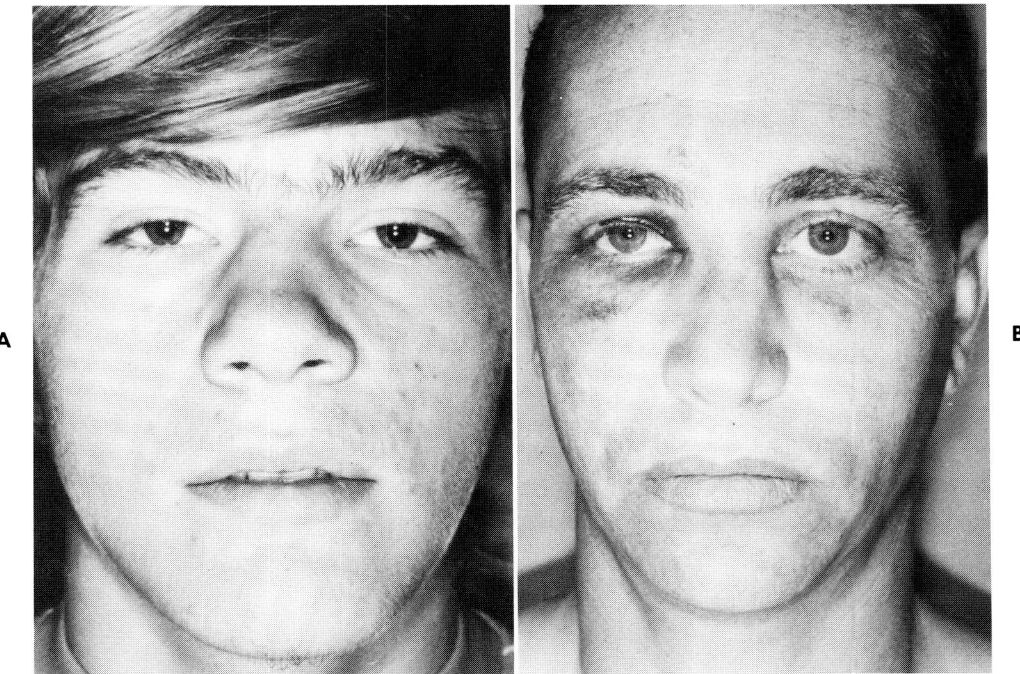

Fig. 12-1. Examples of laterally displaced nasal fractures. **A,** The nasal bones are only minimally deviated to the right. **B,** The nasal bones and septum are moderately displaced to the left.

one zygoma fractured (Fig. 12-2). Again there is a spectrum of nasal deformities from a minimal fracture to such severe comminution that it is not possible to restore the eggshell fragments of the nose into a normal nasal framework. A secondary bone graft to the dorsum of the nose will then be necessary.

Nasoorbital fractures. Because of the thin matchbox-like quality of the nasal septum, ethmoidal air cells, and medial orbital walls, the force of the head striking the shelf-like edge of the automobile dashboard or a rigid steering wheel can break through the weak resistance of the nasal bones and frontal process of the maxilla. The nasal structures are driven into the interorbital space, producing lateral displacement of the interorbital bony structures. The medial palpebral ligaments may be pushed laterally, either attached to a fragment of the lacrimal bone or severed, avulsed, and detached completely. The lacrimal drainage system may be severed at the level of the canaliculi, the common canaliculus, the lacrimal sac, or the nasolacrimal duct. The nasal pyramid may remain relatively intact (Fig. 12-3, *A* to *D*) or may be severely comminuted (Fig. 12-3, *E*).

It is our prediction that the nasoorbital frac-

ture is becoming less common as automobiles in recent years have been designed with an energy absorbing padded convex sheet-metal dashboard (like the convex surface of a tin can) and have been equipped with a lap *and shoulder belt.*

Nasal pyramid fractures without associated injuries. The nasal pyramid (nasal bones and frontal process of the maxilla) or the nasal bones alone can be fractured and comminuted by a frontal blunt force. This frontal trauma may cause any or all of the following fractures: at the midline suture of the nasal bones, at the nasomaxillary suture, at the base of the frontal process of the maxilla, in the septum, at the junction of the thick cephalad half and the thin caudal half of the nasal bones, and at the very caudal tip of the nasal bones. This last-mentioned fracture of the tip of the nasal bones is one that is diagnosed on the lateral roentgenograms, but it is not apparent on physical examination and does not require treatment.

Telescoping of the nasal septum. An uncommon but frequently poorly diagnosed and treated traumatic nasal deformity is the upward displacement of the cartilaginous nasal septum beneath the nasal bones. The cartilaginous septum evi-

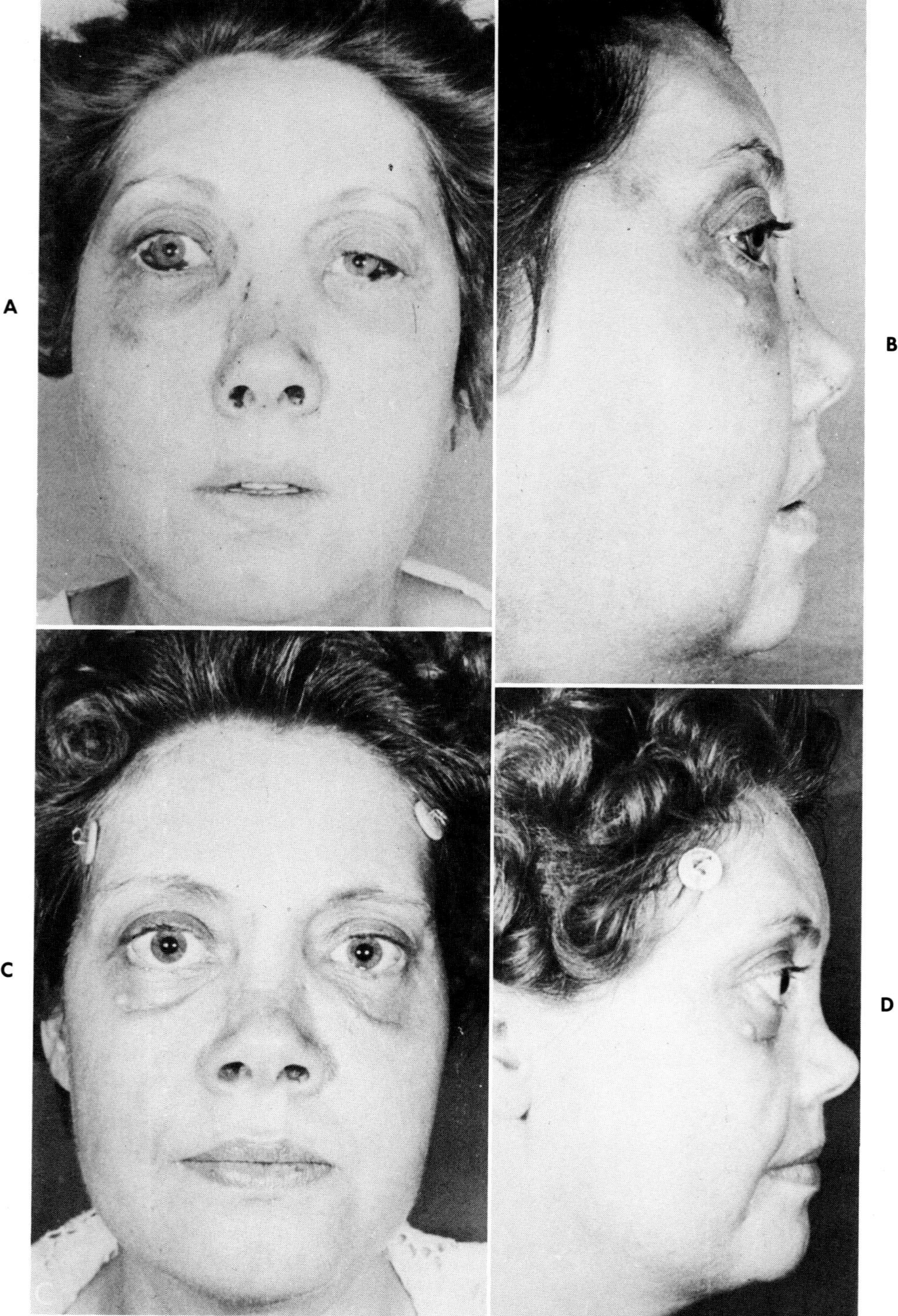

Fig. 12-2. This woman sustained a fracture of the maxilla, left zygoma, and nasal bones in an automobile accident. Fractures of the maxilla and zygoma were treated by open reduction, internal wire fixation, craniofacial suspension, and intermaxillary fixation. Despite closed reduction of the comminuted nasal bones there was not sufficient bone available to restore the normal height of the dorsum of the nose. **A** and **B,** Preoperative photographs. **C** and **D,** Postoperative photographs taken 1 month after treatment.

dently is fractured vertically and separated from the nasal bone at the dorsum of the nose, permitting the caudal segment to be forced beneath the nasal bones.

The diagnosis can be made by palpating the step off at the junction of nasal bones and septum on the dorsum of the nose, with the obvious short-ening of the length of the nose and the increased nasolabial angle on profile.

DIAGNOSIS

History. The type of trauma is of definite significance in the diagnosis and treatment of nasal fractures. As has already been indicated, a lateral

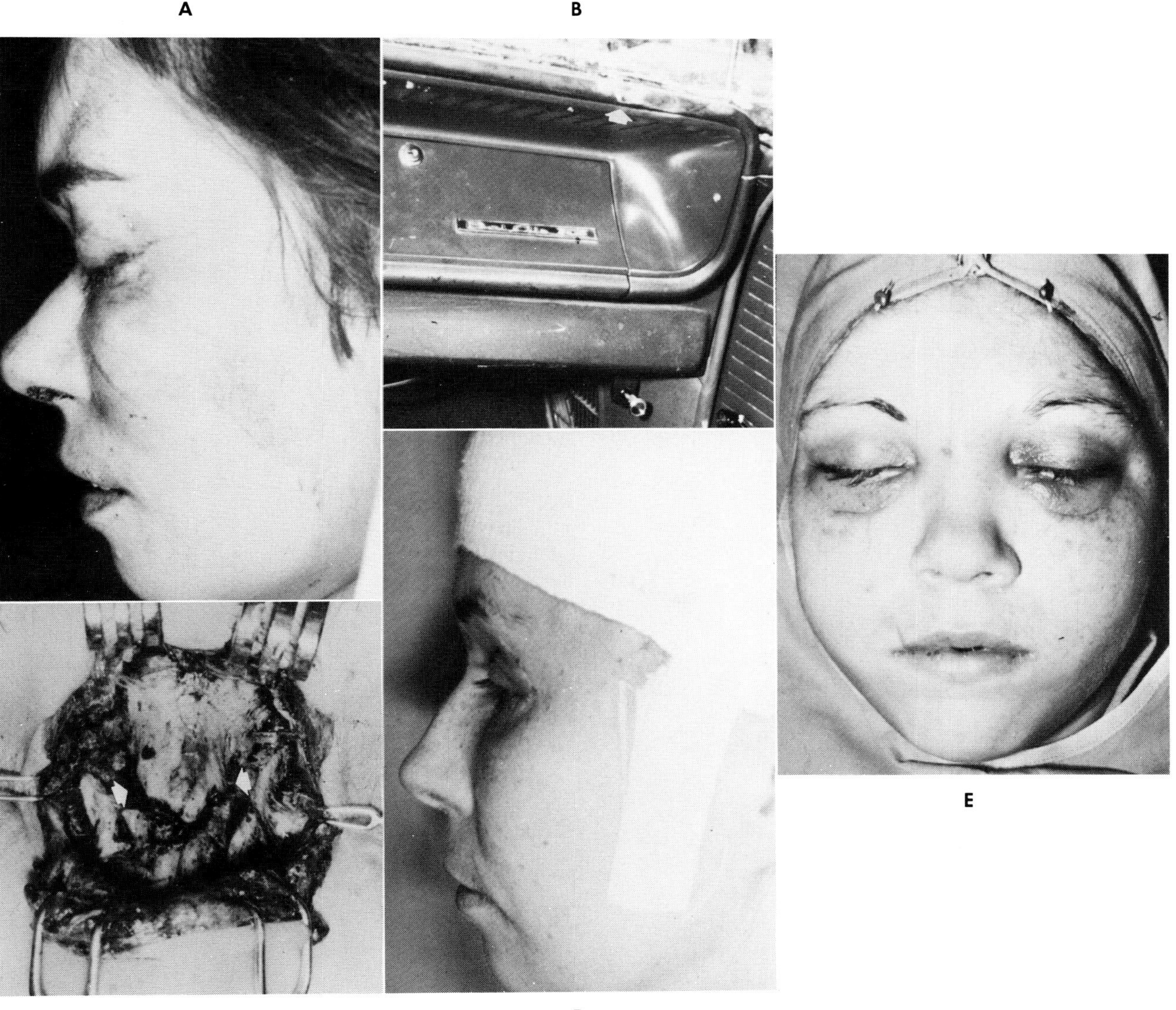

Fig. 12-3. Nasoorbital fracture in which the nasal bones remained intact except for a transverse fracture near the glabella. **A,** Marked inward displacement of the nasal bones. **B,** Arrow points to the shelf-like edge of this dashboard on which the patient who wore a lap seat belt fractured her nose, medial orbital walls, cribiform plate, and orbital roofs. **C,** Arrows point to transverse fractures of nasal bones that were openly reduced and fixed with interosseous wires through the overlying skin laceration. **D,** Patient's appearance 2 weeks after combined reduction of nasal bone fractures and bifrontal craniotomy for reduction of orbital roof fractures and repair of dural lacerations. **E,** Nasoorbital fracture with severe comminution of nasal bones, frontal process of maxilla, and medial orbital walls with detachment of the right medial canthal ligaments.

blow by an elbow, a fist, or a fall produces a lateral displacement of the nasal bones that can be treated by a closed reduction in the emergency department. In contrast, the history of a more severe impact in an automobile accident would cause us to expect a depressed and comminuted fracture.

The length of time between injury and examination is of interest because of the increased swelling of the soft tissues of the nose that may occur as the hours pass by.

The history of previous operations on the nose and especially the appearance of the nose prior to injury are of importance. A number of patients will have a nasal deformity from previous trauma. By having the patient look in a mirror, he can usually tell you if the nasal deviation is the same one that he has had for several years or something new.

Physical examination. *Inspection* of the nose, even from across the room, is often all that is required to make the diagnosis of a nasal bone fracture. As has been previously mentioned, both the physician and the patient should inspect the nose to rule out a malunion of the nasal bones related to prior trauma.

Palpation of the nasal bones between the examiner's thumb and index finger is of great diagnostic assistance in detecting lateral deviation, crepitation, and mobility of the nasal bones. Even when there is marked soft tissue swelling over the nasal bones, gentle but firm palpation can be of assistance in making the diagnosis of a fracture.

Roentgenograms. The need for roentgenograms to make the diagnosis of a nasal bone fracture is greatly overrated. Foman and his colleagues[3] reported that only 50% of patients with a clinically

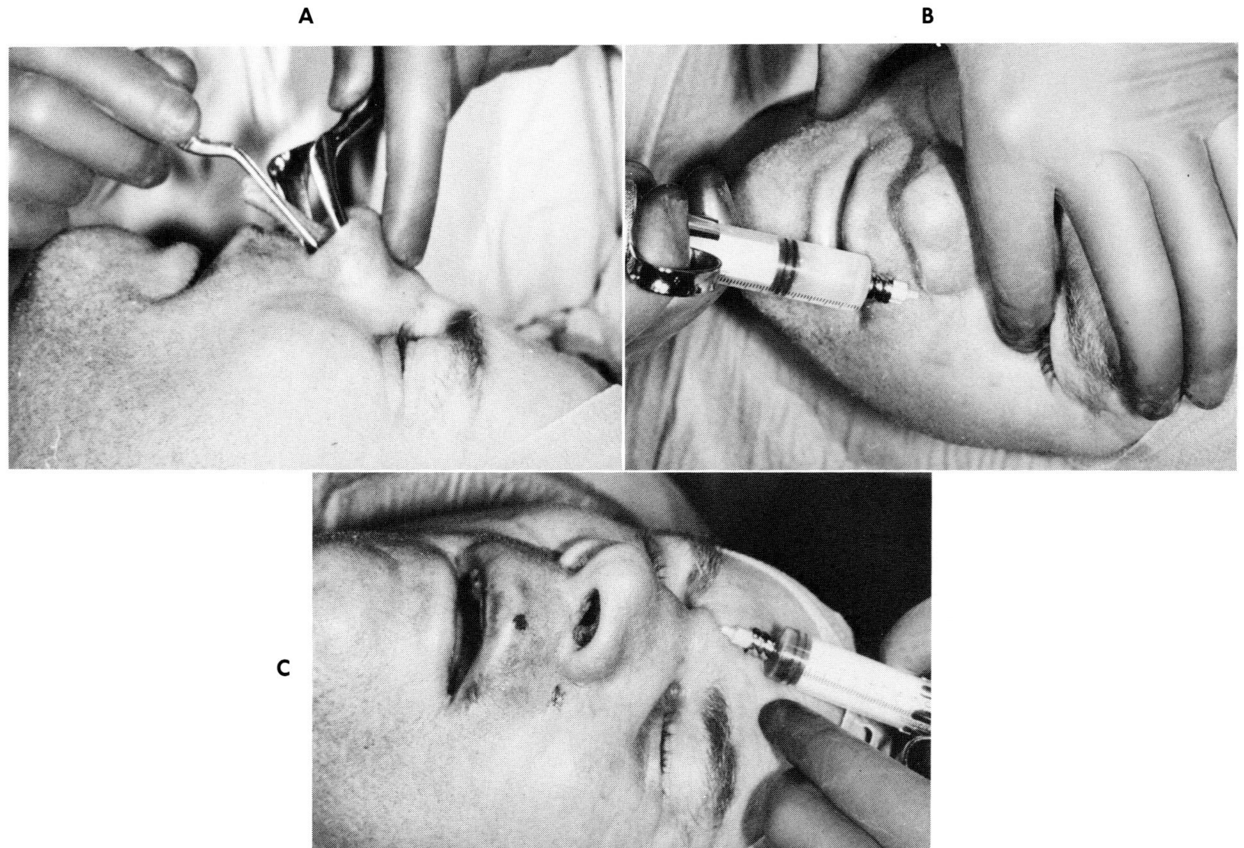

Fig. 12-4. Local anesthesia of the nose can be carried out in the emergency department. **A,** Insertion of three cotton nasal packs with 5% cocaine on each side of the nasal septum. **B,** Bilateral infraorbital nerve block with 1% lidocaine with 1:100,000 epinephrine. **C,** Infiltration beneath the skin over the nasal bones with 1% lidocaine with 1:100,000 epinephrine.

evident nasal fracture had radiologic evidence of fracture.

Though the diagnosis of a nasal fracture can best be made on physical examination, if one wants to increase the chances of visualizing the fracture on roentgenograms, then the lateral view, Water's view, and a view in which the x-ray beam is directed from cephalad downward and onto an occlusal film held under the palate between the patient's incisor teeth should be made.

TREATMENT

Treatment of the various types of nasal fractures will be discussed with emphasis on the laterally displaced nasal fracture.

Treatment of the laterally displaced nasal fracture

Timing. It is easier to reduce a nasal fracture when there is little or no swelling of the soft tissues of the nose. I do not consider a moderate amount of swelling a contraindication to carrying out a closed reduction. Usually it is a matter of finding the time in one's schedule. In most instances I will reduce the fracture on the day of injury or schedule this to be done 1 or 2 days later.

It is usually possible to manually reduce a laterally displaced nasal fracture up to 2 to 3 weeks after fracture. Aloin[4] purposely waits 2 to 3 weeks so that "the repositioning takes place at the same time as nature sets about healing the fracture."

After 3 to 4 weeks, the untreated nasal fracture becomes so solidly healed that a nasal bone osteotomy is required to position the bones in the midline. A 3-mm. wide chisel can be pushed through the skin and used to perform an osteotomy at the junction of the planes of the frontal process of the maxilla and the body of the maxilla. Then with an osteotomy at the suture line between the frontal bone and nasal bone–frontal process of the maxilla, it is possible to bring the nose back into the midline. In older malunion fractures an osteotomy along the nasal dorsum between the nasal bones and the nasal septum may be necessary.

Anesthesia. Children, particularly those under 10 years of age, are admitted to the hospital and scheduled for reduction of the nasal fracture under general anesthesia. Older children and adults are given meperidine hydrochloride, 50 mg. intramuscularly, and then 20 or 30 minutes later the nose is anesthetized with local anesthetic agents. We use intranasal packs with 5% cocaine, fol-

lowed by infraorbital nerve blocks and infiltration over the dorsum of the nose with 1% lidocaine with 1:100,000 epinephrine (Fig. 12-4). From 15 to 20 minutes should be allowed to elapse for the anesthetic to take effect before reducing the fracture.

Operative technique. The operator's fingers are the best tools to reduce a laterally displaced nasal fracture. By placing both thumbs on the side of the nasal bones toward which the nose is deviated, the deformity is overcorrected to the opposite side (Fig. 12-5). *Overcorrection* is the most important single point in the treatment. The nasal bones are then pushed back into the midline or, in severe lateral displacements, they are splinted in the slightly *overcorrect* position. If the nasal bone that was pushed on with the thumbs has been forced too far medially, then a heavy elevator can be inserted through the nostril on that side to elevate the bone into an appropriate position.

For fixation, intranasal packs of Carbozine and an external plaster splint are often used (Fig. 12-5, *D*). These packs or splint will not provide much fixation, though, and many be omitted when the nasal bones are minimally displaced. The splint probably acts principally to protect the nasal bones from additional trauma. The packs are removed in 3 to 5 days and the splint in 1 week.

There is often a tendency for the lateral deviation to recur postoperatively to a less severe degree. Thumb pressure on the nose by the patient several times a day for 2 to 3 weeks to counteract this tendency may provide a happy outcome. All this is related to breaking up the interlocked stresses on one side of the nasal septal cartilage, as has been shown so well by Fry.[5] Breaking the continuity of the cartilage on one side causes it to curve or deviate toward the opposite side. In a postmortem study of individuals with laterally deviated noses, it was found that the septal cartilage had been broken on one surface with deviation of the septum and nasal bone framework to the opposite side.[5]

Many athletes who engage in contact sports have a strong desire to continue playing despite the fact that it will require about 4 weeks for the fractured nasal bones to attain some solidity. As long as the patient understands that his nose may be broken again more easily, there is no reason to prohibit him from playing soon after fracture reduction, preferably with a protective mask. When the nasal fracture occurs within 2 or 3 weeks of the end of the playing season, treatment can be delayed for that period of time.

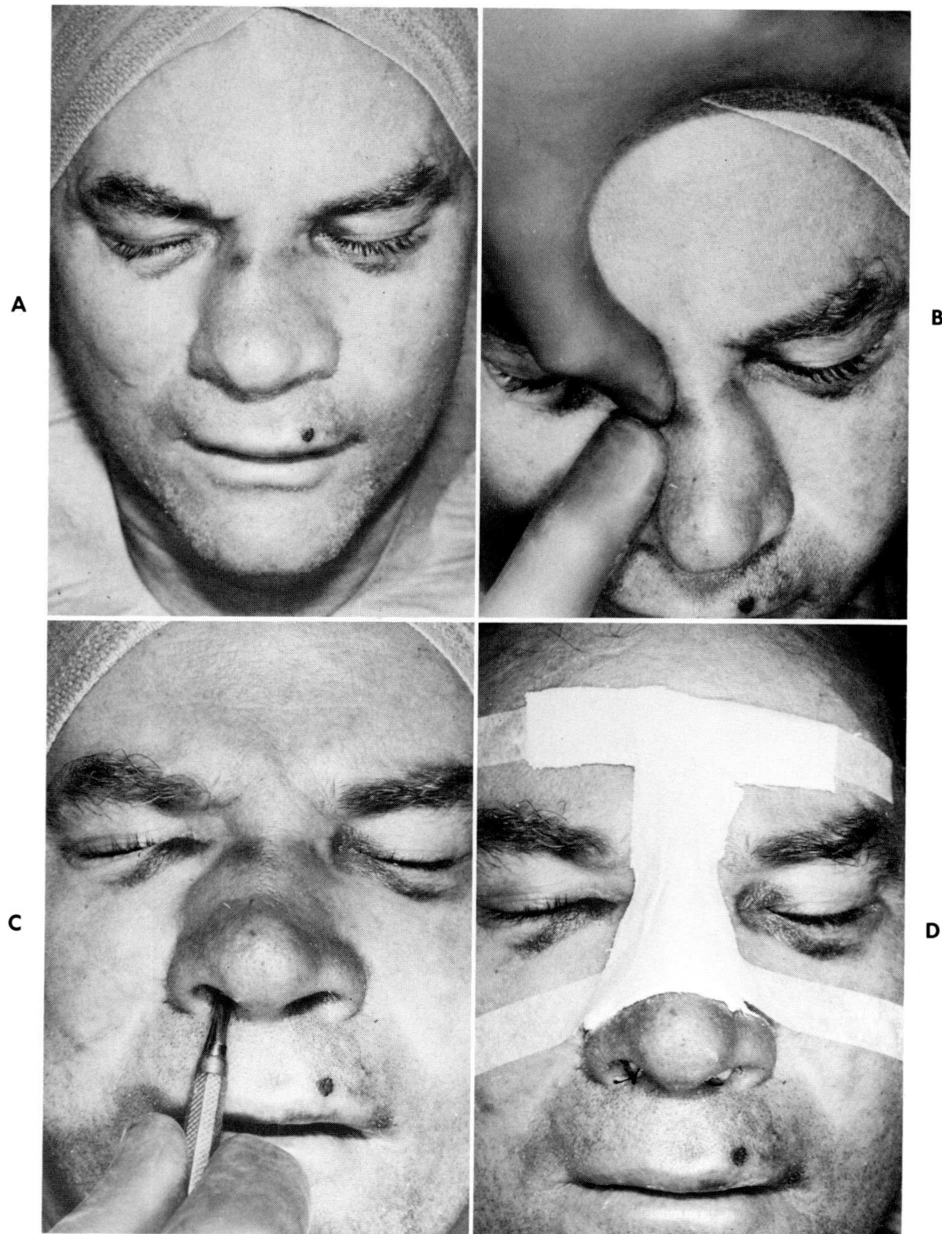

Fig. 12-5. Closed reduction of a laterally displaced nasal fracture. **A,** Appearance of nasal bones after anesthesia has been obtained. **B,** Overcorrection of the lateral deviation by manual pressure. **C,** A heavy elevator is inserted through the right nostril to elevate the right nasal bone, which has been pushed medially in the process of reducing the fracture. **D,** Intranasal packs and the external plaster splint in place.

Treatment of the depressed and comminuted nasal fractures

For the treatment of nasoorbital fractures the writings by Converse and Smith[6] and Dingman, Grabb, and Oneal[7] are recommended.

The treatment of the depressed nasal fracture in association with maxillary and zygomatic fractures involves the elevation of the comminuted fragments with an Asch forceps. In these patients, the nasal packs can be most helpful in immobilizing the fracture fragments.

Treatment of telescoping of the nasal septum

With a towel clip, the nasal septum can be brought down into its normal position. To maintain the septum in position, a 2-0 nylon suture can be passed through the skin over the dorsum of the septum near its junction with the nasal bones. The needle then continues through the cartilaginous septum and out through the skin and is then tied over a metal splint to hold the septum up for 10 to 14 days. The metal splint rests over the nasal bones for support and has two parallel cuts at its dorsal caudal border, over which the suture can be tied.

REFERENCES

1. Schultz, R. C.: Facial injuries, Chicago, 1970, Year Book Medical Publishers, Inc.
2. Kazanjian, V. H., and Converse, J. M.: The surgical treatment of facial injuries, Baltimore, 1959, The Williams & Wilkins Co.
3. Foman, S., and others: Management of recent nasal fractures, Arch. Otolaryng. 55:321, 1952.
4. Aloin, H.: A propos du traitement des fractures du nez, note de technique chirurgicale, J. Franc. Otorhinolarying. 1:302, 1952.
5. Fry, H.: Nasal skeletal trauma and the interlocked stresses of the nasal septal cartilage, Brit. J. Plast. Surg. 20:146, 1967.
6. Converse, J. M., and Smith, B.: Naso-orbital fractures and traumatic deformities of the medial canthus, Plast. Reconstr. Surg. 38:147, 1966.
7. Dingman, R. O., Grabb, W. C., and Oneal, R. M.: Management of injuries of the naso-orbital complex, Arch. Surg. 98:566, 1969.
8. Denecke, H. J., and Meyer, R.: Plastic surgery of the head and neck—corrective and reconstructive rhinoplasty, New York, 1967, Springer-Verlag, Inc.
9. Dingman, R. O., and Natvig, P.: Surgery of facial fractures, Philadelphia, 1964, W. B. Saunders Co.
10. Horton, C. E.: Septal and bone injury of the nose. In Georgiade, N. G., editor: Plastic and maxillofacial trauma symposium, vol. 1, St. Louis, 1969, The C. V. Mosby Co.

Chapter 13

Submucous resection

Herbert Lipshutz, M.D.

HISTORY

The first publication of a septal deviation was by Quelmaltz in 1750.[1] Morgagni in 1762 wrote on septal deviation.[2] From then to about the middle of the nineteenth century there were very few writings until Langenbeck (1843),[3] Dieffenbach (1845),[4] and Chassaginac (1851),[5] all of whom advocated total or partial removal of the cartilaginous septum. There were further contributions to the literature, but among the early writings Adams (1875)[6] proposed moving the septum into the midline either by fracturing or by crushing. Burckhardt,[7] in 1875 wrote of septal surgery submucously. Killian (1904),[8] however, was the first to perform a submucous resection of the cartilaginous septum and strongly advised leaving the dorsum and the anterior end of the cartilage for support. Asch[9] in 1890 was doing surgery on the septum by cutting and repositioning, but not submucously. Curiously, Freer[10] (1902) stated the entire septal cartilage could be removed without creating any deformity. Trendelenburg in 1898 was probably the first to correct a deviated, angulated septum.

From the middle and late nineteenth century and the early twentieth century to 1929, when Metzenbaum[11] described replacement of the angulated or dislocated end of the septum and not resection, there was very little reported in the literature. Peer (1937)[12] advocated resection of the angulated end of the septum and, if necessary, a free septal columellar graft was to be used for support.

There has been relatively little added to the basic techniques; changes were mostly placing incisions and different methods of straightening the cartilage and shifting the septum. The swivel knife (Ballenger)[13] has facilitated the simple submucous resection. There are still differences of opinion as to the necessity of septal surgery submucoperichondrially. Brown and MacDowell[14] agree that septal work can be done through the mucosa.

ANATOMY

The septum (Fig. 13-1) is divided into an osseous and cartilaginous portion. The osseous portion is formed by the perpendicular plate of the ethmoid, the vomer, the frontonasal spine of the frontal bone, the rostrum of the sphenoid, and the crests of the nasal, maxillary, and palatal bones. Practically, we are concerned with the perpendicular plate of the ethmoid, the spine of the maxilla, and the vomer.

The cartilaginous portion is formed by the septal cartilage, the vomeronasal cartilages, and the medial crura of the alar cartilages. The septal cartilage is irregularly quadrilateral in shape and separates, anteriorly, the right and left nasal fossae. It represents the ventral extremity of the primordial cartilaginous cranium. It is attached to the perpendicular plate of the ethmoid posteriorly, to the vomer and maxilla inferiorly, and to the nasal bone superiorly and extends between the alar cartilages anteriorly.

The so-called vomeronasal cartilages are two long narrow strips attached to the septal cartilage and the vomer. They may be impossible to distinguish from the septal cartilage.[15]

SEPTAL DEFORMITIES

Deformities of the septum may be congenital or acquired, and when acquired they are almost always traumatic. Patients, therefore, clinically have symptoms of airway obstruction. In many instances the septal deviation is grossly visible either from a direct dorsal view or examination of the nose from below with the obstructing septum visible in the nostril.

Although it is impossible to classify septal deformities, there are some deviations that seem to be more commonly seen than others. Generally, we can divide deformities into three categories: (1) anterior septal deformities, (2) middle, upper, or posterior nasal septum deviations, and (3) complete displacement of the anterior end of the septum from the nasal spine (Fig. 13-2, *A*). The angulated or deviated lower end of the septum may be only part of the problem, and further examination will reveal subsequent difficulty more proximal or posterior.

When the septal deformity results from severe injury, there may be marked thickening caused by old fracture and scar as well as angulation.

Middle and upper deformities may be one of angulation that may be separate or combined with lower deviation (Fig. 13-2, *B* and *C*). There may be bulging of the septum in a large curve (Fig. 13-3), and the septum may touch either the middle or the inferior turbinates, causing breathing obstruction. Many of the upper obstructions are congenital rather than traumatic in origin.

Finally (Fig. 13-4), we do see an intact septum that has been displaced in its entirety from its nor-

mal position on the vomer bone and nasal spine either laterally or medially.

Sometimes the deformed upper middle or posterior cartilage will be noticed to be continuous with the displaced bony perpendicular plate of the ethmoid. Under most conditions the vomer bone itself is uncommonly displaced or even injured.

Surgery. Operative procedures and techniques for correction of the previously described deformities have not varied much since the late nineteenth and early twentieth centuries. There is still discussion regarding whether or not submucous resection should be done as a separate procedure or per-

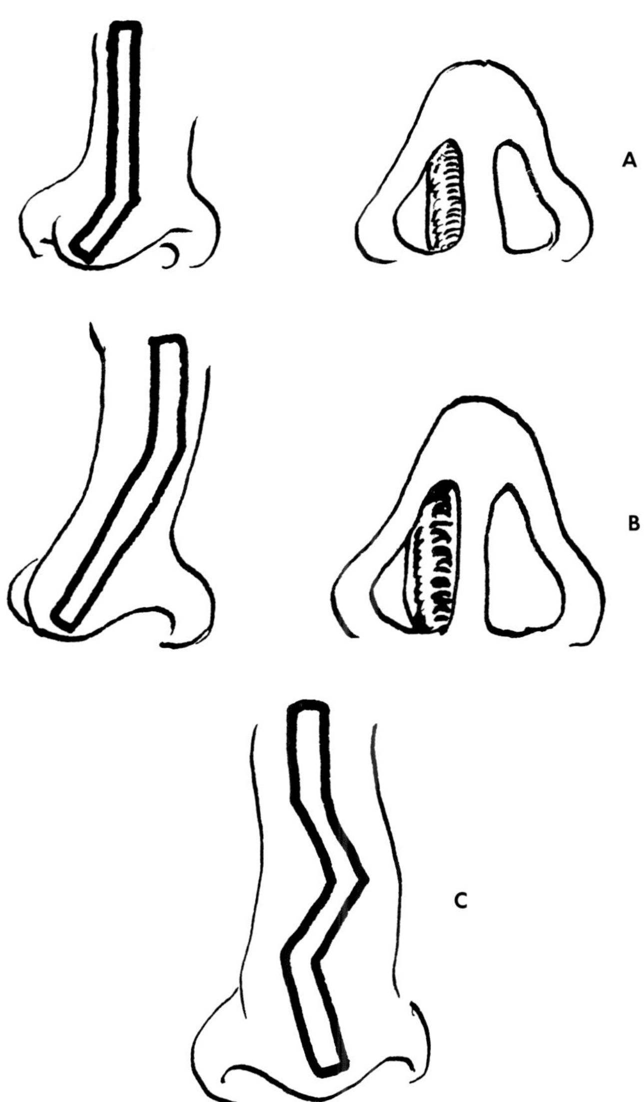

Fig. 13-2. **A,** Lower cartilaginous septal angulation. **B,** Upper cartilaginous septal angulation. **C,** Upper and lower septal angulations combined.

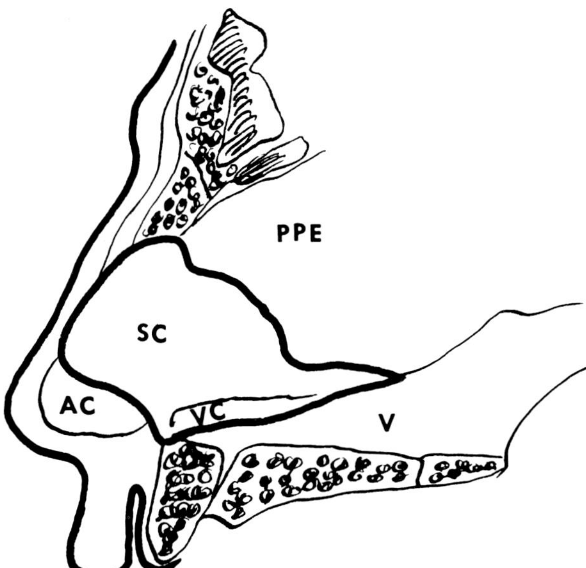

Fig. 13-1. Anatomy of the cartilaginous and bony septum. *PPE*, Perpendicular plate of the ethmoid; *SC*, septal cartilage; *AC*, alar cartilage; *VC*, vomer cartilage; *V*, vomer.

formed in a combination with a rhinoplasty when necessary.[16] There is also discussion among surgeons as to whether or not it is necessary to do mucoperichondrial dissection for septoplasty.[14, 15] In our experience, it has been found unnecessary in selected instances to perform the surgery submucously.

Procedures. If only a submucous resection on septoplasty is to be done, most of the anesthesia can be adequately obtained by initially spraying the nose with cocaine 10% to 20%, waiting approximately 2 minutes, and then packing the nose with umbilical tape or selvage gauze soaked in 10% or 20% cocaine with 1:1,000 epinephrine added (Fig. 13-5). If it is to be a combined operation then the nose itself is infiltrated. As a standard solution for nasal correction of the external deformity, we have found that a basic solution of 20 ml. of 1% xylocaine containing 150 units of hyaluronidase and about 10 drops of 1:1,000 epinephrine is more than satisfactory. Under ordinary conditions no more than 10 ml. of this local anesthesia solution is necessary. It has been stated that in a combined procedure the best time to perform the submucous resection is just after the hump is removed[16]; however, in those instances where more room is necessary, performing the lateral osteotomy first gives greater access to the nasal vault since the bones and upper lateral cartilage can be easily retracted.

To perform a simple submucous resection, an incision is made in the mucous membrane on the convex side of the nasal septum parallel to the

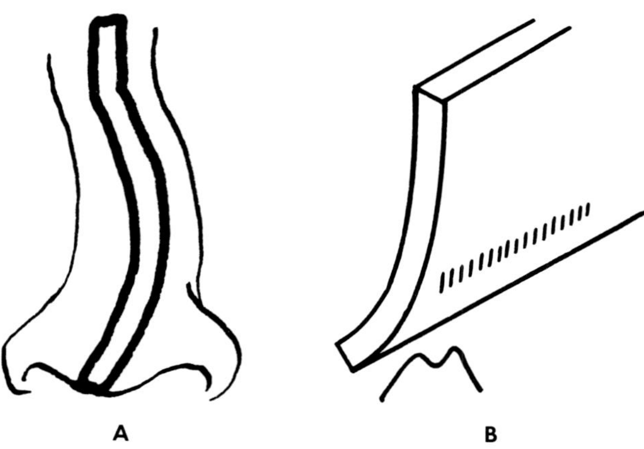

Fig. 13-3. **A,** Bulging septal deformity. **B,** Bulging septal deformity with displacement.

Fig. 13-4. Intact, displaced, angulated septum.

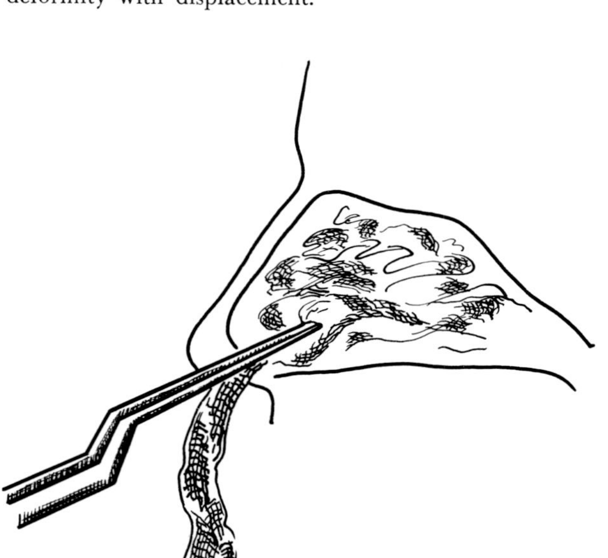

Fig. 13-5. Packing nose with epinephrine and cocaine–soaked gauze.

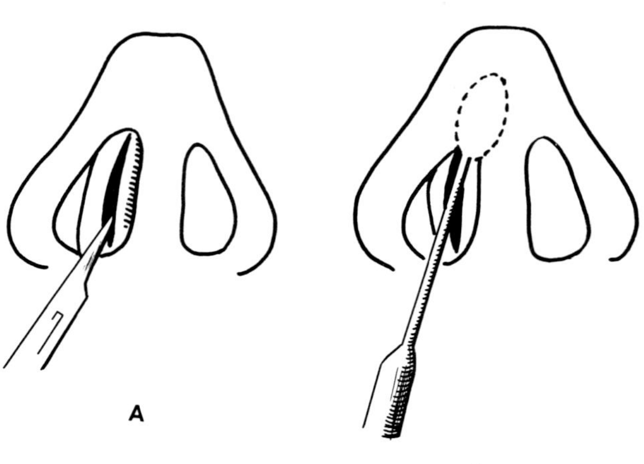

Fig. 13-6. **A,** Left incision in anterior end of protruding septum. **B,** Right insertion of perichondrial elevator to raise mucosa.

columella and extending to the superior portion of the anterior end of the septum down to the nasal spine (Fig. 13-6). Sharp dissection separates the mucoperichondrium from the edge of the cartilage, then a perichondrial elevator is inserted and the mucoperichondrium raised. An incision is made through the distal end of the cartilage (Fig. 13-7), leaving a small rim anteriorly, and a swivel knife is inserted.[13] This is then passed in a circular fashion completely around the bulging cartilage but leaving the rim of the septal cartilage in all directions for support. The incision in the mucous membrane can be closed with catgut or silk as desired. The nose is usually packed with Xeroform gauze.

While there has been discussion on first whether it is necessary to elevate the mucous membrane on both sides, we do not advise it under most circum-

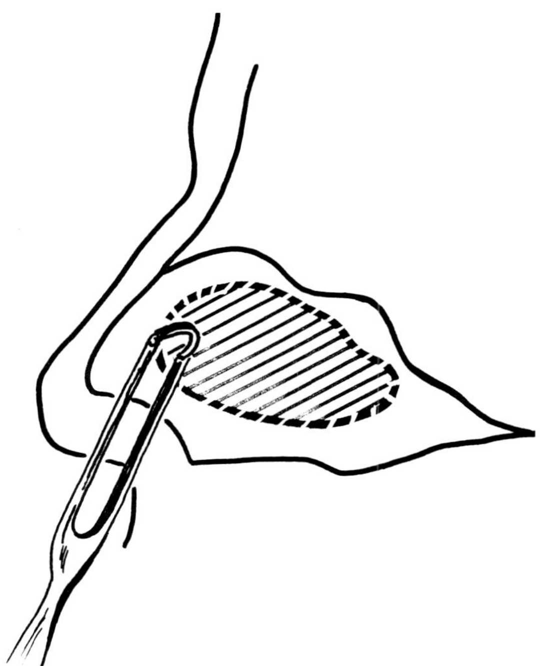

Fig. 13-7. Insertion of swivel knife to remove central bulging septum.

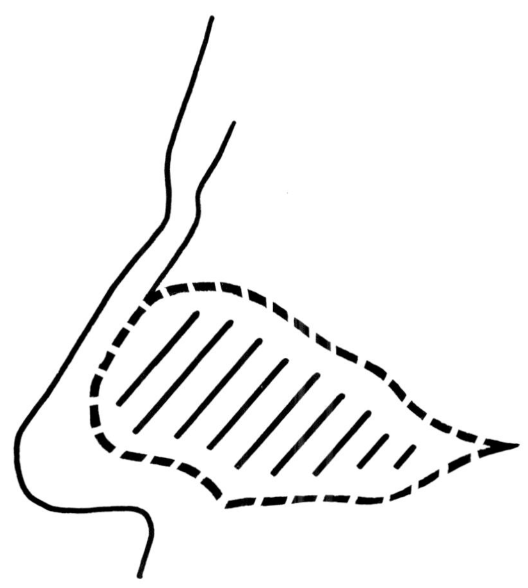

Fig. 13-8. Scoring length of septum.

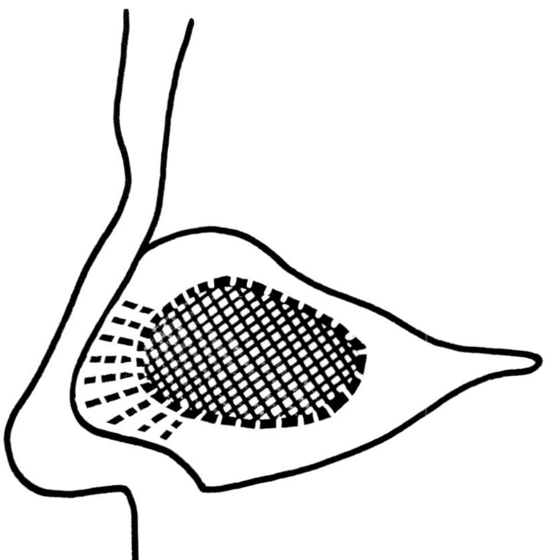

Fig. 13-9. Removal of septum plus septal scoring.

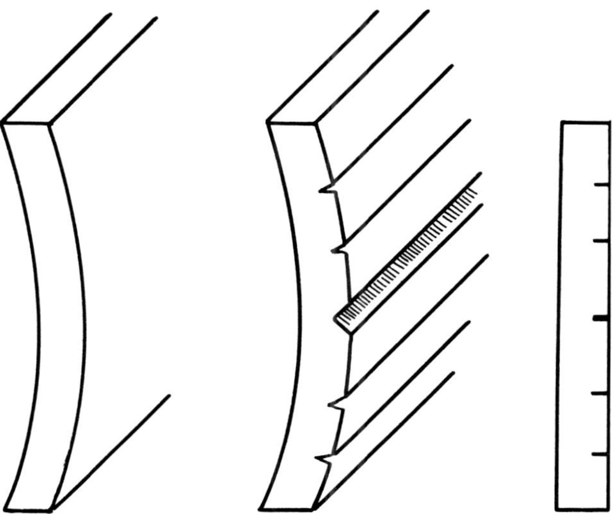

Fig. 13-10. Horizontal scoring plus V excisions.

stances, since leaving the mucous membrane on one side preserves the blood supply to the cartilage. Also controversial is whether or not it is best to dissect the mucous membrane from the convex side.[17] In our experience this has been found to be the easiest side on which to elevate the mucous membrane. This is a direct approach, and the septum is visible grossly throughout. Finally, there will be less of a tendency to perforate the septum in dissecting off the mucous membrane, especially in a scarred septum.

Bulging septum. A septum with bulging (Fig. 13-8) may be treated by scoring the septum along its entire length in multiple incisions completely through and through the cartilage.[18] This may be performed through the mucous membrane on one side without having to dissect it free.

The anterior end of the septum may be scored (Fig. 13-9) after removing a septal bulge in order that the dorsal curved portion can be straightened, but the scoring must not completely go through the dorsal edge.[19]

Another method that has been advocated (Fig. 13-10) for correction of the bulging septum is horizontal scoring and V excisions without going through the concave side. The largest V excision should be in the middle.[17] If necessary, a strip can be removed along the vomer so that the septum can be reset into the midline. This method does not require suturing of the cartilage and the position can be maintained by packing alone.

Septal angulations. This type of septal deformity extends from the simplest to the most complicated. The simplest is tip angulation (Fig. 13-11), which in some cases, if it is small enough, can be excised without leaving any deformity. When the

distal fragment is too large to disregard, a V excision on the convex side will allow the distal end to be straightened, and this may be held in place by direct suture or by packing.

When the entire septum is angulated (Fig. 13-12) it may be simply straightened by incising the cartilage along its inferior border and extending the incision along the perpendicular plate of the ethmoid. It is possible that a portion of the vomer may have to be removed so that the septum may be able to be set in the midline. Upon occasion, a portion of the perpendicular plate of the ethmoid may have to be removed to help correct the deviation. There is unanimity of opinion that when the posterior bone is deflected, it should be removed; otherwise it will only sequester, and it eis not needed functionally.

High angulations (Fig. 13-13) are actually handled in a manner similar to lower angulations, with V excisions on the convex side. If there is considerable scarring (Fig. 13-14) at fragment sites

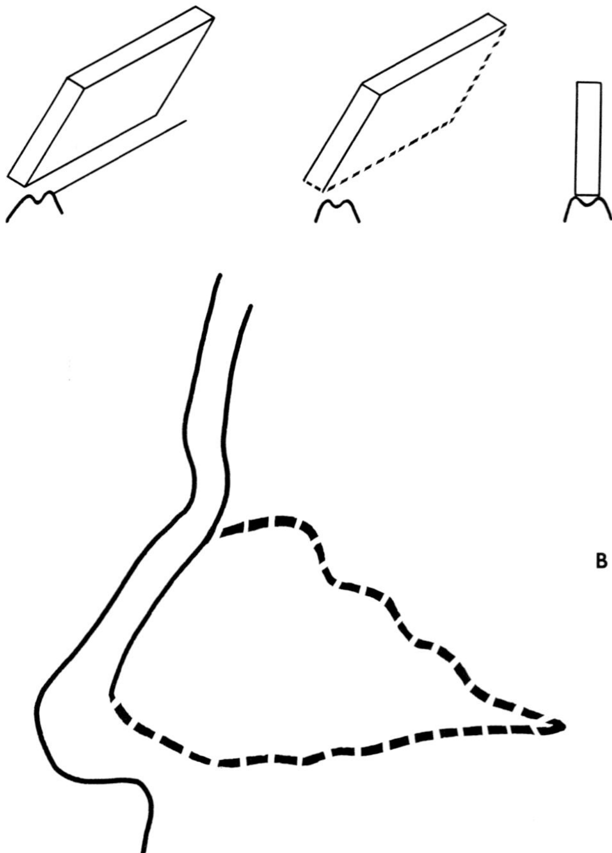

Fig. 13-12. **A,** Angulation of entire septum. **B,** Incision for entire septal angulation.

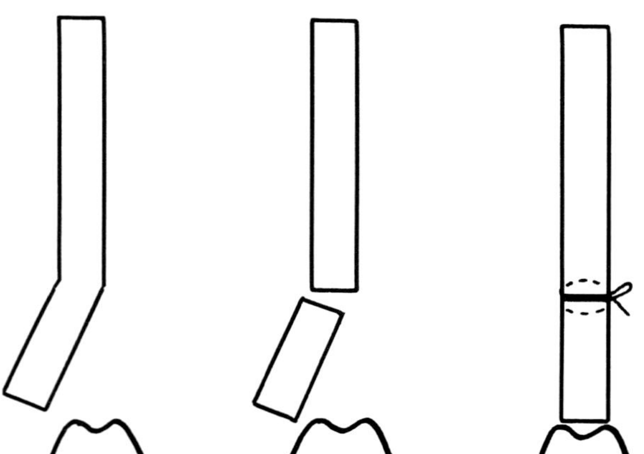

Fig. 13-11. Correction of tip angulation.

it may be necessary to excise more than the **V** in order to straighten the septum. This may have to be done in several places to obtain complete correction. Immobilization is accomplished by sutures or by packing or both.

In some instances when there has been fracture of nasal bones, lateral osteotomies may have to be performed to straighten the nose. In severe crushing injury to the nose the septum may be short and thick, causing obstruction. This type actually is very difficult to handle, and only the actual thinning down of the thickened septum would be satisfactory. The mucous membrane has to be dissected free with considerable care since it is usually markedly adherent and scarred, and a perforation may result.

Septal perforations are probably the most common complication of septal surgery and are almost always traumatic in origin, but they also may be caused by syphilis or tuberculosis. In most instances, small perforations may be left alone. Bleeding from

the crusted edges and excessive amounts of air through the perforation may require repair by mucous membrane flap rotation or advancement or by a flap of mucous membrane from the inferior turbinate. Also described is the use of an obturator to close the septal defect. When there is more tissue loss from tumor resection, tissue may have to be brought from a distance via pedicle flap or tube.

REFERENCES

1. Quelmaltz, S. T. In Haller's Disputat ad morborum historian, Paris, 1757; quoted by Denecke, H. J., and Meyer, R.: Plastic surgery of head and neck, vol. 1. Rhinoplasty, New York, 1967, Springer-Verlag, Inc.
2. Morgagni, G.: De sedilius et causis morbarum, Padua, 1762.
3. Longenbeck, B. V.: Hondluech der Anatomie, Gottingen, 1843.
4. Dieffenbach, J.: Die operative cherurgie, Leipzig, 1845.
5. Chassaignac, C.: Gazette des hopetaux, Paris, 1851.
6. Adams, G.: Brit. Med. J. 13:421, 1875.
7. Burckhardt, N.: Bericht uber die chirurgische Abteileing der Ludwigshospital—Charlottenhilfe wohrend der Jahre, 1885-1887.
8. Killian, G.: Die submucose Fensterresiktion der Nosenscheidewand, Arch. Laryng. Rhinol. (Berlin) 16:362, 1904.
9. Asch, M.: A new operation for deviation of the nasal septum with a report of cases, New York, 1891, D. Appleton & Co., p. 76.
10. Freer, O.: The correction of deflections of the nasal septum with minimum of traumatism, J. Am. Med. Ass. 38:636, 1902.
11. Metzenbaum, M.: Replacement of the lower end of the dislocated septal cartilage versus submucous resection of the dislocated end of the septal cartilage, Arch. Otolaryng. 9:282, 1929.
12. Peer, L.: An operation to repair lateral displacement of the lower border of the septal cartilage, Arch Otolaryng. 25:475, 1937.
13. Ballenger, H. C.: Otology, rhinology, and laryngology, Philadelphia, 1943, Lea and Febiger.
14. Brown, J. B., and McDowell, F.: Plastic surgery of the nose, St. Louis, 1951, The C. V. Mosby Co., p. 240.
15. Schaeffer, J. P.: The nose, paranasal sinuses, Nasolacrimal passageways and olfactory organ in man, Philadelphia, 1920, P. Blakiston's Son & Co., pp. 78-83.
16. Dingman, R. O.: Local anesthesia for rhinoplasty, and the nasal septum in rhinoplastic surgery, Plastic Reconstr. Surg. 28:251-260, 1961.
17. Dingman, R. O.: Correction of nasal deformities due to defects of the septum, Plast. Reconstr. Surg. 18:291-304, 1955.
18. Becker, O. J.: Principles of otolaryngologic plastic surgery, ed. 2, Rochester, Minn., 1958, American Academy of Ophthalmology and Otolaryngology.
19. Converse, J. M.: Plastic and reconstructive surgery, vol. 2, Philadelphia, 1964, W. B. Saunders Co., p. 740.

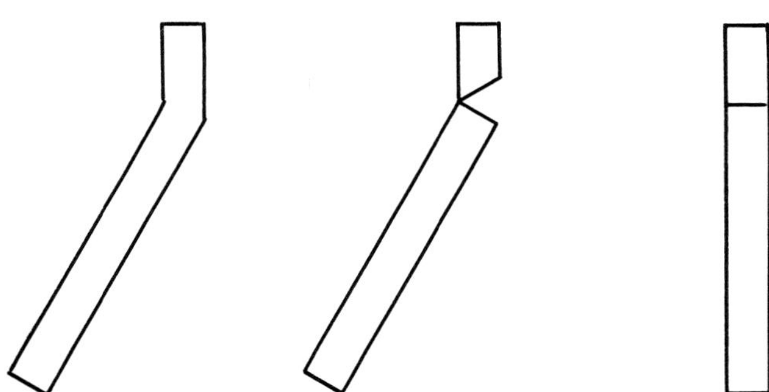

Fig. 13-13. High angulations of septal cartilage.

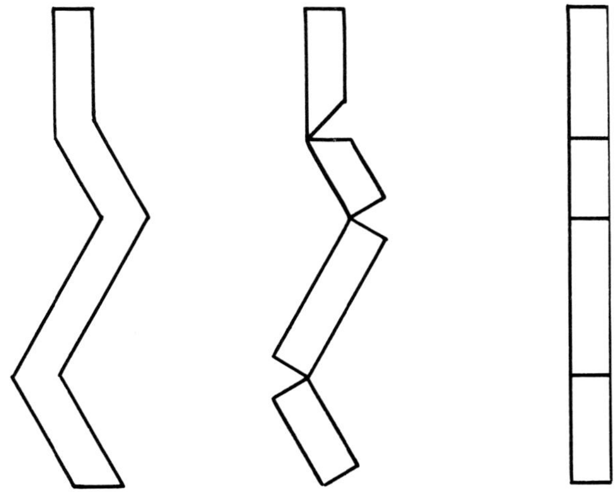

Fig. 13-14. Correction of multiple angulations of septum.

Chapter 14

Combined septoplasty and rhinoplasty

Charles E. Horton, M.D., F.A.C.S.

Although most patients who consult a plastic surgeon are concerned primarily with the appearance of their nose, the function of the nose must also be considered. Nasal obstruction can be caused by many things:

1. The size and configuration of the nostrils as they direct the air current flow
2. Narrowing by the valvular action of the upper lateral cartilages, which also direct the laminar airstream
3. Excessive width of the columella
4. Shortness of the columella and abnormally small openings of the nose
5. Abnormal anterior portions of the inferior and middle turbinates, which direct the turbulence and air current direction on inspiration
6. Abnormal configuration of the posterior ends of the inferior turbinates, which control the air flow on expiration
7. The erectile tissue of the turbinates, which expands and decreases in relation to atmospheric, emotional, and other stimuli
8. Tumors and foreign bodies
9. Trauma, which deviates the normal structures of the nose so that, with bending of the nasal bones and of the septum, the airways are diminished

The mucosa on the inside of the nose is covered with cilia, which actively sweep dust and foreign particles posteriorly toward the pharynx. Ciliary action can be interrupted by excessive heat or cold, by scarring, allergy, or infection, and by numerous chemicals. When ciliary activity is absent, crusts form in the nose and nasal air flow resistance occurs. At the attachment of the upper lateral carti-

lage to the septum, interference with respiration may occur when the upper lateral cartilage does not move with respiration (the so-called valve area). If this area is too wide, too much air may enter the nose and dryness may occur. This may be interpreted by the patient as a condition similar to nasal obstruction. If the nose is too narrow at this point, impedence of air flow will occur and an obstruction in breathing will result. Collapse of the valve area or of the alar area may occur on inspiration, particularly when the muscles that motivate the nasal structures have been injured, scarred, or are absent. The compressor nares and dilator nares as well as the levator labii superioris alaeque nasi help move the upper and lower lateral cartilages and both open and close the alar rims and the upper lateral cartilage area.[1]

For many years, the simple procedure of submucous resection has been in disrepute. It has been recommended to treat every condition from headache to sinusitis and emphysema, and when such chronic conditions did not respond, the procedure gained a bad general reputation. Many authorities believe that when a difficult submucous resection is performed, the cartilaginous structures of the nose are so weakened that a simultaneous performed osteoplastic rhinoplasty could be hazardous. Numerous other nasal surgeons feel a combined procedure is unnecessary in the usual rhinoplasty with minimal septal deviation. We feel that submucous resection is indicated if a dorsal hump is removed in rhinoplasty and it is necessary to reduce the width of the nose by infracturing the nasal bones.

By simple pressure on the outside of one's own nose at this area, it is easy to see that the diminution of the valve area by even 1 mm. produces nasal air flow resistance. Therefore, even though the septum is virtually straight at this point, if the nasal bones and upper lateral cartilages are to be moved medially, a submucous resection will decrease the thickness of the median partition of the nasal cavity and will help prevent the complication of increased nasal airway resistance postoperatively.

The septum is not straight in most noses. A curve high in the septum may produce internal pressure on mobilized nasal bones and will not allow the nasal walls to be moved medially into the position desired for external symmetry. A curvature of the septum inferiorly along the vomerine-palatine groove and crest can cause severe nasal obstruction after the rhinoplasty surgery. The anterior septum may be deviated to one side or the other of the midline and attached eccentrically to the side of the nasal spine, which also may be displaced. These distortions of the septum inevitably cause external twisting of the tip of the nose unless they are corrected. Adequate septal release and repositioning can only be obtained by doing an adequate septoplasty or submucous resection before or at the time of the osteoplasty rhinoplasty.[2] Previous septal surgery may produce scarring and may be inadequate to correct septal curvatures when subsequent rhinoplasty is performed. An initial submucous resection may also be too radical, and struts needed for rhinoplasty support at later surgery may be removed. In our hands, a combined procedure is preferable.

RHINOPLASTY

To perform the combined operation, it is important for the operator to have in mind the desired shape and size of the nose (Fig. 14-1, *A*). Brilliant green dye is used to mark the new height of the nasal profile, to indicate the area and the amount of the upper lateral cartilage to be excised, to demarcate the trimming of the lower lateral cartilage, to indicate whether or not the columella should be thinned, and to show the level of the cut along each maxillary crest to separate the nasal bone from the maxilla (Fig. 14-1, *B* and *C*). These markings are also useful to emphasize the desired deepening of the glabellar angle and the changes required at the septolabial angle. A mark is used on the cheek to indicate the elevation desired for the nasal tip and whether the base or the tip of the septal tip is more or less shortened. Nasal marking can also indicate whether or not a hidden columella

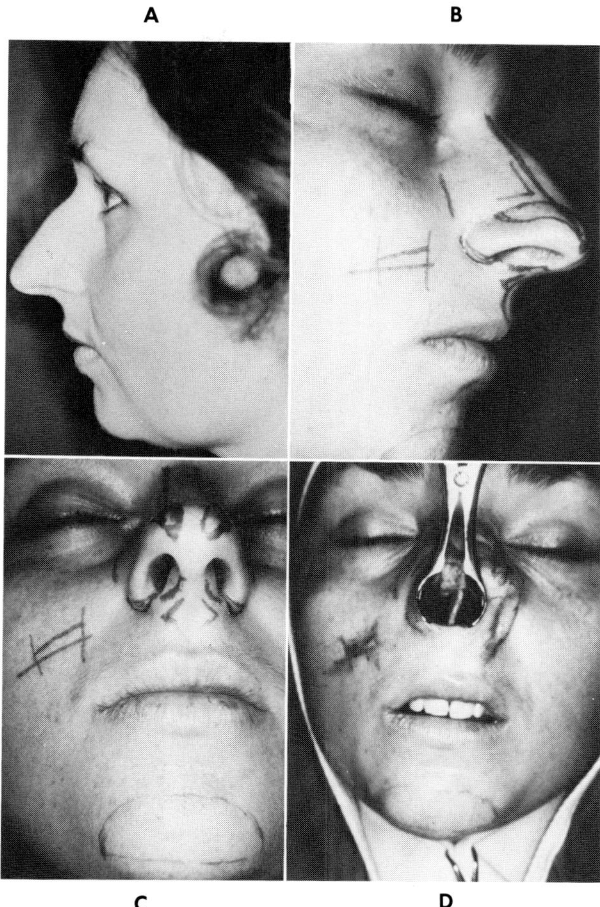

Fig. 14-1. A, Preoperative view. **B,** Preoperative markings, profile. **C,** Preoperative markings, front. **D,** Intranasal approach to septum.

is present and whether less upper lateral cartilage should be resected to proportion an overhanging columella. Marking is also useful if the nose is twisted and lengthened on one side when more lateral cartilage resection is required on the long side.

Each operator should select the type of anesthesia that creates the best personal and patient environment in which to operate. Many rhinoplasty surgeons will not operate if the patient will not accept local anesthesia. Many patients will not have nasal surgery unless they are given a general anesthetic. Certain operators state that excessive bleeding is encountered in general anesthesia for nasal surgery. We prefer to operate with local-thiopental (Pentothal) anesthesia technique. However, when patients request, we do not object to operating with local anesthesia alone or with general anesthesia. When a general anesthetic is given, we continue to inject lidocaine 1% with 1:100,000 epinephrine. Using this technique, we have had little more bleeding with general than with local anesthesia. Because of the allergic complications of cocaine, lidocaine can be substituted for nasal packing medication if desired.[3]

A through-and-through incision is made in the columella sufficiently high so that it is not visible from a lateral view. This incision should extend from the nasal spine to the dome of the nose and then between the upper and lower lateral cartilages. It is important that this incision be extended laterally and inferiorly for proper mobilization. The skin of the nose is undermined through this incision. The upper lateral cartilages are divided at the top of the septum. The dorsum of the distal septum is sectioned back to the nasal bones at the level previously marked on the skin. If the upper lateral cartilages are higher than this previously determined level, the upper edges of these cartilages are also cut along the desired level. Using an osteotome, each nasal bone is divided at the proposed nasal height, allowing removal of the nasal hump and the upper portion of each upper lateral cartilage. A triangular portion of the nasal diploe is then removed on each side of the septum at the glabella. Excess soft tissue is removed from the outside of each upper lateral cartilage. The dorsum of the nose is smoothed with a rasp. If the lower ends of the upper lateral cartilages protrude into the vestibule, they are trimmed. If the columella is hidden and the lateral alar rims are to be elevated, a larger amount of the lower end of each upper lateral cartilage is trimmed to include both the membrane and the cartilage. The lower lateral cartilages are then contoured by incising at the posterior and superior margins of the dome at the top of each lateral cartilage to within 1 to 2 mm. of the desired anterior edge of the dome. At that point, the incision is carried laterally, leaving a 1- to 2-mm. strut of the lower lateral cartilage to preserve continuity of the medial and lateral crus. This cut produces a flap of lower lateral cartilage that is hinged laterally and that can be readily brought out from the nasal cavity for easy exposure. While the operator holds either the cartilaginous or mucous membrane layer, the assistant grasps the other, and these layers are separated to allow resection of the upper three fourths of the lower lateral cartilage. When the tip of the nose is very thin, it is not necessary to remove a large amount of the cartilage. However, in noses that

are thick and bulky, the majority of the lateral crus of the lower lateral cartilage can be excised. When columellar height, or the height of the tip of the nose, is to be reduced, a hockey stick–shaped section of cartilage is removed. At this time, the septal tip may be resected to elevate the tip to the desired level. If the columella was "hidden" preoperatively, no resection is done; however, if a drooping columella is to be corrected, a larger resection of cartilage and soft tissue is necessary.

Once the previous procedures are completed, a submucous resection should be performed. An adequate dorsal and inferior septal strut must be preserved. Therefore it is essential to do the hump and profile lowering and the septal tip resection prior to the submucous resection. If the submucous resection is done prior to the hump and tip removal, the struts might be weakened by unanticipated later profile cartilage resection, and inadequate support of the nasal tip might result.

SUBMUCOUS RESECTION

By retracting the columella, the surgeon can expose the tip of the septum. Sharp pointed scissors dissection facilitates identification of the proper plane between the mucoperichondrium and the cartilage. When the proper plane has been identified, he uses a small blunt dissector to elevate the mucosa, keeping well below the top of the cartilaginous septum. After a superior submucosal tunnel has been made, a blunt hockey stick elevator is introduced to the back of the septum and the dissection is continued from posteriorly to the anterior floor of the nose. The elevator should be held with firm pressure against the septum at all times during the dissection to avoid penetration of the septal mucosa. The mucosa should be elevated on both sides. A speculum blade is inserted on each side of the tip of the septum (Fig. 14-1, *D*). A cut can be made at the base of the septal cartilage approximately 1 cm. behind the inferior edge of the lower part of the septum and extended superiorly to within 1 cm. of the top of the nasal septum. A swivel knife is introduced in the cartilaginous incision and the midportion of the septum is removed. By inserting a nasal speculum the surgeon can identify the remaining deviations and spurs of the septum. Usually, most deviations will be found inferiorly along the floor of the nose.

Posterior to the inferior strut at the level of the palatal crest, the swivel knife will not have removed all of the cartilaginous septum. By incising

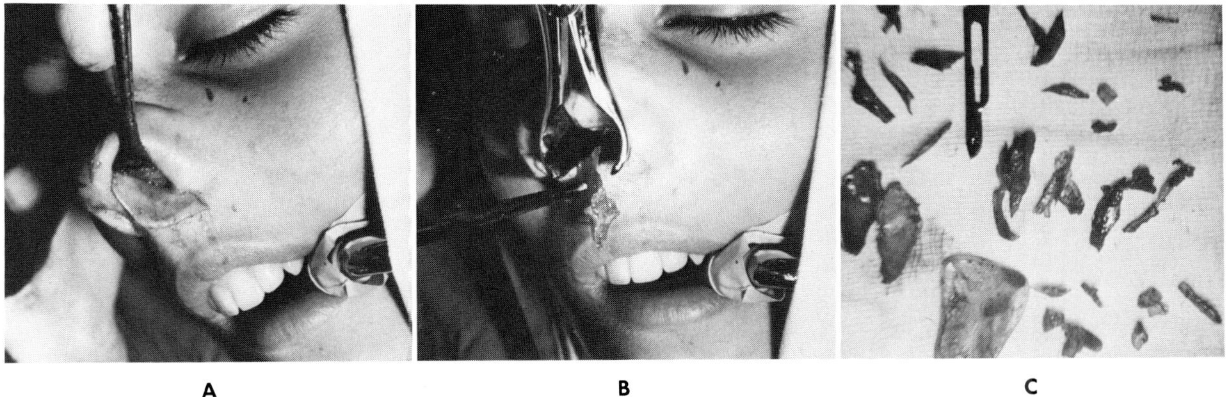

A   B   C

Fig. 14-2. A, Mobilization of anterior septal strut. **B,** Removal of bone spur. **C,** Usual septal resection.

this cartilage behind the anterior strut, the surgeon can use an elevator to raise the remaining cartilage along the floor of the nose, dislodging it from the vomerine-palatal groove. With a punch, he removes the remainder of the deviated septum as far posteriorly as necessary to correct the deviation. Careful inspection to make certain that no septal deviation persists superiorly is necessary, for, if present, this septal deviation will push outward against the mobilized nasal bones. Mucosa can be elevated from the concave site of a distorted strut and the cartilage scored to release the tension of the cartilage. Occasionally it is necessary to score both sides; however, care should be taken not to weaken the dorsal strut excessively. If the mucous membrane is totally freed and drops from its superior septal attachment, it should be resutured.

Dissecting at the nasal spine with the points of small scissors, the surgeon can divide the fibrous tissue holding the inferior cartilage strut and free the anterior cartilage strut (Fig. 14-2, *A*). Moving posteriorly, he can elevate the mucous membranes on each side of the bony palatal crest, by exploring with the point of the scissors to find the exact site where the bony deviations occur and by elevating the soft tissue with the point of the sharp scissors. Once the proper plane has been established, a hockey stick type of elevator can be used to further elevate the mucosa from the inferior bony spurs. When the spurs are adequately freed of mucosa (Fig. 14-2, *B*), a chisel is introduced behind the nasal spine and all abnormal bone resected (Fig. 14-2, *C*).

To complete the rhinoplasty, the surgeon introduces a chisel at each piriform angle and separates the nasal bones from the maxilla. The nasal bones are then moved to the desired position in the midline, keeping the bases of the nasal bones in contact with the maxillary bone laterally and making certain that the inferior edges of the nasal bones are not pushed too far medially.

DISTORTED INFERIOR NASAL STRUT

If the lower nasal strut remains deviated at the tip, it is necessary for the surgeon to score the cartilage to release the tension in the manner described by Fry.[4-9] If an acute angulation occurs, cuts should be made on the concave side and a mattress suture used to hold the septum into a better position. If the nasal spine is eccentrically located, a chisel or punch can be used to remove the bone on one side or the other, leaving a small portion in the midline to which the cartilaginous septal strut can be sutured. Occasionally, if the septum is excessively weakened or mutilated, a portion of the cartilage previously removed from the central portion of the septum can be used as a graft to the tip or to the dorsum of the nose. Silastic and polyethylene plates have been placed inside the septal mucous membrane flaps to replace distorted cartilage. These are frequently extruded. If the septal tip strut is inadvertently destroyed or unusable, an autogenous bone graft can be done at the time of the rhinoplasty—submucous resection procedure, or grafting can be performed as a secondary operation. The use of plastic sheets placed on each side of the distorted septum has been reported by Fredericks. These struts are left in place for 1 to 2 weeks and removed after healing.

Excessive columellar width results in many instances not only from soft tissue but from inferiorly situated aberrant cartilages at the base of the nostril openings. This cartilage is not continuous with the medial crus of the lower lateral cartilage and is extremely thin, lying adjacent to the skin. To reduce columellar width it is necessary to remove not only soft tissue but also the aberrant columellar cartilage in many cases.

If the septum has been excessively deviated to one side or the other, the middle and inferior turbinates will usually have hypertrophied to enlarge into the space so created, to prevent an excessive air flow on the enlarged side. It is occasionally necessary to perform a submucous resection or a partial resection of the turbinate.

Occasionally, the maxillary bone at the piriform angle will be deformed. This sharp edge of bone can be removed to give more space to the base of the nose to allow a better airway opening.

Following adequate septal resection and repositioning, the columellar incisions are closed with 4-0 plain catgut. A transfixion suture of 4-0 chromic catgut may be used inferiorly to add stability to the strut. Petrolatum gauze packing is placed in each side of the nostril and a dressing consisting of tape and a metal splint is applied. We do not routinely give antibiotics or antiinflammatory medication.

The septal cartilage and bone removed at rhinoplasty can be used as graft material to augment the chin (Fig. 14-1), orbital floor, and tip strut and occasionally as a free graft to the dorsum. Septal cartilage has also been utilized as filler material to raise the alar rim base in cleft lip–nasal deformities.

Weir excisions at the alar bases are used when necessary to reduce the nasal tip height and the size of the nostril openings.[10]

Complications of submucous resections are usually minimal. A permanent septal perforation can be avoided by careful dissection and by making certain that all adjacent tears in the mucous membrane are sutured and/or carefully packed. Many operators make an inferior unilateral opening in intact septal mucosa after septoplasty to allow dependent drainage to prevent septal hematoma. Perforations of the palate have been reported following septal resections. Cerebrospinal fluid leak is uncommon. The most frequent complication following a combined rhinoplasty and septoplasty is secondary obstruction caused by minor curvatures of the remaining cartilage struts. Because most of the cartilage has been excised with the technique described, obstruction is usually minimal and can frequently be corrected by a small secondary pro-

cedure with minor revisions of the remaining deviated septum. In our hands, septoplasty by using vertical or transverse incisions and removing selected small areas of septal cartilage has not proved to be satisfactory. If adequate struts are preserved, resection of cartilage is superior to repositioning, and fewer postoperative airway problems are noted.[11, 12]

DISCUSSION

In a review of approximately 1,000 postoperative rhinoplasty patient charts, only eighteen were noted to have complained of septal nasal obstructive symptoms following combined rhinoplasty and submucous resection. In the majority of these patients, a simple further selective resection of the remaining struts of the septum or further surgical correction of the enlarged turbinates proved beneficial. Other patients who were suspected to be allergic were treated by referral to an allergist. In contrasting this with experiences reported by other surgeons who state that the postrhinoplasty airway obstructive problem is frequent, we feel that the combined operation, even though little septal deformity is present, has much to offer to prevent diminution of the airway. When the septum is involved in the deviation problem, the combined operation is mandatory. A submucous resection performed "elsewhere" is usually not satisfactory when rhinoplasty is to follow. Superior septal deviations may persist or inadequate septal cartilage for strut reconstruction may be present. Secondary septal surgery is difficult because of difficult identification or proper tissue planes.

Simultaneous surgery of submucous resection and osteoplasty rhinoplasty has been essential in preventing postrhinoplasty breathing difficulties.

REFERENCES

1. Ogura, J. H., and others: Nasal obstruction and the mechanics of breathing, Arch. Otol. **83:**135-150, 1966.
2. Perks, E. R., and Rozner, L.: A method of measuring nasal obstruction, Arch. Otol. **80:**200-205, 1964.
3. Thomas, W. C.: Submucous resection of the nasal septum; special reference to anesthesia, Southern Med. J. **53:**1407-1413, 1960.
4. Fry, H.: The importance of the septal cartilage in nasal trauma, Brit. J. Plast. Surg. **20:**392-402, 1967.
5. Byars, L. T.: Surgical correction of nasal deformities, Surg. Gynec. Obstet. **84:**65-78, 1947.
6. Fry, H.: Interlocked stresses in human nasal septal cartilage, Brit. J. Plast. Surg. **19:**276-278, 1966.
7. Fry, H.: Cartilage and cartilage grafts: the basic properties of the tissue and the components respon-

sible for them, Plast. Reconstr. Surg. **40:**426-439, 1967.

8. Fry, H.: Nasal skeletal trauma and the interlocked stresses of the nasal septal cartilage, Brit. J. Plast. Surg. **20:**146-158, 1967.

9. Fry, H.: The distorted residual cartilage strut after SMR of the nasal septum, Brit. J. Plast. Surg. **21:**170-172, 1968.

10. McDowell, F., Valone, J. A., and Brown, J. B.: Bibliography and historical note on plastic surgery of the nose, Plast. Reconstr. Surg. **10:**149-185, 1952.

11. Horton, C. E.: Spetal and bone injury of the nose. In Georgiade, N. G., editor: Plastic and maxillofacial trauma symposium, vol. 1, St. Louis, 1969, The C. V. Mosby Co.

12. Lamont, E. S.: Reconstructive surgery of the nose in congenital deformity injury and disease, Am. J. Surg. **65:**17-45, 1944.

Chapter 15

Posttraumatic rhinoplasty

Robert M. McCormack, M.D.

Posttraumatic rhinoplasty differs from the usual elective cosmetic rhinoplasty in that the great variety of more severe anatomic deformities requires meticulous analysis for each individual case. Rarely are any two problems identical. There are two major types of posttraumatic nasal deformities.

LATERAL DEVIATION OF BONY AND CARTILAGINOUS STRUCTURES

The more common posttraumatic nasal deformity is lateral deviation of bony and cartilaginous structures of the nose, particularly of the bony and cartilaginous septum (Fig. 15-1). This lateral deviation is often more complex than a single convex or concave curved axis and involves a double curve with the anterior end of the septum protruding obliquely into the nares of the side opposite to the major deviation. Careful preoperative intranasal examination will illustrate the marked asymmetry of the structures. The most difficult part of the planning is the clinical judgment as to location of the incisions in the cartilaginous and bony structures to try to produce the anticipated symmetry after thorough mobilization of these anatomic structures (Fig. 15-2). This frequently requires cutting the dorsal structures in an asymmetrical manner because that is the way they have healed. This basic principle of asymmetrical incisions of the bony and cartilaginous framework to create ultimate symmetry after mobilization forms the basis of posttraumatic rhinoplasty planning.

The bony and cartilaginous septum of the laterally deviated nose also shows a variety of deformities that frequently interfere with the airway on one or both sides. The original trauma may have fragmented the septum so that the fragments heal with displacement and at times, of course, there may be an actual traumatic perforation of the septum because of loss of cartilage and mucosa. Again the aim should be to mobilize the bony and cartilaginous septum by proper incisions. It is this thorough mobilization through these incisions that is more effective in attaining reasonable symmetry of the septum than is mere submucous resection of displaced cartilage. Often, of course, an element of submucous resection, particularly at sharp angular portions of the anterior cartilaginous septum, must be performed for adequate mobilization. This septal mobilization in the marked deviation of the posttraumatic nose requires more than one incision perpendicular to the dorsal profile line hinging these fragments on the intact mucosa of the opposite side. In severe cases, we have not hesitated to carry this incision all the way through to the dorsum of the septum. This has not resulted in saddle deformity of the distal cartilaginous septum. The columellar incisions over the anterior end of the septum must be made carefully, and again the actual incision on one side may have to be asymmetrical as compared to the other in anticipation of mobilization of the anterior end of the septum for proper symmetry and refitting of the columella. It is beneficial in posttraumatic rhinoplasty to extend the columellar incisions slightly into the floor of the nostril, actually exposing the anterior nasal spine of the maxilla and the abutment of the cartilaginous septum on this spine. This exposure aids in careful mobilization of the anterior end of the septum if needed. At times an additional incision along the vomer base parallel to the superior surface of the hard palate is needed for complete mobilization.

POSTERIOR DISPLACEMENT

The second common type of posttraumatic nasal deformity is the posteriorly displaced sunken-in nose resulting from a frontal blow (Fig. 15-3, *A* and *B*). The secondary mobilization and reduction of the bony and cartilaginous nasal structures becomes even more difficult in this instance since the healing following severe impaction of the displaced fragments often makes it impossible to bring these structures forward completely. A very careful incision with a very low lateral osteotomy and thorough mobilization in the glabellar area as well as thorough mobilization of the bony and cartilaginous septum can lead to reasonable narrowing and

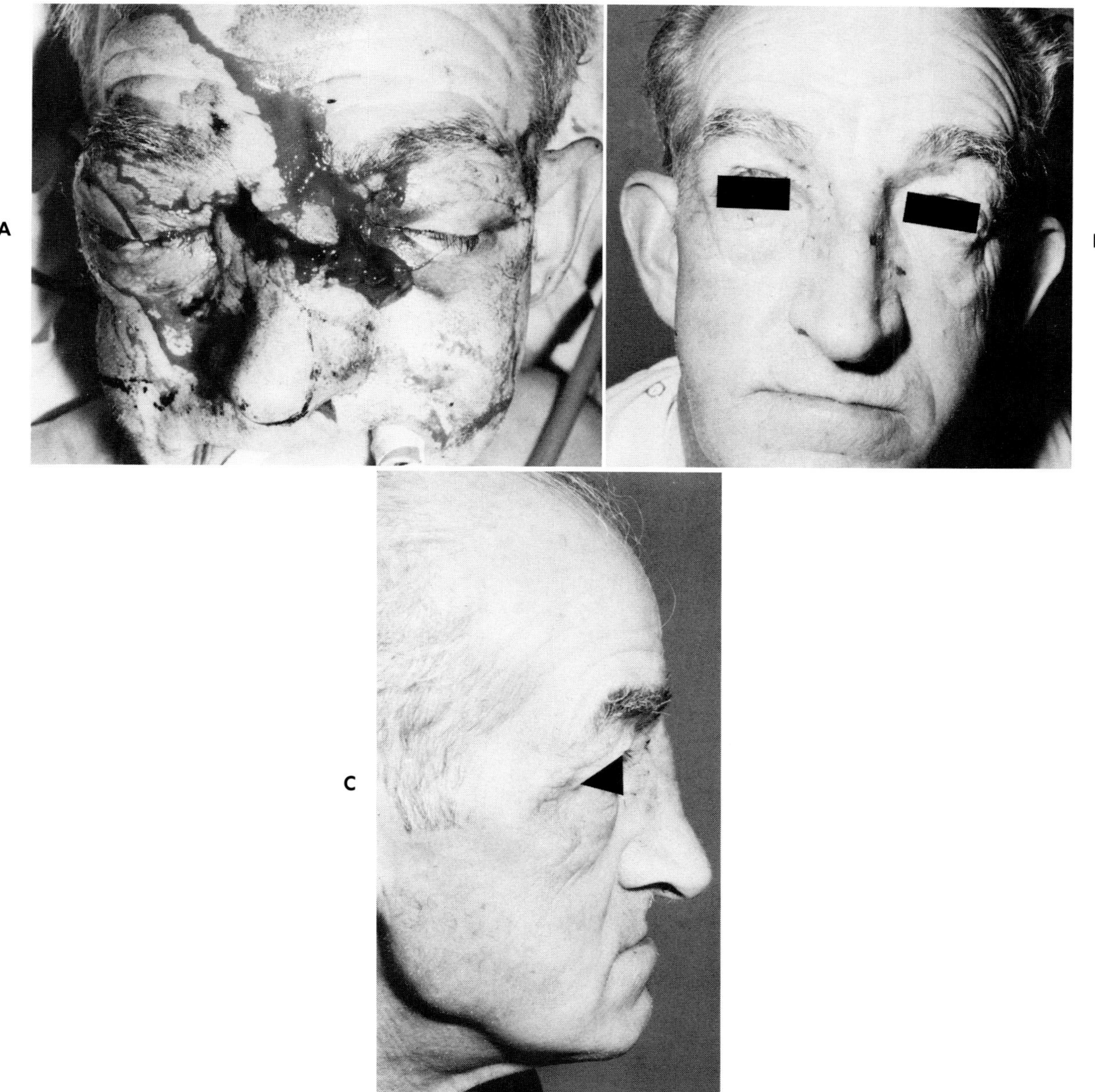

Fig. 15-1. A, Severe compound fracture of bony and cartilaginous structures of nose with marked displacement. These structures were mobilized by meticulous soft tissue repair including suturing of mucosa, absorbable suture fixation of fracture of cartilaginous septum, and external transverse wire fixation over bolus dressing of lateral bony structures. **B,** Frontal view 2 weeks after open reduction. **C,** Lateral view 2 weeks after open reduction.

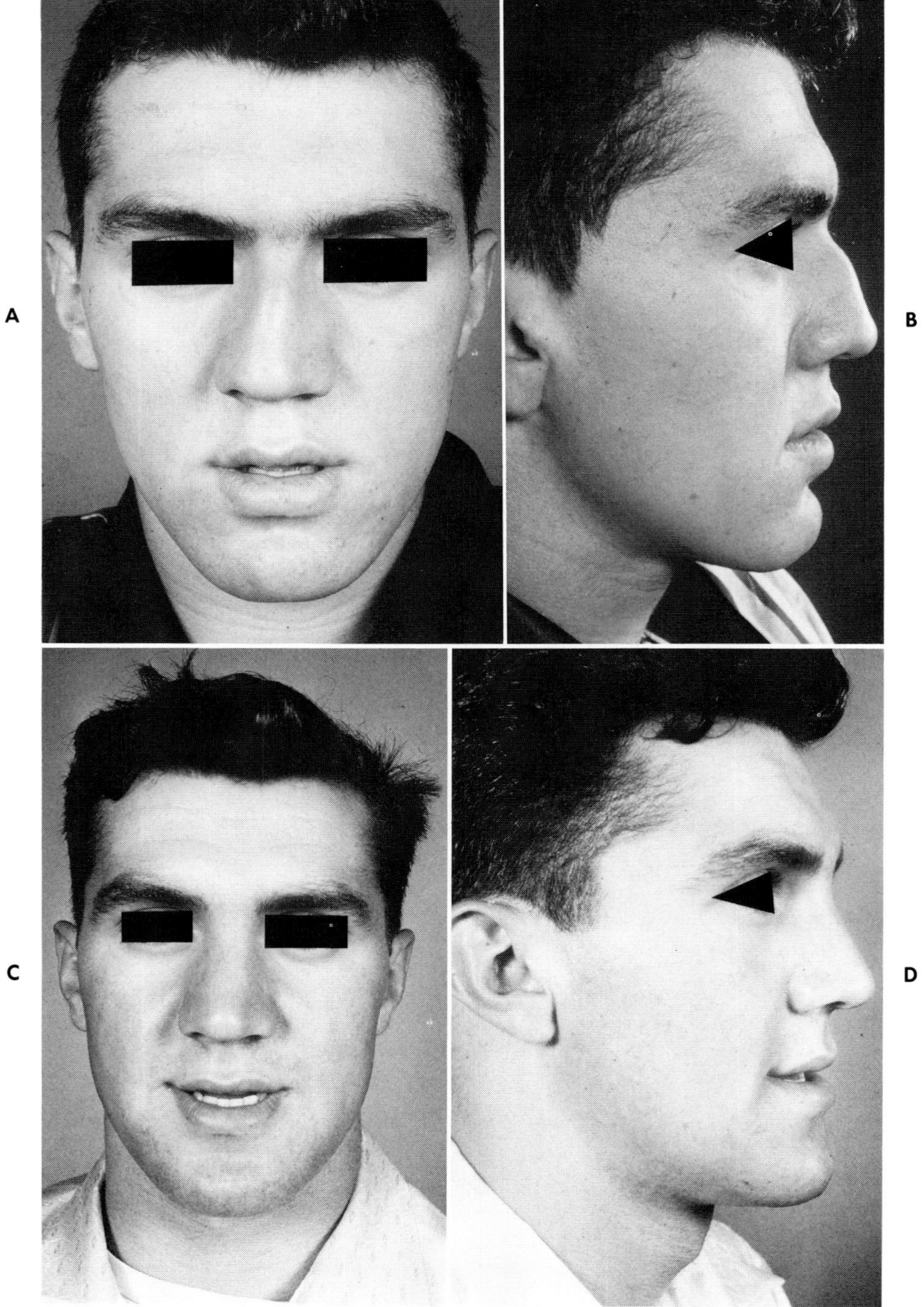

Fig. 15-2. A, Preoperative frontal view with deviation of cartilaginous and bony structures secondary to blunt trauma. **B,** Preoperative lateral view. **C,** Postoperative frontal view after posttraumatic rhinoplasty with extensive mobilization of all displaced structures. **D,** Postoperative lateral view.

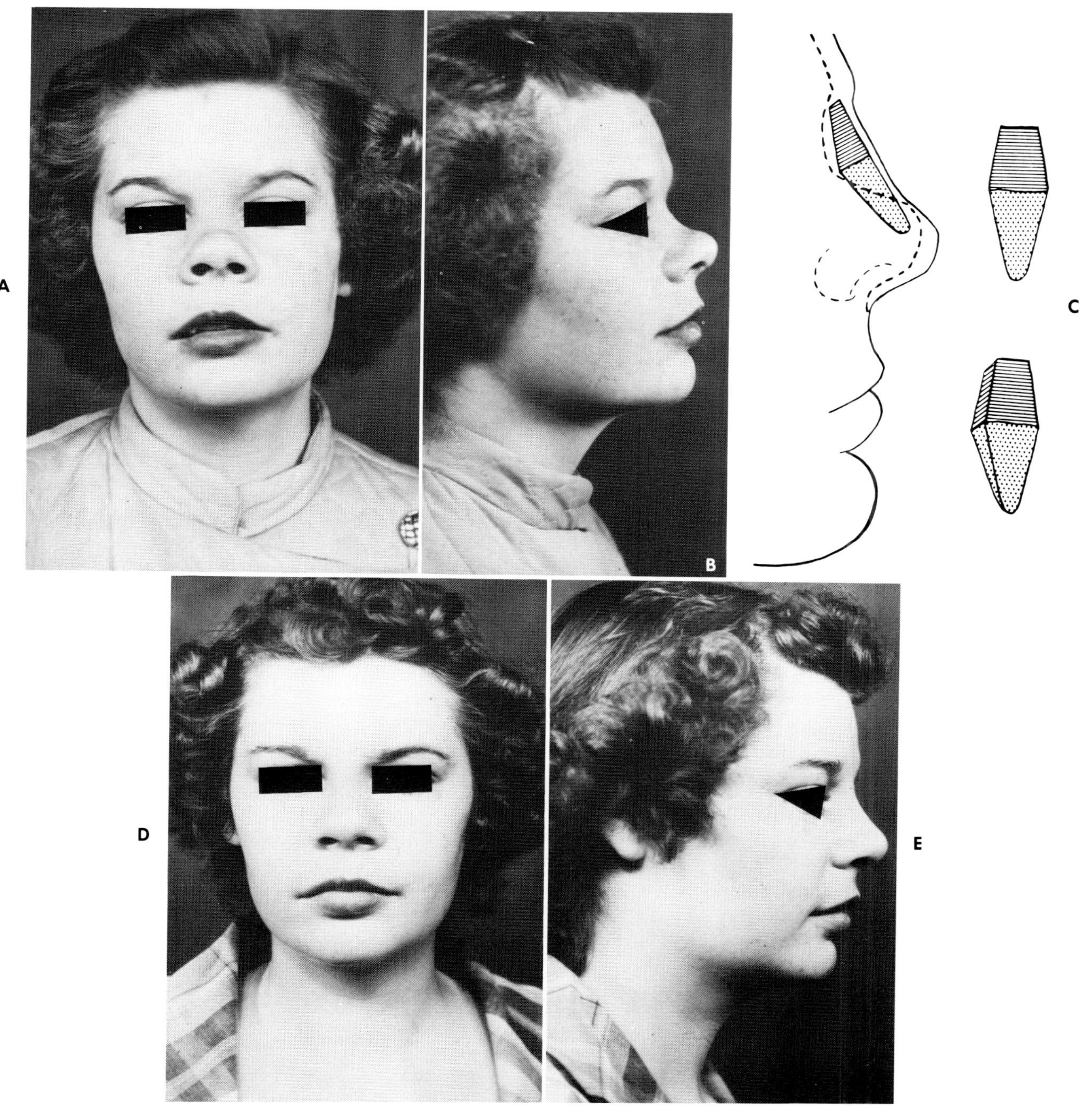

Fig. 15-3. A, Preoperative frontal view with severe posterior displacement and impaction of nasal maxillary fracture. **B,** Preoperative lateral view illustrating severe posterior displacement and shortening of nasal profile. **C,** Composite osteocartilaginous graft taken from the costochondral junction of the rib. The profile outlines are traced from the preoperative and postoperative photographs. This type of dorsal nasal graft allows bony union in the superior portion. The lined areas are bony and the dotted areas cartilaginous. **D,** Postoperative frontal view after the costochondral nasal graft. **E,** Postoperative lateral view.

forward reduction. This reduced position can serve as a foundation for possible later additional bone or cartilage grafting to the dorsum of the nose, if that is indicated. If the posterior displacement has resulted in the nostrils tilted up so that they can be viewed from in front, a thorough mobilization of the cartilaginous septum and vomer is often required in order to displace the septum dorsally and inferiorly. If a second stage dorsal graft is required we have preferred, over the years, a rib graft taken at the costochondral junction so that the superior portion is bone and the inferior portion is cartilage. The inferior portion is usually slightly longer and the piece is carved in a generally fusiform shape so that a cantilever effect supports the tip of the nose, being careful not to make the cartilaginous portion too long to avoid pressure necrosis and/or erosion of soft issues (Fig. 15-3, *C* to *E*).

In our experience general endotracheal anesthesia is often preferable to local anesthesia for the extensive mobilizations that are required in the posttraumatic rhinoplasty. This aids our ability to do extremely thorough mobilizations that often require fracture and osteotomies by small osteotomes in a variety of locations along the dorsum, the lateral border of the bony nose, the glabellar area, and the vomerine ridge of the septum. These extensive osteotomies plus carefully planned incisions in the cartilaginous septum, the anterior end of the septum perpendicular to the dorsal profile, plus careful submucous resection, if indicated, provide thorough and adequate cartilaginous mobilization. Only then have we been able to attain adequate reduction. The general endotracheal anesthesia is supplemented by injections of lidocaine with 1:100,000 epinephrine, for hemostasis.

Another important factor in adequate mobilization and reduction in posttraumatic rhinoplasty is careful use of extranasal and intranasal palpation. After mobilization of the structures, one can palpate carefully with the little finger through the nares the osteotomy lines described before, the bony and cartilaginous septum (including the actual incision of the septal structures), the glabellar area (to detect any residual fragments that have not been adequately mobilized), the vomerine ridge, and the anterior nasal spine, as well as the anterior end and the dorsum of the septum, including the

dorsal edge of the displaced lateral nasal bones and the bony septum. This systematic palpation has been of help in evaluating and ensuring that the mobilization has been extensive enough to allow adequate reduction. The intranasal palpation often gives more accurate information than extranasal palpation or initial inspection through the nasal speculum. After the structures are reduced, one can palpate the dorsal profile line, the anterior end of the septum, and the airway bilaterally to ensure that these are in satisfactory positions and that a functional airway is present.

Postoperative care is similar to that of elective cosmetic rhinoplasty, except that one must be doubly careful to avoid skin pressure, particularly on any external scars in the skin of the nose, since necrosis may result. Also, the packs are usually left in the nose longer, often leaving one pack in the nostril that required extensive airway correction for a day or two longer than the opposite side. Occasionally transverse wire fixation is required to maintain the reduction of the extensively mobilized nasal structures.

In summary, the basic principles of posttraumatic rhinoplasty involve:

1. Careful analysis of the anatomic deformity presented by the severely laterally deviated or posteriorly displaced bony and cartilaginous structures
2. Adequate anesthesia and reasonable hemostasis, usually under general endotracheal anesthesia
3. Extensive mobilization of the displaced structures by thorough osteotomies and cartilaginous and mucosal incisions
4. Careful intranasal palpation as a major aid to analysis of the deformity under the anesthesia and for checking the adequacy and accuracy of the mobilization after reduction of the structures, including satisfactory symmetry and functional airways
5. Care in postoperative splinting and nasal packing to ensure retention of the symmetry gained

For posttraumatic deformity of the nose, the operative term *rhinoplasty* is probably not descriptive enough and a term like *osteotomy and open reduction of the nose* would be more accurate.

Chapter 16

Saddle nose deformity

James W. Smith, M.D.
George T. Craig, M.D.

The term *saddle nose* is used to describe a central collapse of the nasal bridge. Etiologically, it may be classified as:

1. Posttraumatic (which now constitutes the greatest number of cases)
2. Congenital, genetic, ethnic (for example, Negroid nose)
3. Postsurgical
4. Postinfectious (specific, such as lues, leprosy, yaws, and the like, or nonspecific, such as from a pyogenic septal abscess)

While the external deformity can look the same in most cases, there may be striking differences in the underlying nasal bones, septal and lateral cartilages, and membranous lining, depending on the etiology (Fig. 16-1). These alterations necessarily modify the treatment plan and its effectiveness. For example, Gillies in a classic monograph[1] describes the unique nature of the syphilitic nose and its correction by the epithelial inlay method. A multiplicity of other procedures, and especially implant materials, dating from ancient times, has been employed to correct the deformity. These materials have included precious metals, ivory, autogenous, homologous, and heterologous bone and cartilage, and, in the modern era, alloys such as tantalum and vitallium and most recently the various alloplastics, the current favorite being silicone rubber. This fascinating and lengthy story has been admirably and exhaustively documented by Denecke[2] and by McDowell.[3]

One is led to the presumption on the basis of this same long history and multiplicity—as in all such instances—that no one method or material has been entirely satisfactory. Safian reported success with ivory and later turned to silicone rubber.[4] Millard feels this latter material offers the simplest approach and is reasonably safe.[5] Stark prefers an L-shaped cancellous iliac bone graft.[6] Foman used this material but never in the L-shape.[7] Barsky likewise prefers iliac bone,[8] as does Converse, who calls it the most popular implant material right now.[9] Denecke says that the results with bone grafts are "contradictory" and that he prefers cartilage.[2] Brown and McDowell did not seem to express strong preference for one material over another.[10]

There seem to be few reviews of results of correction of the saddle deformity with implants in the literature. Farina has lately reported on 169 patients treated with osteoperiosteal tibial grafts with a failure rate of 9.5%.[11] However, ninety-

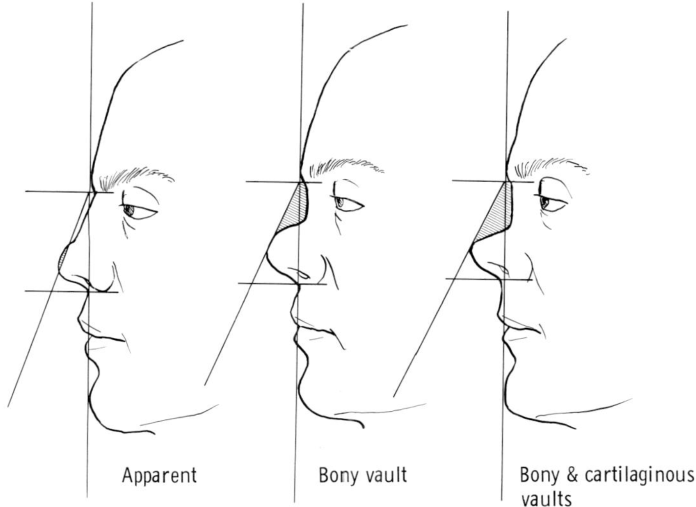

Fig. 16-1. Different types of saddle nose deformities.

eight of these were patients with Hansen's disease, which is quite a different problem from the post-trauma cases commonly encountered in the United States. Campbell reported a series of 100 autogenous iliac bone grafts in 1969.[12] He cited three failures and five grafts that later shifted. Ohmori reported ten cases in 1965 treated with a modification of the Gillies epithelial inlay.[13] Bruck reported on 156 cases treated with iliac crest grafts in 1969, but seemingly without further specifics on results.[14]

Against this background, we were prompted to review our experience with implants for saddle nose deformity at the New York Hospital from 1932 to the present. The records of 148 patients who received 213 implants were analyzed. Of these patients, 102 received one implant and 46 received multiple implants. Implant materials included:

Cartilage autograft	77
Cartilage homograft	61
Bone autograft	54
Bone homograft	3
Alloplastics	17
Dermal autograft	1

All of the bone autografts were cancellous iliac. Most of the alloplastics were silicone rubber. Most of the implants were done at the New York Hospital, with some patients being referrals after treatment failures elsewhere.

Analysis of success or failure can rest on several bases, with varying degrees of objective precision, for example, patient and/or surgeon satisfaction, third party approval, x-ray evidence of "take," frank dissolution of the graft, mobility of the graft, slough of skin and exposure of implant, infection, and so on. Since most attention and detailed follow-up—quite naturally—were available in the cases of those patients who received multiple implants, and since the percentage breakdown of etiology in these patients paralleled that of those who received only one implant (Table 2), it seemed that this group could yield the most useful information. Especially did it seem significant to know what material proved to be the "final" implant material, or, expressed in another way, in how many cases did a given implant have to be removed and/or another graft inserted for whatever reason. Ultimately, this seemed the valid criterion by which to judge success or failure.

Table 3 summarizes the findings in this multiple implant group. An outline of specific complications in the single and in the multiple implant group is indicated in Table 4. As might be ex-

Table 1. Etiology of the deformity by number of cases and approximate percentage

Cause of deformity	Number of cases	Percentage of total
Trauma	50	34%
Congenital	37	25%
Prior surgery	37	25%
Infection	10	7%
Cleft lip–nasal deformity	7	5%
Syphilis	5	3%
Radiation to nose	2	1½%
Total patients	148	

Table 2. Patients with one implant

Cartilage autograft	44
Bone autograft	25
Cartilage homograft	31
Bone homograft	1
Alloplastic (Silastic)	1

Table 3. Patients with multiple implants

	Number of grafts	Number of patients	Final procedure
Cartilage autograft	33	27	12
Bone autograft	29	23	16
Cartilage homograft	30	22	5
Bone homograft	2	2	2
Alloplastic	16	12	7
Other	1	1	1

*Forty-six patients with 111 implants.

Table 4. Complications

	Cartilage autografts	Cartilage homografts	Bone autografts	Bone homografts	Alloplastic
Warping	13	7	0	0	0
Loss of skin	0	3	2	0	2
Infection	3	3	1	0	3
Displaced graft	2	0	2	0	1
Resorbed	7	8	9	1	0
Total number of grafts	77	61	54	1	17

pected, most of these complications occurred in the multiple group and most resulted in the eventual loss of the implant.

CONCLUSIONS

Analysis of the preceding statistics leads us to make several conclusions and judgments.

1. A large number of all patients who are going to have an implant placed for purposes of

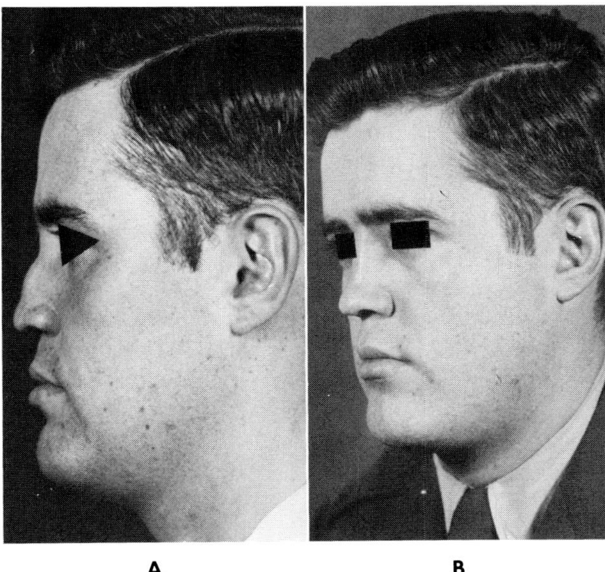

A   **B**

Fig. 16-2. This patient originally was injured in an airplane accident. A traumatic saddle nose deformity resulted. **A,** Results after the initial correction was done with a cartilage graft; it warped. **B,** The cartilage was removed and an iliac bone graft (autograft) was substituted. The patient is shown here 2 years after surgery.

correction of a saddle deformity are going to go on to having more than one implant placed (forty-six of 148 patients had multiple implants) before they and/or the surgeon feel that the desired or the final result has been achieved. Put in another way, the success rate can be expected to be no more than about 70%.

2. Cartilage, both autogenous and homologous, is clearly the most unsatisfactory material by a wide margin (Fig. 16-2). This is partly because of its well-known propensity to "warp," which might be alleviated by its being carved along certain lines of tension in the manner of Gibson[15] and as further clarified by Fry.[16] However, we also found a large number that were attendant with other complications of infection, erosion through skin, and displacement. A recent survey of French plastic surgeons indicates that cartilage has been progressively discarded for the correction of saddle nose deformities.[17]

3. The success-failure rate with alloplastic materials is approximately 50%, but the total number of cases and implants is probably too small to draw very firm conclusions beyond this statement of fact.

4. Bone homografts were used in only three instances, so no firm conclusions are drawn.

5. The most suitable material is autogenous bone. Even here, however, the success rate as measured by the number of "final" implants in the multiple implant group did not approach the good results reported by others, most lately by Campbell.[11] For the reasons stated previously, we did not assess for success or failure the twenty-five patients who received a cancellous iliac bone graft as their only implant—a much more favorable group—which might have improved our overall statistics.

Finally, we make a conclusion and judgment not really based on the above statistics, as it cannot be, but on a "clinical impression" from a perusal of the comments in these same clinical records. It seems that those patients who have the greatest sensitivity about their personal appearance are the most prone to have unsatisfactory results. Whether this results from something like their coercive insistence on the surgeon so that he does "a little bit more" in these patients and thereby jeopardizes the results or from something as simple as their manipulation of their noses postoperatively has to be entirely speculative. In any case, this factor considered together with the statistical analysis already presented should encourage cautious patient selection accompanied by frank and realistic preoperative counseling in these saddle nose deformed patients.

REFERENCES

1. Gillies, H. D., and Millard, D. R.: The principles and art of plastic surgery, Boston, 1957, Little, Brown & Co., pp. 103-111.
2. Denecke, H. J., and Meyer, R.: Plastic surgery of head and neck, vol. 1, New York, 1967, Springer-Verlag, Inc., p. 141-194.
3. McDowell, F., Valone, J. A., and Brown, J. B.: Bibliography and historical note on plastic surgery of the nose, Plast. Reconstr. Surg. 10:149-85, 1952.
4. Safian, J.: Progress in nasal and chin augmentation, Plast. Reconstr. Surg. 37:446-52, 1966.
5. Millard, D. R.: Corrective rhinoplasty and augmentation mentplasty. In Grabb, W. C., and Smith, J. W., editors: Plastic surgery—a concise guide to clinical practice, Boston, 1968, Little, Brown, and Co., p. 509.
6. Stark, R. B.: Plastic surgery, New York, 1962, Harper and Row, Publishers, p. 350.
7. Foman, S., and Bell, J.: Rhinoplasty—new concepts, evaluation and application, Springfield, Ill., 1970, Charles C Thomas, Publisher, pp. 45-47, 178.
8. Barsky, A. J., Kahn, S., and Simon, B. E.: Principles and practice of plastic surgery, ed. 2, New York, 1964, McGraw-Hill Book Company, pp. 267-268.
9. Converse, J. M., and others: Deformities of the nose. In Converse, J. M., editor: Reconstructive plastic surgery, Philadelphia, 1964, W. B. Saunders Co., pp. 745, 752.

10. Brown, J. B., and McDowell, F.: Plastic surgery of the nose, Springfield, Ill., 1965, Charles C Thomas, Publisher, pp. 180-217, 408-423.

11. Farina, R., and Villano, J. B.: Follow-up of bone grafts to the nose, Plast. Reconstr. Surg. **48:**251-255, 1971.

12. Campbell, R. M., and de Cordier, M.: A study of 100 autogenous bone grafts to the nose. In Robbett, W. F., editor: Proceedings of the Centennial Symposium of the Manhattan Eye, Ear & Throat Hospital, vol. 2: Otolaryngology, St. Louis, 1969, The C. V. Mosby Co., pp. 223-228.

13. Ohmori, S.: Surgical treatment of saddle nose, Plast. Reconstr. Surg. **36:**117-123, 1965.

14. Bruck, H. G.: Fifteen years experience in the reconstruction of saddle nose by means of autologous bone grafts, Mschr. Ohrenheilk, **103**(12):548-553, 1969; reported in Excerpta Medica—Plastic Surgery **1**(12):185, abstract 1290, 1970.

15. Gibson, T., and Davis, W. B.: The distorsion of autogenous cartilage grafts: its cause and prevention, Brit. J. Plast. Surg. **10:**257-274, 1957-1958.

16. Fry, H.: Cartilage and cartilage grafts: the basic properties of the tissue and components responsible for them, Plast. Reconstr. Surg. **40:**426-439, 1967.

17. Peck, A., and Jausseran, M.: Study of implants in saddle nose, Ann. Chir. Plast. **15:**38-43, 1970; reported in Excerpta Medica—Plastic Surgery **2**(9):170, abstract 1186, 1971.

Chapter 17

Considerations in the management of premalignant and malignant lesions of the nose with emphasis on cryosurgery

Robert M. Goldwyn, M.D.

Setrag A. Zacarian, M.D.

Considerations in the treatment of premalignant and malignant lesions of the nose are patently appropriate to a symposium on aesthetic surgery. The saying "It's as plain as the nose on your face" emphasizes the fact that a nasal deformity, either congenital or acquired, cannot be easily hidden. With respect to cutaneous growths of the nose, our goal here as elsewhere should be curing without deforming; admittedly, we cannot always attain this objective.

The first practical measure to avoid deformity is to avoid unnecessary surgery. Before undertaking an excision of a nasal lesion whose diagnosis is doubtful, obtain a biopsy. Microscopic findings may obviate surgery. Any incision in the nose with removal of more tissue than a small-sized biopsy should not be done in a cavalier fashion. Nasal skin is precious; scars may heal unkindly and are difficult to mask, particularly in the male whom society has denied the use of makeup.

For many premalignant lesions, such as senile keratoses, good alternatives to classical excision and closure are available: surgical shaving, cytotoxins, and cryosurgery. For malignant lesions, such as certain epitheliomas, cytotoxic agents, cryosurgery, and, occasionally, irradiation are also useful.

SURGICAL SHAVING

Surgical shaving removes a lesion, like a keratosis, in the same manner as one takes a split graft; it allows a specific diagnosis and prompt re-epithelialization with minimal scarring.[1]

CYTOTOXIC AGENTS

Of the cytotoxic agents, the most widely used is a 5-fluorouracil for premalignant actinic keratoses resulting from long exposure to sunlight or other sources of ultraviolet radiation and also for premalignant keratoses associated with late x-ray dermatitis. Although it has also been employed in the treatment of some superficial basal cell carcinomas and intraepithelial squamous cell carcinomas, at the moment this substance has not been approved for such use by the Food and Drug Administration. 5-Fluorouracil is not indicated for treatment of seborrheic keratoses or senile warts.[2]

Fluorouracil is a modified pyrimidine similar to uracil and thymine, with whose metabolism it competes. It acts by blocking the methylation of deoxyuridylic acid to thymidylic acid, thereby interfering with the synthesis of deoxyribonucleic acid. Its specific action in keratoses is not known, but it may achieve its effect by immunologic means.

Preparations of the drug recently marketed under the name Efudex (Roche) include a 5% 5-FU cream and a 2% or 5% 5-FU solution in propylene glycol. Fluoroplex (Herbert Laboratories), a 1% 5-FU solution in propylene glycol, works very satisfactorily. It can be applied to the lesion twice daily with a small cotton swab for 2 to 6 weeks. If a growth is close to the eyes, extreme care should be used. Many older patients cannot see well and therefore should not be allowed to apply this medication. For them, a cream or no cytotoxic substance would be better.

Topical 5-fluorouracil causes marked erythema in the area treated; there keratoses become crusted and fall off. Scarring is minimal. For actinic keratoses of the hands and arms, 5-FU is not as effective as for lesions on the face. This substance has the additional advantage of causing the disappearance of premalignant lesions too small to be detected clinically.

The intensity of this agent's action is increased by an occlusive dressing and by ultraviolet light, as in sunlight. Patients experience pruritus, burning, and phototoxicity and should be warned about the inflammatory response in the areas treated. Patients should be seen every 2 weeks or sooner to evaluate the drug's effectiveness. Healing may not be complete 1 or 2 months after therapy has stopped. If irritation is severe, cold water compresses or topical corticosteroid preparations are useful. Occasionally, the drug causes an allergic skin reaction.

It is probably wise not to have a pregnant patient use this solution, because there might theoretically be some effect on the fetus even though this has not been established. Before using 5-fluorouracil, one might wish to biopsy a lesion, but certainly if the growth persists after a satisfactory therapeutic trial, then surgery or some other form of treatment should be employed.

IRRADIATION

This is not the place to detail the advantages and disadvantages of irradiation for cutaneous carcinomas of the nose. However, there are certain patients in whom irradiation is the procedure of choice. These are often patients who are elderly, with large extensive carcinomas, and who refuse other forms of treatment. Sites where irradiation is valuable are the inner canthus of the eye and the immediately adjacent nasal skin. Modern techniques of x-ray delivery and improved methods of shielding should now make us less reluctant to enlist the aid of the radiotherapist. The effect of irradiation can be potentiated by topical 5-fluorouracil.

SURGICAL EXCISION

When surgical extirpation is necessary, the degree of difficulty for the surgeon and the extent of deformity for the patient depend upon the nature of the lesion, its size, and its location. Areas of the nose where closure is less difficult (not "easy") are the root, the nasolabial region, and, in certain instances, the middorsum. Where closure almost always presents a problem is on the tip, the columella, the nostril rim, and the inner canthal area. As mentioned previously, for epitheliomas in these difficult locations, cryosurgery and irradiation should be considered.

It would be inadvisable to present a tedious catalogue of procedures that can be employed to reconstruct the nose following excision of a cutaneous malignancy. Instead, a few helpful techniques will be emphasized.

Simple yet useful is the subcuticular suture— 4-0 nylon or monofilament wire for closing defects on the dorsum of the nose. The wire reduces wound tension on skin sutures and can be left 7 to 14 days, thus permitting earlier removal of skin sutures (5-0 & 6-0 nylon) and occasionally even obviating the need for them.

For moderately sized lesions, local rotation flaps are indispensable. Some of these have been well summarized by Elliott.[3]

So useful is the nasolabial flap that one might wonder if that tissue existed to aid plastic surgeons. It can be employed to reconstruct the nostril rim and the columella and for nasal lining. The donor site is easily closed. Occasionally another operation is required to make the point of rotation at the base of the flap less conspicuous.

Local flaps can be combined with grafts, either split or full thickness, placed in a less obtrusive position at the side of the nose where the glasses rest rather than on the dorsum, which is more obvious. Local flaps have the obvious advantages of carrying their own blood supply and providing excellent color and contour. If a lesion is particularly invasive, or if it is recurrent, it is wise to obtain frozen sections during the procedure to make certain that the margins are clear before covering with a local flap. Although distal flaps are admittedly helpful and often the only solution, they should be avoided whenever possible. The liberal use of the nasolabial flap in bold dimensions or a combination of local flaps often eliminates the need for a forehead flap, for example.

Delayed closure should not be forgotten as a very helpful technique to cover large defects where bare bone is exposed, where there is insufficient tissue for local flap, and where one wishes to avoid the disfigurement and inconvenience of a distal flap. The would can be left to granulate for 7 to 12 days and a split graft from the supraclavicular region can then be applied, even on an out-patient basis. The take is almost always excellent and the ultimate contour very satisfactory.

CRYOSURGERY

The cryosurgical approach for the eradication of benign and particularly premalignant lesions of the skin had its beginnings at the turn of the century.[4-6] Liquid air was the first refrigerant and, for a short period, carbon dioxide came into use. With technologic improvements, the commercial production of liquid nitrogen has rendered this cryogen the most universal and useful for cryosurgery. The early attempts[7-8] to destroy malignant tumors of the integument were abandoned, and for the most part cryosurgery was limited to benign and precancerous growths. The method of application has been essentially the liquid nitrogen-soaked cotton swab, firmly pressed on the respective lesion for a matter of several seconds.

In the past 40 years, with a clearer understanding of the biologic and physicochemical alterations of cells and tissues subjected to sub-zero temperatures, cryobiology and cryogenics have offered the clinician a keener interpretation of therapeutics with this modality. Modern cryosurgery entered a new phase with the pioneering work of Irving Cooper.[9] Technologic barriers were overcome by the engineers and more sophisticated instrumentations were designed for the cryosurgical approach to cutaneous,[10-12] oral,[13] and visceral[14] neoplasms.

For the effective eradication of malignant cutaneous tumors certain principles must be strictly adhered to.

1. *Delivery system* must be versatile and mobile with controlled direct spray of the cryogen (liquid nitrogen).
2. *Freezing time* will have to vary from one to several minutes, depending upon the size of the tumor and the achievement of a minimum temperature of $-25°C$.
3. *Temperature monitoring* is essential for larger tumors. A microthermocouple needle should be inserted below the tumor to register the temperature on a pyrometer.
4. *A double freeze thaw cycle* may be necessary for effective cryonecrosis.
5. *Freezing* must extend beyond the visible margins of the tumor.

When these parameters are followed, cryosurgery is both a refined and a simple procedure. The

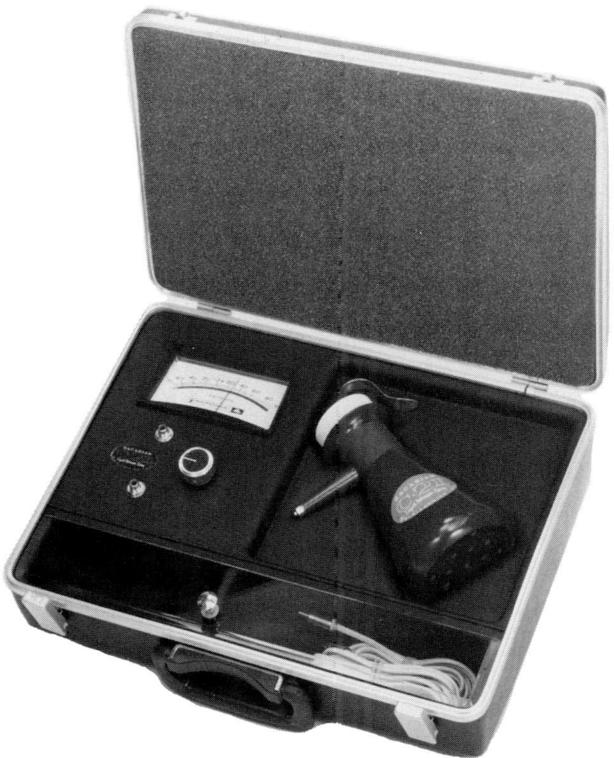

Fig. 17-1. The cryosurgical instrumentation of the C-21 unit including the microthermocouple needle and pyrometer. (Courtesy Frigitronics, Inc., Shelton, Conn.)

direct spray of liquid nitrogen has been most useful; a probe or disk with circulating liquid nitrogen can be readily attached to the C-21 unit (Fig. 17-1) for both oral lesions and the free margin of the eyelid. In the past 7 years, one of us (Zacarian) has cryosurgically treated 730 patients with some 1,240 malignant tumors of the skin. With the exception of sixty-eight epidermoid carcinomas and forty-two basosquamous cell carcinomas, the remaining for the most part were basal cell carcinomas. It is important to note that of the 730 patients, 220 presented their lesion on the nose. This constitutes approximately 30% of all facial cancers. A further observation revealed that the most common site of involvement of the nose, some 20%, was the ala nasi. This area, along with the tip and nasolabial folds, is indeed a therapeutic challenge to any cancer therapist. Cryosurgery for tumors of these regions is particularly rewarding from a cosmetic point of view. We all appreciate the thin skin overlying the nose and the ears and we should have high regard for the underlying cartilage. Chondronecrosis following cryosurgery is extremely rare, as are hypertrophic scars and keloids.

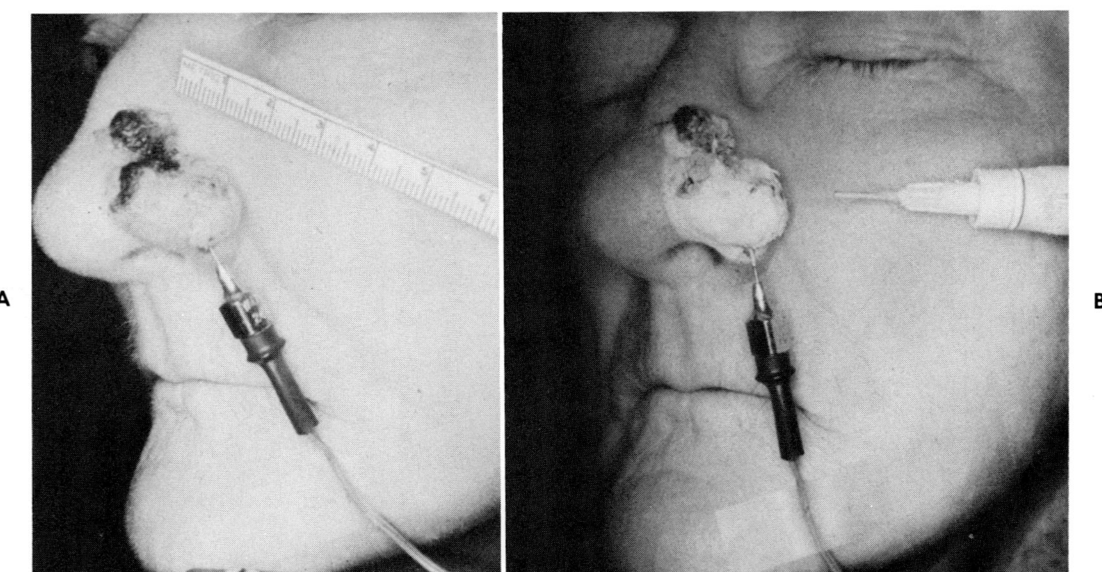

Fig. 17-2. An elderly patient with a large fungating basal cell carcinoma of the left nostril. **A,** The microthermocouple needle is positioned below the tumor to underlying cartilage. **B,** The tumor was frozen for 4 minutes with a double freeze thaw cycle and a monitored temperature of −30°C.

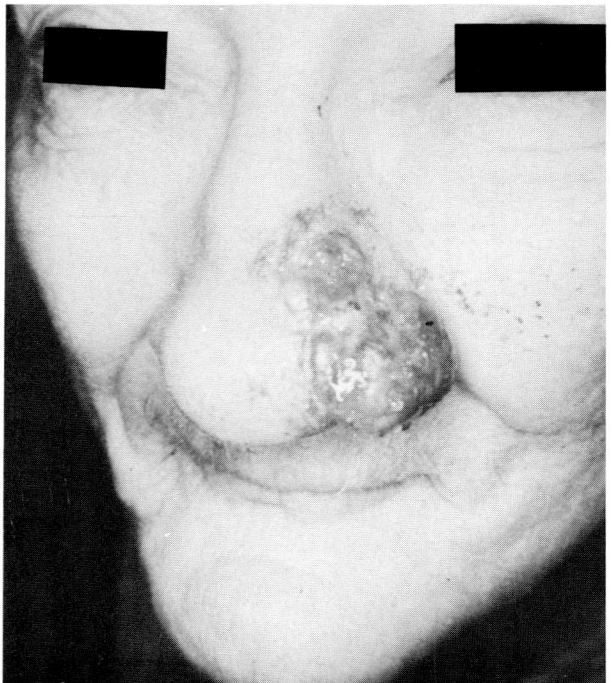

Fig. 17-3. Same patient in Fig. 17-2 24 hours after cryosurgery. Note the marked edema and serous exudate.

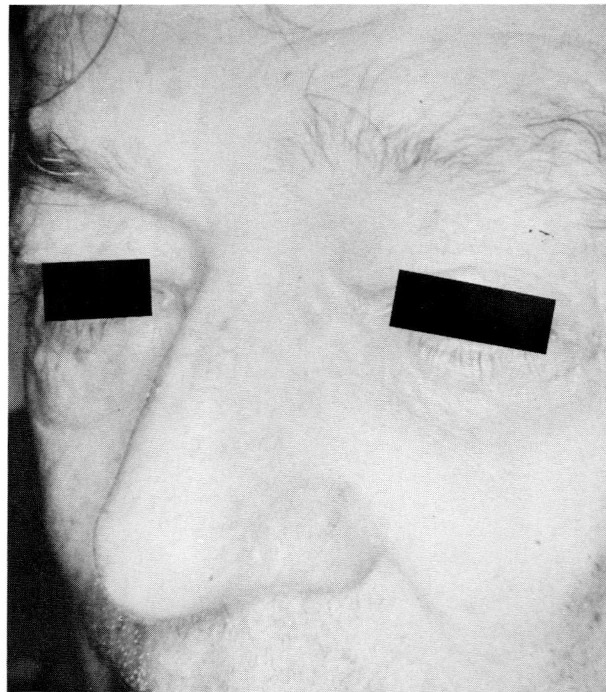

Fig. 17-4. Same patient as in Fig. 17-2 4 weeks after cryosurgery. Complete clinical eradication of the tumor and absence of hypertrophic scars. This patient has been followed for 18 months without a recurrence.

Cryosurgery is therefore especially useful for lesions whose removal by standard excision would leave considerable deformity.

Another indication for cryosurgery is the recurrent skin cancer following irradiation and/or surgery when further surgery or irradiation is not feasible or advisable.

The essential technique has been outlined previously.[10, 11] The importance of monitoring the temperature of the ice front subcutaneously or below the tumor during the freezing procedure should be stressed (Fig. 17-2). The only failures were because the thermal gradient was not properly monitored or the freezing did not extend sufficiently beyond the visible margins of the tumor. Local anesthetic can be used when inserting the thermocouple needle and a double freeze thaw cycle will enhance cryonecrosis.[15] Immediately following the initial freezing, the tumor should be allowed to return to ambient temperature and to freeze again to $-25°$ to $-30°C$. The total cryosurgical time may vary from 2 to 4 minutes. Care must be taken to prevent the spray of liquid nitrogen to adjacent vital areas as the eye or lips, if freezing the tumor approximates these areas. The initial pain during freezing disappears but returns during the thawing period for several minutes. This momentary discomfort is well tolerated by most patients. A hemorrhagic bulla along with edema and serous exudate may last for several days (Fig. 17-3), and this is followed by an adherent eschar for several weeks. This will slough within 3 to 4 weeks. Healing time is slow but indeed a small price to pay for so minimal a scar that results from this procedure (Fig. 17-4).

Slight depigmentation may be noted for a while and very infrequently there is a hypertrophic scar. In our experience a keloid has not been observed.

Time and space do not allow a discussion in depth regarding pathogensis of cryonecrosis. Suffice to say that sub-zero temperatures of $-25°$ to $-30°C$. destroy in part both normal and malignant cells, but more importantly, the thrombosis and complete permanent occlusion of the microvessels of the integument produce ischemic necrosis and final destruction of both normal and malignant tissue.[16-17] The reason for superior wound healing following cryosurgery still remains an enigma. There is no doubt, however, that every cryosurgeon will attest to the minimal deformity and superior cosmetic end result from cryogenic

surgery. The cure rate with this modality appears equal to other existing techniques. In 7 years, 1,240 malignant tumors of the skin have been treated by one of us (Zacarian) and only twenty-seven recurrences noted, a cure rate of 97%.

Since cryosurgery of malignant cutaneous tumors is a comparatively new field, it must wait for a more general endorsement. This will demand greater use, careful follow-up, and further investigation.

REFERENCES

1. Goldwyn, R. M., and Arndt, K. A.: Useful adjuncts for improving the aging face, presented at the Symposium on Aesthetic Surgery, November 13, 1970, Scottsdale, Arizona. In press.
2. The Medical Letter **13**:45-46, 1971.
3. Elliott, R. A., Jr.: Rotation flaps of the nose, Plast. Reconstr. Surg. **44**:147-154, 1969.
4. White, A. C.: Liquid air, its application in medicine and surgery, Med. Records **56**:109-114, 1899.
5. White, A. C.: Possibilities of liquid air to the physician, J.A.M.A. **36**:426-428, 1901.
6. Pusey, W. A.: The use of carbon dioxide snow in the treatment of nevi and other skin lesions, J.A.M.A. **49**:1354-1357, 1907.
7. White, A. C.: Liquefied oxygen and x-ray treatment of malignant growths, Interstate Med. J. **9**:657-660, 1902.
8. Whitehouse, H. H.: Liquid air in dermatology, its indications and limitations, J.A.M.A. **49**(5):371-377, 1907.
9. Cooper, I. S.: Cryogenic surgery of basal ganglia, J.A.M.A. **181**:600-604, 1962.
10. Zacarian, S. A.: Cryosurgery in dermatology, Int. Surg. **47**(6):528-534, 1967.
11. Zacarian, S. A.: Cryosurgery of skin cancer and cryogenic techniques in dermatology, Springfield, Ill., 1969, Charles C Thomas, Publisher.
12. Goldwyn, R. M., and Roseoff, C. B.: Cryosurgery for large hemangiomas in adults, Plast. Reconstr. Surg. **43**:605-611, 1969.
13. Gage, A. A., and others: Cryotherapy for cancer of the lip and oral cavity, Cancer **18**:1646-1651, 1965.
14. Cahan, W. G.: Cryosurgery of malignant and benign tumors, Fed. Proc. **24**(2):S241-S248, 1965.
15. Stone, D., Zacarian, S. A., and Peri, C.: Comparative studies of mammalian normal and cancer cells subjected to cryogenic temperatures *in vitro*, J. Cryosurg. **2**:43-52, 1969.
16. Zacarian, S. A.: Histopathology of skin cancer, following cryosurgery, Int. Surg. **54**(4):255-263, 1970.
17. Zacarian, S. A., Stone, D., and Clater, M.: Effects of cryogenic temperatures on microcirculation in the golden hamster cheek pouch, Cryobiology **7**(1):27-39, 1970.

Chapter 18

Reconstruction of the nose with local flaps

James H. Hendrix, Jr., M.D.

Whenever the nose is lost in whole or in part, whether full thickness or partial thickness, the search for materials for reconstruction is guided by many factors. Among the factors that must be considered are the condition of the patient, the causative agent or disease process, the availability of tissue near and far, and the steps or stages necessary to execute the job.

Partial losses of the nose can often be repaired immediately, thus minimizing the length of time the victim needs to bear his deformity. Very early restoration of missing parts is desirable except in extreme mitigating circumstances. Circumstances or conditions that might dictate that repair be delayed are disease—local or generalized—infection, concomitant injuries to other parts, injuries to adjacent tissues or parts that might not be usable initially but that could be used after their recovery.

Skin coverage that matches best in color, texture, and thickness is that brought in from near at hand—the nearer the better. Also, fewer stages are usually required to accomplish a major job of coverage when flaps of adjacent tissue are used.

DEFECTS AND TYPES OF FLAPS

Skin defects of the lower half of the nose can usually be conveniently repaired by flaps from the nasofacial groove or a large rotation of skin from higher up on the nose (Fig. 18-1). These flaps have excellent blood supplies and usually heal promptly. The skin of the nasofacial groove is often spared in injury and by cancer. The pedicles of these flaps can be short and narrow or, if need be, can be denuded and used as island pedicles based on the lateral nasal artery and vein. An adventitial or subcutaneous flap can be ap-proximately 1 cm. wide and fairly thick. It can be extended by dissection deep into the nasofacial groove. The flap can be tunnelled under or placed under a portion of the incised skin in order to place it in its desired bed. Defect repair is no problem and leaves minimal deformity and scarring.

Full-thickness defects of the lower portion of the nose, such as a notch, can be handled by designing a flap from the adjacent ala. The missing tissue in the new defect at the upper edge of the advanced alar cartilage can be covered with a free graft from the ear (Fig. 18-2).

Skin defects of the upper half of the nose lend themselves to repair by flaps from the glabella region and the midline of the forehead (Fig. 18-3).

When losses of skin coverage of the nose extend beyond the nasofacial groove, complicating or eliminating the use of flaps from that area, a combination of flaps may be necessary to give the desired result. A combined flap repair might consist of a rotation advancement flap from the cheek or cheek and neck plus a forehead flap (Fig. 18-4).

In complete loss of the nose, restoration of framework is necessary. This can be done after coverage and lining have been provided. However, bony support is more often provided at the time soft tissue coverage is swung into position. The elimination of procedures cuts down on scar contracture and shrinkage. Millard[1] has promoted faster and more direct repairs of this nature.

LINING OF FLAPS

When there is a through-and-through defect or complete loss of the nose, lining of good quality mucous membrane or hairless skin is necessary.

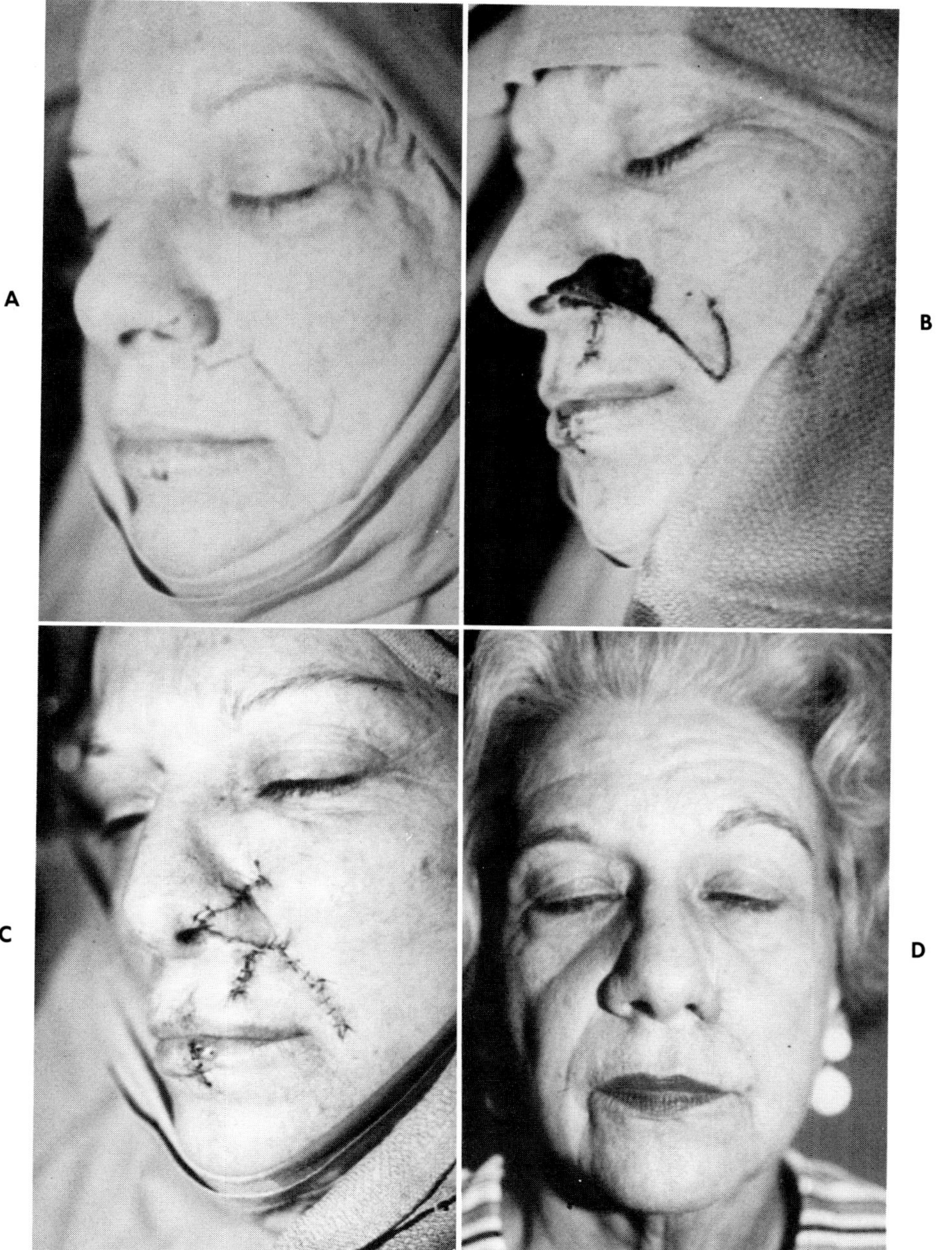

Fig. 18-1. A, Carcinoma of left ala, cheek, and upper lip. Excision and flap marked. **B,** Defect after excision. Flap incised. **C,** Closure with flap in position. **D,** Final result.

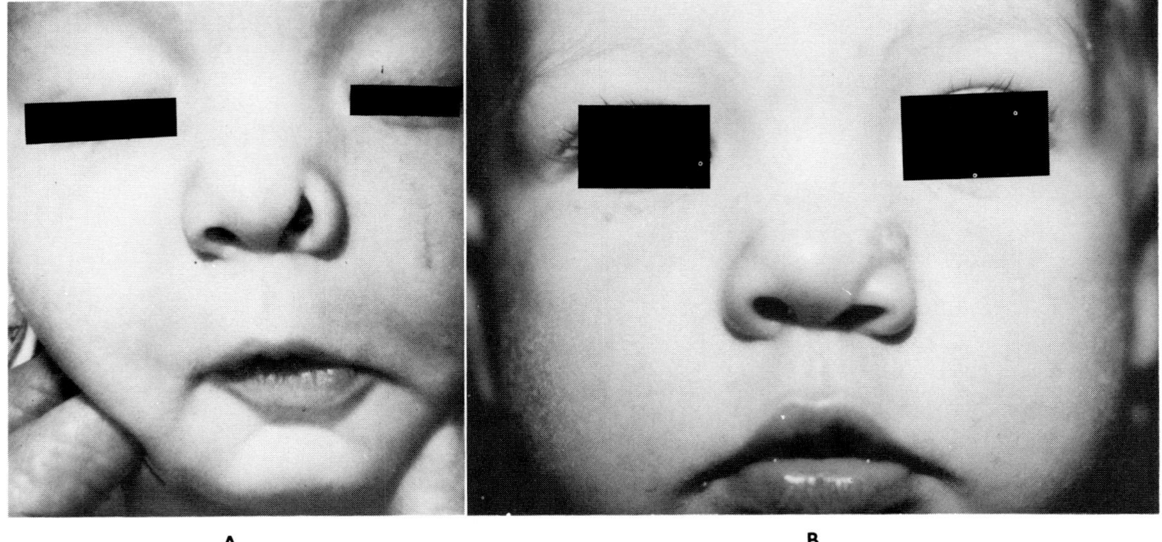

Fig. 18-2. **A,** Congenital notch deformity of ala. **B,** Correction by alar flap and full-thickness skin graft.

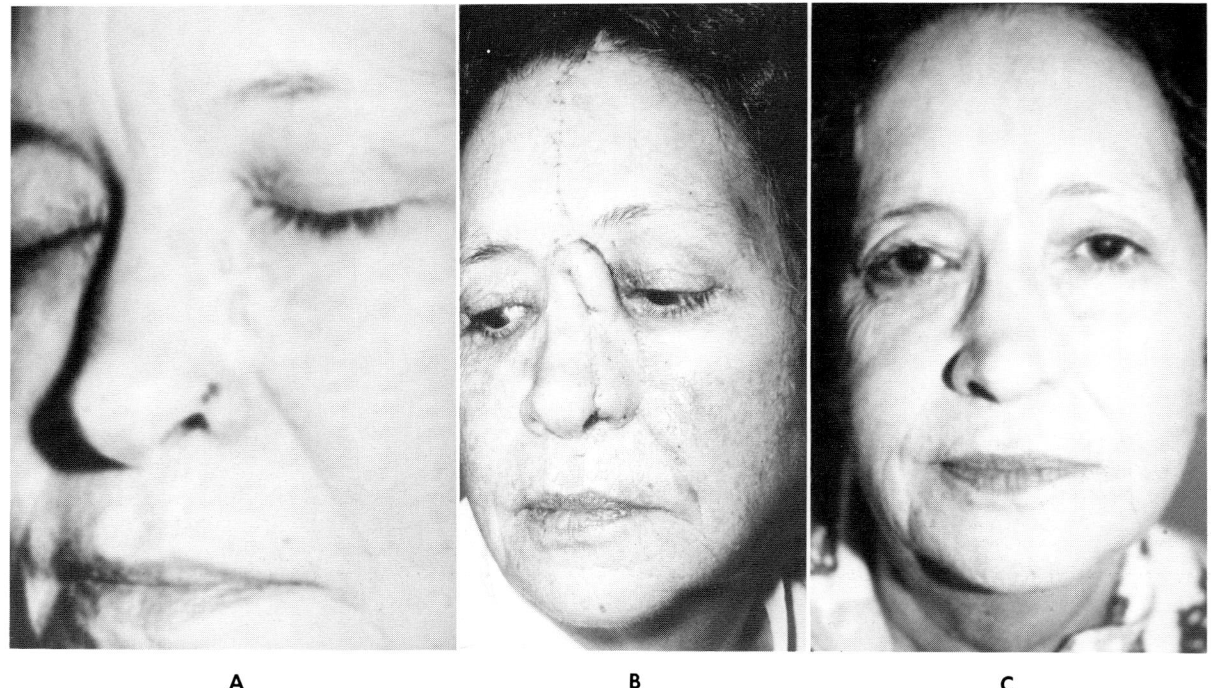

Fig. 18-3. **A,** Carcinoma on alar margin of nose. Thin atrophic scar of nasofacial groove with beginning breakdown secondary to radiation therapy. **B,** Forehead flap in position. **C,** Flap set in. After a few months, lateral edge of flap will be trimmed and cheek skin advanced to nasofacial groove.

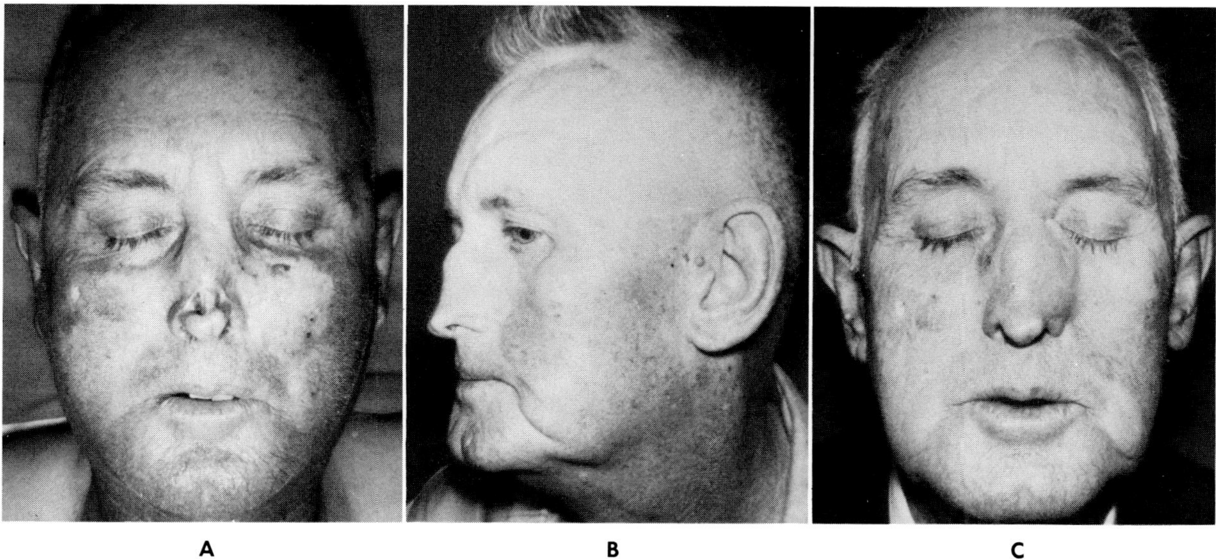

Fig. 18-4. A, Through-and-through defect of lower nose. Squamous cell carcinoma of nose, left lower lid, and cheek. **B,** After resection, repair was with forehead flap plus the rotation of cheek flap. Split skin graft to forehead and preauricular area. **C,** Front view of postoperative result.

Local mucous membrane flaps are considered first, but when the size of the defect rules them out, local turnover flaps are a possibility. In a through-and-through defect, one may suture a free graft across the defect like the skin over a drum frame prior to dropping the flap in place. Lining may be supplied by applying a graft to the underside of a flap approximately 2 weeks prior to transfer of the forehead coverage flap. An example of this is the chondrocutaneous flap described by Millard[2] and others.

CONCLUSION

Whenever possible, reconstruction of loss of coverage of a portion of the nose should be performed using local tissue. This provides coverage of more nearly normal color and texture. The contour is usually more nearly normal. Flaps rarely need delay, and in many instances reconstruction can be accomplished in one procedure. These procedures can frequently be done under local anesthesia, and there is little resultant deformity from the donor site, since these defects can usually be closed to create a fine line scar. This, in turn, can frequently be hidden in the normal facial grooves or along the normal lines of tension.

REFERENCES

1. Millard, D. R.: Eyelid repair with a chondromucosal graft, Plast. Reconstr. Surg. **30:**267, 1962.
2. Millard, D. R.: Hemirhinoplasty, Plast. Reconstr. Surg. **40:**440, 1967.

Chapter 19

Surgery of the alar wings and base

Robert Pool, M.D.

The aesthetic goal of an attractive nose through a corrective rhinoplasty is an artistic challenge that requires surgical perfection in the execution of the repair in all nasal anatomic regions. The shape, height, and location of the nasion or root of the nose determines its character, but the architecture of the tip and alae establishes its aesthetic appeal (Fig. 19-1). Nowhere is surgical perfection more needed than in properly balancing the ala and tip to the rest of the nose and to the face on which they reside. And nowhere is the choice more exquisitely delicate.

External excisions at the alar base were among the earliest utilized in rhinoplasty, being described by Weir in 1892, modified by Joseph, subsequently fertilized by Aufricht, and artistically expanded by Millard[1] to include the alar rim as well as the base. This was followed by a long period of disuse that is hard to explain except for the exaggerated fear of external scars. These fears have largely been unwarranted.

It is legitimate to question the validity of alar base excisions in reference to:

1. Whether the need for these excisions can be anticipated preoperatively
2. Whether they should be done primarily or secondarily
3. How the surgeon can calculate the size and geometric shape of the area to be excised

DEFINITION OF TERMS

In order to communicate effectively and simply on this subject, it is appropriate that we define some of the terms utilized in describing the alar configurations.

Alar "flare" is that external rotation or shift of the ala and its rim from the midline (seen in exaggerated fashion in the cleft lip). It may be present preoperatively or may be accentuated by the surgical lowering and narrowing of the radix and/or the alar cartilages, presenting to view the inner nostril instead of the more attractive curve of the rim (Fig. 19-2). Lowering and narrowing of the root with radical alar cartilage resection will produce greater tendency to alar flare (especially in a female with thin overlying skin). There will also be a tendency to develop the "knock-kneed" alar appearance (Fig. 19-3),

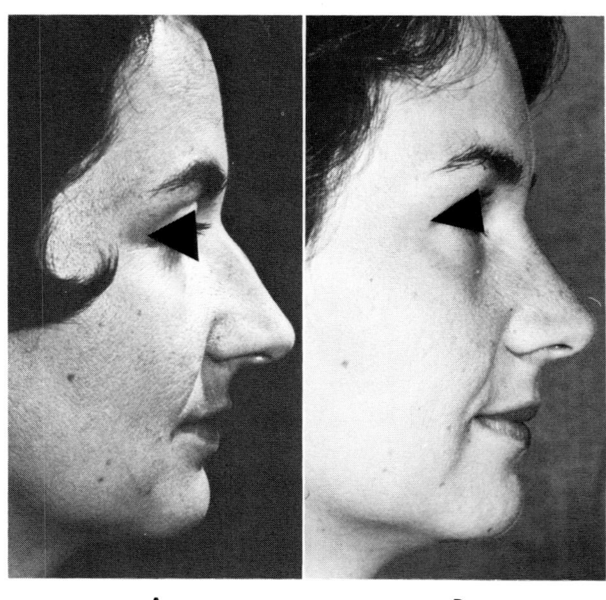

Fig. 19-1. A, Preoperative view of tip needing delicate refinement. **B,** Postoperative view showing that nostril attractiveness is not only related to facial angle but to its pleasing curve from columella to cheek.

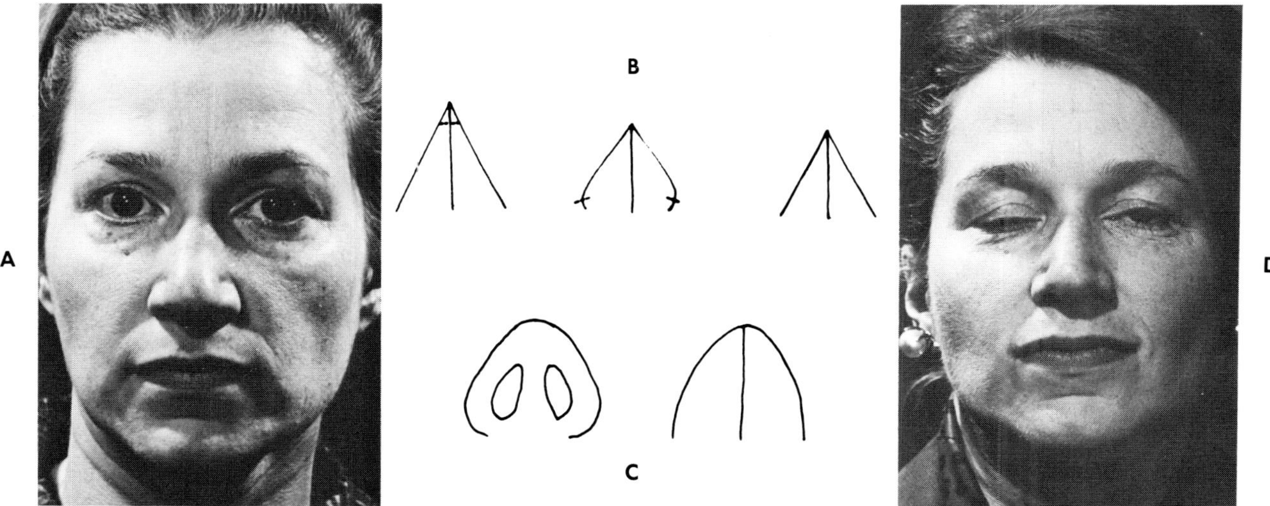

Fig. 19-2. A, Patient who had three previous procedures and uncorrected alar flare and widening. Sagging sidewalls are disproportionate to inner canthal width and upper nasal width. **B,** Schematic drawing showing alar widening after reduction of tip height. This will occur if cartilage rim is intact or skin is thick, or both. Base excision restores pyramid at tip. **C,** Schematic drawing showing totally wide tip with only central reduction in height, as in **A. D,** Postoperative appearance after alar base excision, giving alar narrowing.

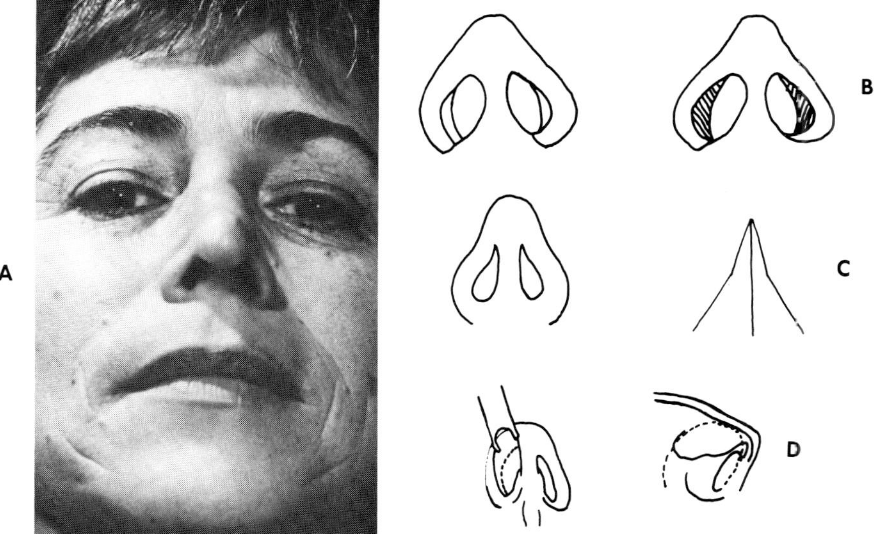

Fig. 19-3, A, "Knock-kneed" nostril and alae caused by loss of alar cartilage support and lowering of central tip support. **B,** Schematic drawing illustrating alar flare, with exposure of less attractive inner nostril. **C,** Schematic drawing of unattractive nostril contour that may appear later (6 to 18 months) postoperatively. **D,** Pernicious intranasal circumferential incision giving pinched tip, aggravating alar flare, and sometimes internal obstruction from scar web.

quite often the assassin of a natural nasal appearance. Both of these latter conditions may not be apparent until 9 months or later after surgery. Interruption of function of the nasalis muscle may aggravate this condition, as does the use of uninterrupted, circumferential, intranasal incisions from the base of the columella across the dome to the piriform margin laterally (Fig. 19-3, *D*). It is well to remember that the distal nose in most humans is highly mobile. This motion can aggravate the degree of intranasal scarring, which eventually can pinch the supratip area and further evert the alae.

Alar width refers to its greatest transverse dimension (Fig. 19-4, *A*). It will widen when the nasal tip must be surgically reduced in height or width by alterations in the alar cartilage. The more radical alar cartilage alterations produce greater widening. The skin of the nose is a drape, and when its supporting framework is altered it may droop, sooner or later. This is precisely where alar base excisions can be pivotol.

In so doing we must clarify the Weir and Joseph excisions. The Weir excision (Fig. 19-5, *A*) is a basilar wedge excision extending superiorly in the nasolabial angle. The Joseph "modification"

is a wedge excision at the very nostril base. To clarify this it may be helpful to use the term *internal circumference* to describe the internal nostril dimension. If this is excessive, tissue must be removed from the internal rim and can best be done at the base as described by Joseph (Fig. 19-5, *C*).

External nostril circumference is a term that would apply to the outer skin dimension. If this line is lengthened, the wedge excision must include more external skin laterally and/or superiorly. When it extends superiorly it is generally called a Weir excision. In practice these definitions may be somewhat artificial since aesthetic demands of the face may require a subtle blending of the two techniques.

In general, if the rim is flared outward or the nostril is "knock-kneed" (instead of oval), tissue must be removed primarily from the rim. If the ala is too wide because of an excessive external skin dimension, the excision must remove more of the lateral skin. This brings us to the aesthetic evaluation of alar width and shape.

Great attention has been paid to the height, shape, and angle of the nasal profile and much less consideration to the frontal visage. This as-

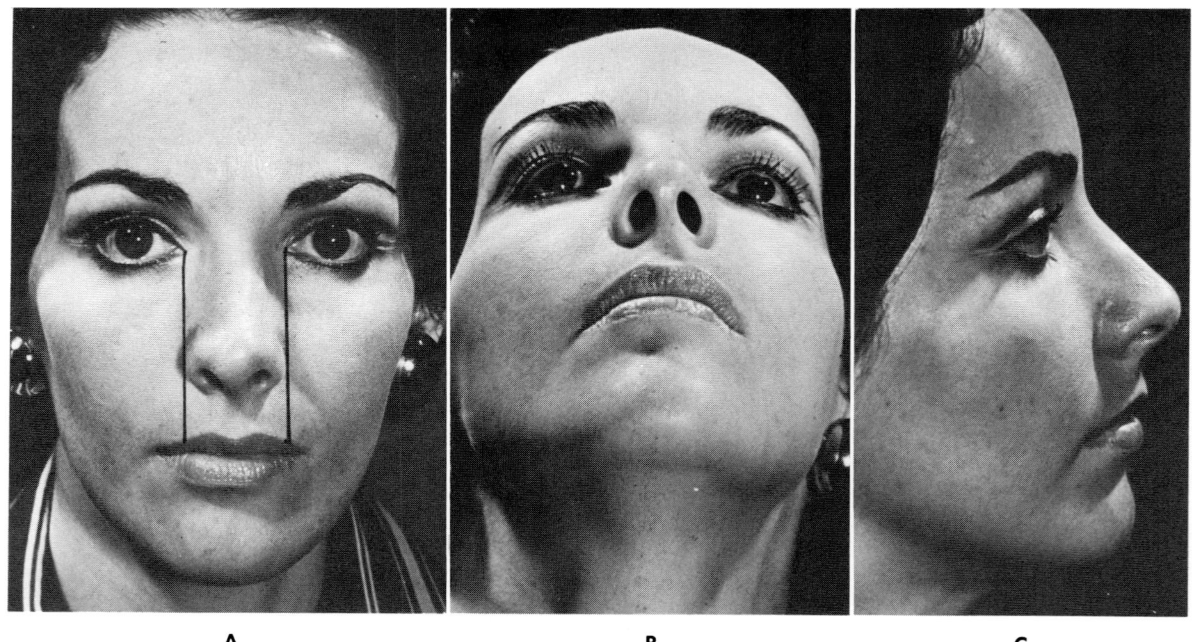

A   B   C

Fig. 19-4. A, Harmonic anterior guidelines. Nostril width should be proportionate to inner canthal width. This is a helpful guide preoperatively as well as during surgery. **B,** Attractive oval nostrils and alar width. **C,** Lateral view. Alar rim shows attractive arching and smooth curves, which are feminine in their appeal.

pect must be evaluated at rest and in motion with an artistic analytic three-dimensional eye.[3, 4] Preoperative study of the elemental components of the root and tip and their relationship to the attachments of the lip and cheeks will steer the sensitive surgeon in the correct direction, if he charts the course dictated by facial balance.

AESTHETIC GUIDELINES

One of the useful landmarks in evaluating the anterior harmonic proportion of the eyes and nose is to compare the lateral external alar width of the nose with a line dropped vertically from the medial canthus of the eye. If the external border of the ala is lateral to this line preoperatively or after tip reduction, it is safe to assume that narrowing would be aesthetically helpful (Fig. 19-4, *A*). Also, if the nostril has dropped into the "knock-kneed" posture or is flared, an alar base resection is indicated. If tip height is reduced at

the medial crus of the alar cartilage, leaving the juncture between the lateral margin of the alar cartilage and alar skin intact, there is less tendency for alar widening or nostril collapse.

Converse[5] has emphasized the value of keeping this juncture intact to minimize this alteration in nostril shape. The type of alar cartilage resection performed depends on a multiplicity of factors not to be considered here but which must be decided before alar base revisions are considered.

SURGICAL CORRECTION AT THE ALAR BASE

This surgical correction is the last step in the rhinoplasty and begins with a determination of the best camouflaged incisional site at the nostril floor. This area should be lightly scored with a blade, ensuring symmetry with caliper measurements supported by a visual survey from several viewpoints. Incision is made completely through

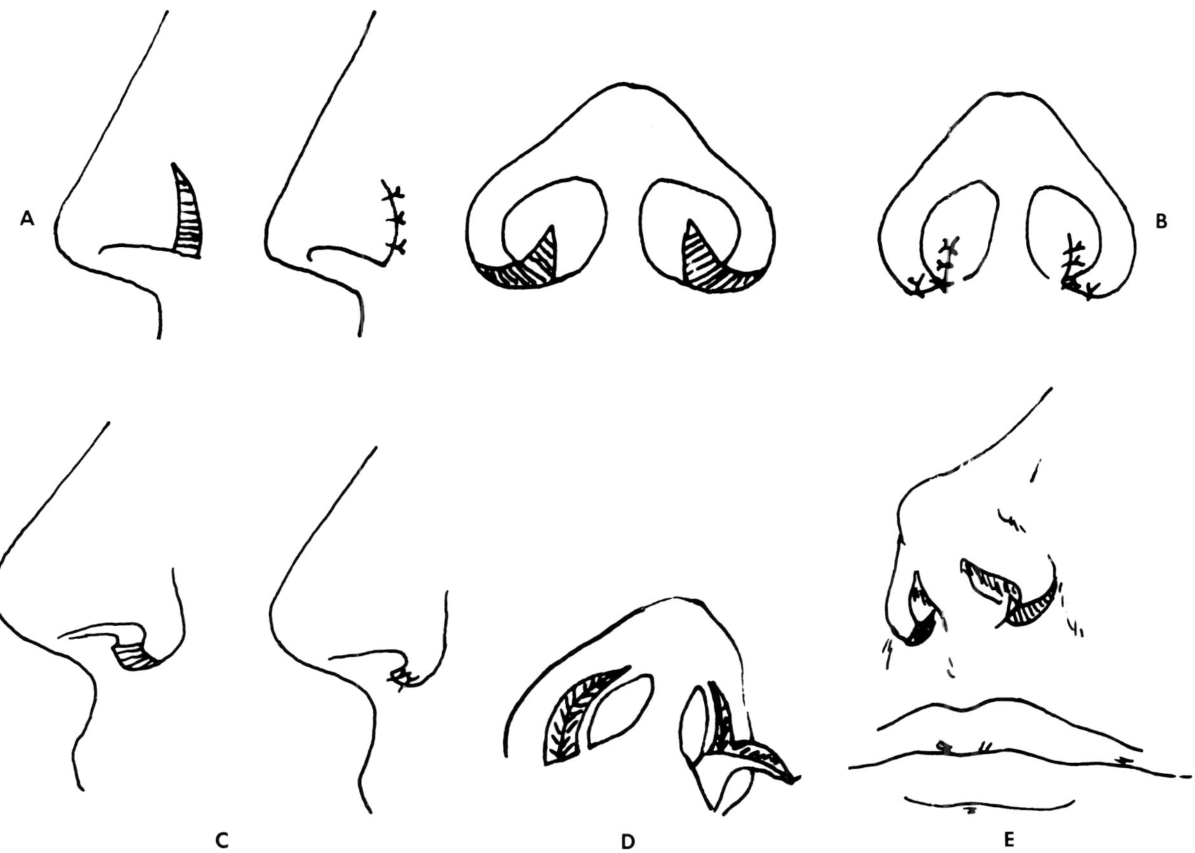

Fig. 19-5. **A,** Weir alar base excision, as illustrated by Joseph. Lateral view, showing its purpose in lowering sidewalls. **B,** Weir excision, anterior view, area to be excised and after closure. (From Joseph.) **C,** Joseph modification, showing it to be primarily an alar base excision. In practice a subtle blend of **A** and **B** may be necessary. **D,** Millard alar rim thinning. **E,** This may be combined with alar base excision.

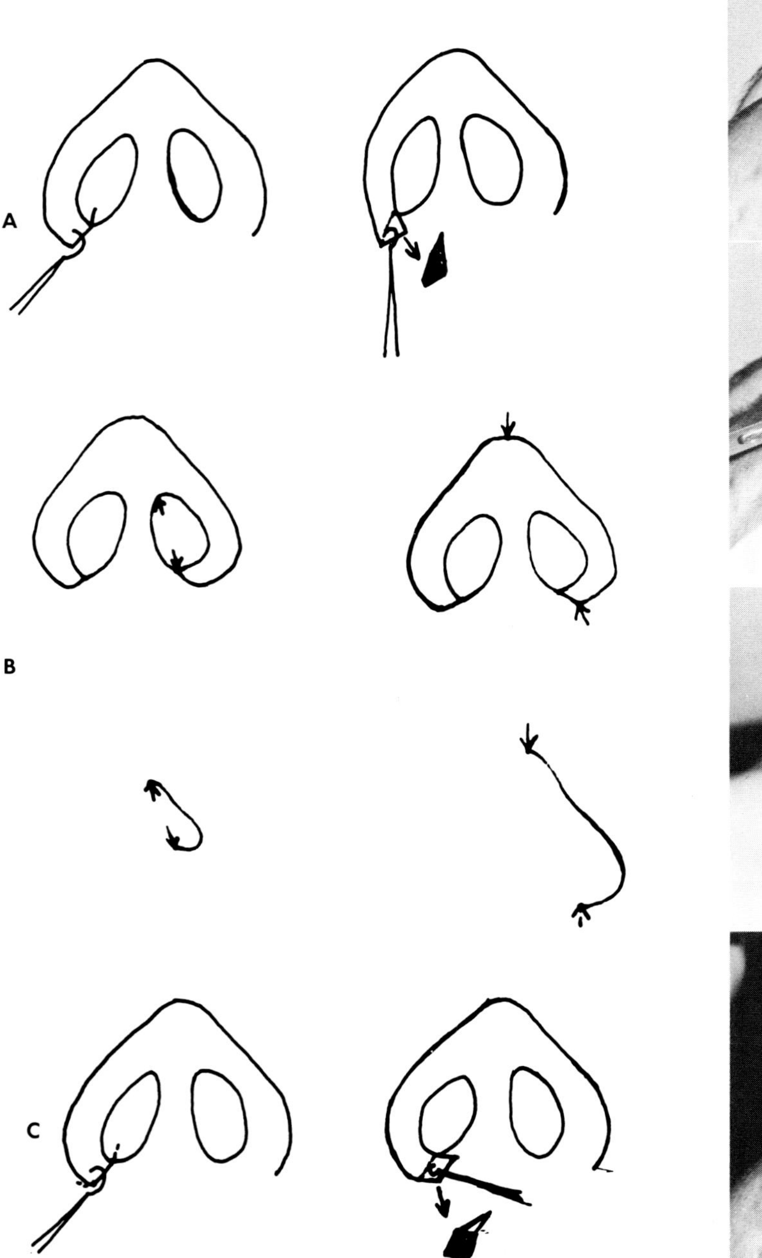

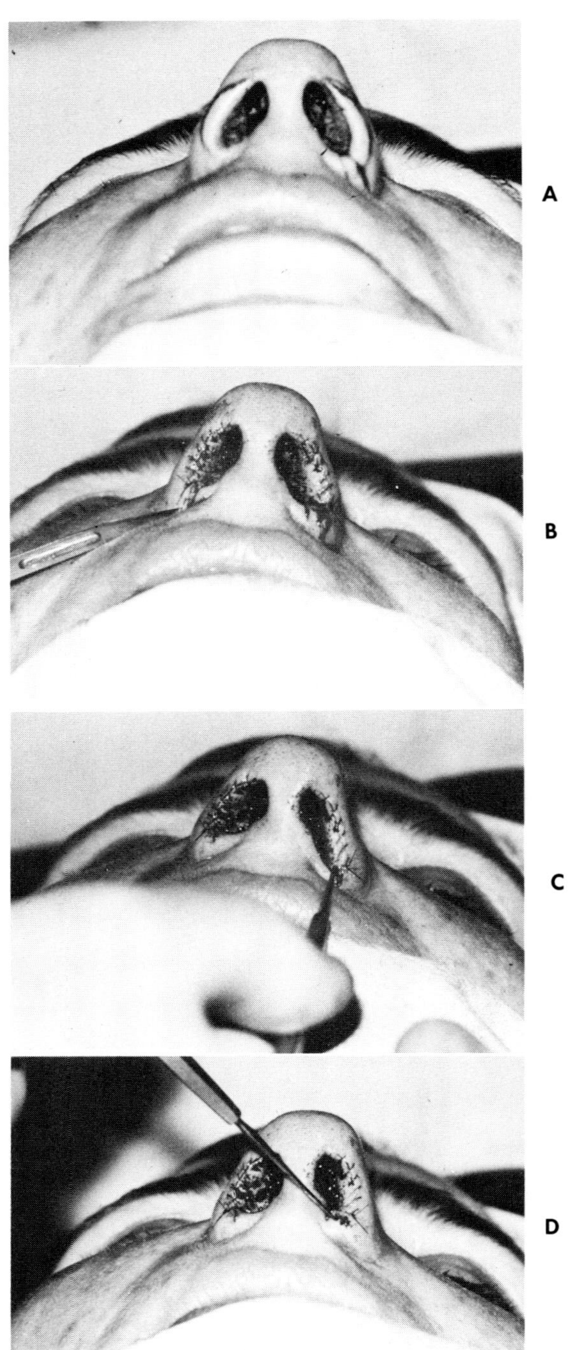

Fig. 19-6. **A,** Schematic drawing after dividing alar base and inserting hook for downward traction to produce oval nostril on the right. Notice shape of overlapped area to be excised. **B,** Compare internal and external nostril dimensions. These determine nostril shape. Excision of an equal amount of both makes nostril smaller and narrower. **A** and **C** show how to make the nostril more oval or round by shape of area excised. **C,** Medial advancement gives a smaller, more round nostril shape, compared with **A.**

Fig. 19-7. **A,** Operative details, rim markings. (Note that tip is swollen from operative procedures.) **B,** Excision completed to thin and shorten sidewall. Alar base excision to be tailored. Note site of basilar excisions scored with blade. **C,** Downward traction is a guide to nostril shape after rim excision in that it produces *a more oval nostril.* **D,** Medial advancement produces *a rounder nostril* and narrows alar width.

the alar rim at this juncture, extending only 3 to 4 mm. into the nostril floor. A hook is placed in the incised lower rim and repositioned downward (caudad) and/or medially as the contour dictates (Fig. 19-6). Downward correction reduces alar rim flare, converting the nostril to a more pleasing oval shape. Delicate placement of the rim sutures is the final key that assures nostril shape.

The amount of external tissue to be removed is measured by exerting hook tension laterally until the desired external contour is achieved. (Fig. 19-7). The excess is removed with a No. 11 blade and closed with fine sutures. As little of the nostril floor should be excised as necessary, and suspending a heavy ala on the delicate columella by attempting excessive medial advancement should be

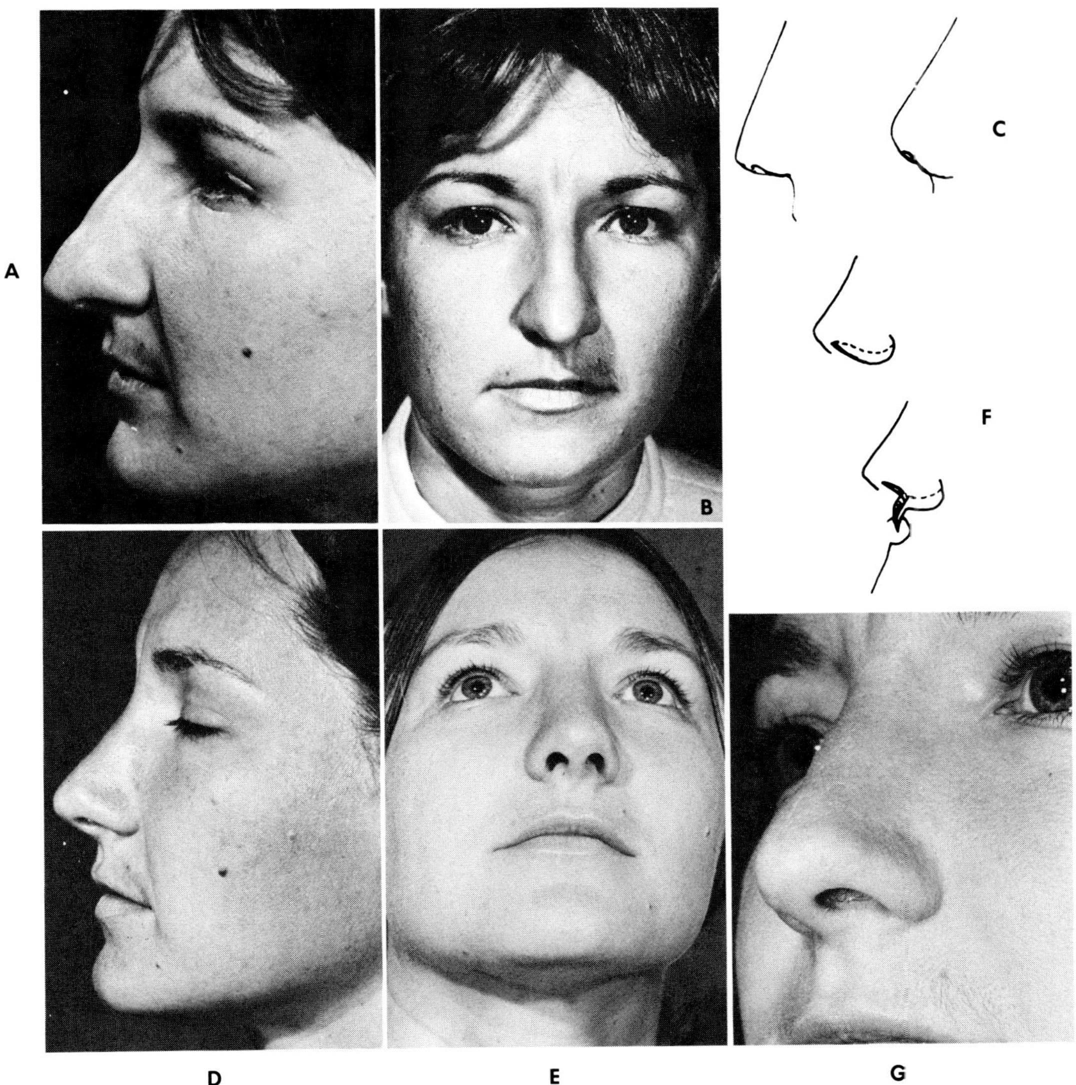

Fig. 19-8. A, Long nose centrally and laterally, with thick skin and unattractive alar rims and nostril shape. **B,** Anterior visage: long nose, narrow radix, wide tip, and thick distal skin resulting in harsh unfeminine architecture. Compare with Fig. 19-4, *A.* **C,** Central shortening would give hidden columella, alar overhang, alae widening, and an unattractive nostril contour. **D,** Postoperative lateral view: total shortening, Millard alar rim excisional shortening plus alar base excision. **E,** Postoperative anterior view. Nostril shape is attractive and proportionate to inner canthal width. **F,** Alar rim excision to shorten sidewall droop after central shortening. (After Millard.) **G,** Oblique view showing inconspicuous scars at rim and base with feminine nostril shape.

avoided. An unattractive round nostril may replace one that was originally a pleasing oval if internal nostril circumference is compromised. Preoperative calculations are a helpful guide but do not supplant artistic judgment at the operating table.

Millard,[1] Rees,[6] and Falces[7] have given excellent descriptions of correction in the non-Caucasian nose, emphasizing these excisions in conjunction with that newer technique, the alar rim excisions.

ALAR RIM EXCISIONS

Germination of this unique adjunct in rim and nostril sculpturing must be credited to Millard, who has given us two excellent articles on its use[1, 2] (Fig. 19-5, *C*). The strip excision at the rim as shown in Figs. 19-5, *C*, and 19-8, *F* will:

1. Thin bulky alar rims
2. Carve a delicate flaring curve in an overhanging sidewall, exposing the columella in profile
3. Correct anatomic irregularities in the rim, congenital or otherwise

This procedure should immediately come to mind in the long large nose with thick bulky skin that requires a large reduction in height and length (Fig. 19-8, *A*).

Unappealing redundancy of the alar rim may be present after shortening of the central nasal length, in spite of cephalad advancement of the residual alar cartilage and columella. The upper alar rim near the septum is more apt to advance superiorly than is the nostril rim near the alar base, with its attachment to the cheek.

In the occasional preoperative patient the sidewalls will present themselves flush with the columella on profile view, the "double barrel shotgun nose." An attractive nose requires an attractive nostril shape, preferably with a visible columella and a gentle upward curve of the rim (Fig. 19-1). The overhanging nostril can be improved with a proper resection. But what about the external scars so produced?

The Oriental and Negroid noses commonly have flush alar and columellar profiles and can be successfully treated by rim sculpture without formation of keloid scar.[7]

Experience has shown that nonconstricting rim sutures that are removed before the fifth postoperative day leave inconspicuous scars (Fig. 19-8, *G*). Care must be taken to avoid visible eversion of the vibrissae, which reside 1 to 5 mm. from the alar rim. Shortening in excess of these measurements is best done by sufficient thinning of the lateral rim to allow approximation without tension on the thinner nostril lining.

Rim excisions can be anticipated preoperatively in a few patients, but more commonly they are better evaluated after central shortening. If the alar rim overhang is severe it can be corrected at the primary procedure.

SUMMARY

Alar base and rim procedures are valuable adjuncts in producing an artistic appearance to the nostril and in exerting control over nostril size and shape in the anterior, lateral, and oblique views. Guidelines in the choice and execution of these procedures have been given.

REFERENCES

1. Millard, D. R.: Alar margin sculpturing, Plast. Reconstr. Surg. **40:**337-342, 1967.
2. Millard, D. R.: Adjuncts in augmentation mentoplasty and corrective rhinoplasty, Plast. Reconstr. Surg. **36:** 48-61, 1965.
3. Seghers, M. J., Longacre, J. J., and deStefano, G. A.: The golden proportion and beauty, Plast. Reconstr. Surg. **34:**382, 1964.
4. Aufricht, G.: Rhinoplasty and the face, Plast. Reconstr. Surg. **43:**219, 1969.
5. Converse, J. M.: Reconstructive plastic surgery, vol. II, Philadelphia, 1964, W. B. Saunders Co.
6. Rees, T. D.: Nasal plastic surgery in the Negro, Plast. Reconstr. Surg. **43:**13, 1969.
7. Falces, E., Wessu, D., and Gorney, M.: Cosmetic surgery in the non-caucasian nose, Plast. Reconstr. Surg. **45:**317, 1970.

Chapter 20

Surgery of rhinophyma

Michael M. Gurdin, M.D., F.A.C.S.
Gene A. Carlin, M.D., F.A.C.S.

Rhinophyma—commonly called "potato nose" but also enjoying such esoteric names as "whiskey," "copper," and "rum" nose, and, in recent American usage, "W. C. Fields nose" after that famous wit—is a skin lesion more commonly involving the lower half of the nose. It may involve the entire nose and spread to the adjacent skin of the cheeks and generally is associated with acne-rosacea of the surrounding skin of the face.

This process is much more common in men than in women, and it is slow growing and painless. Relief is sought only when the cosmetic deformity bothers the patient or breathing is impaired by obstruction of the nares.

The etiology is unknown. Although generally associated with acne-rosacea, said by some to be a precursor of this disease, this relationship has not been proved. Alcohol, spicy foods, and exposure to the sun and wind have long been related to the disease, but careful evaluation reveals these to be exacerbating factors rather than actual causes.

The gross clinical picture (Fig. 20-1) is one of a large, boggy misshapen mass, violaceous in color, with numerous dilated vessels throughout. Palpation reveals numerous enlarged sebaceous glands from which sebum can be expressed.

A classic description of the microscopic pathology was written by Wende and Bentz in 1904.[1]

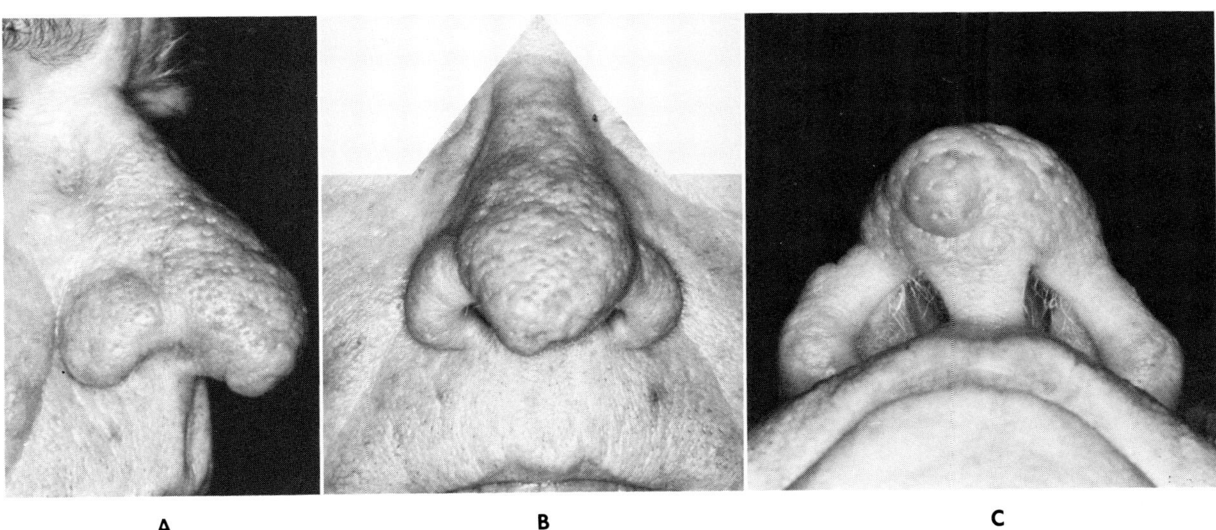

A   B   C

Fig. 20-1. Preoperative photographs. **A,** Oblique view; **B,** front view; **C,** base view.

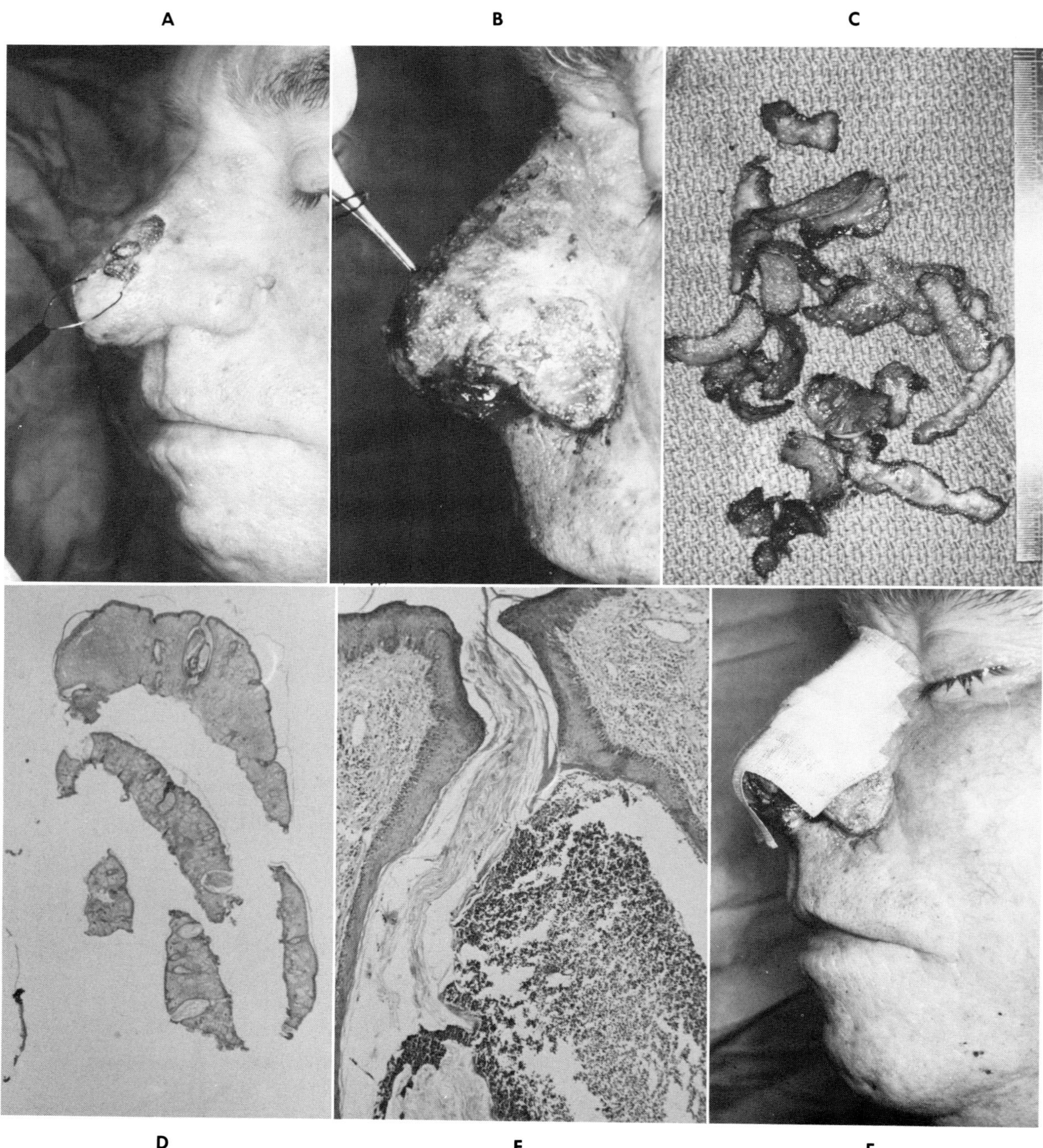

Fig. 20-2. **A,** Sculpting with wire loop and cutting current. Note absence of active bleeding. **B,** Large vessel is picked up with fine-tooth forceps and coagulated. Note irregular base left. These are the depth of the sebaceous glands from which reepithelialization will occur. No cartilage is exposed. **C,** Tissue removed at time of surgery. **D,** Microscopic section (×10). **E,** Microscopic section (×100). **F,** Primary dressing.

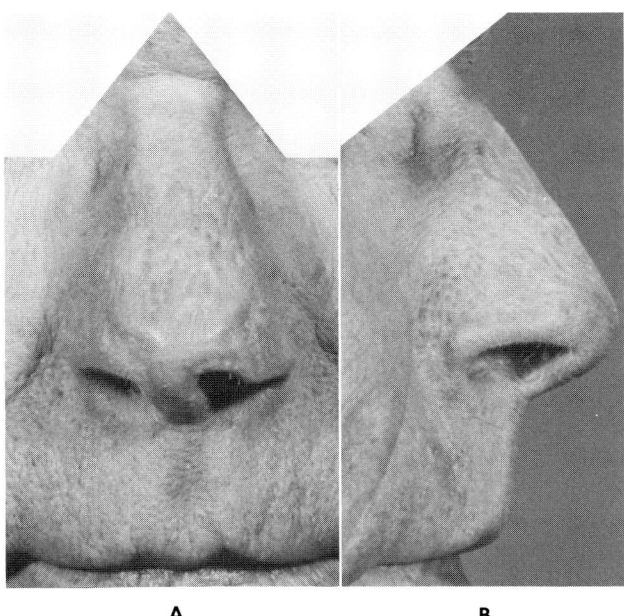

A   B

Fig. 20-3. A, Three-month postoperative photograph (front view). Note reappearance of texture of skin and absence of line of demarcation. **B,** Three-month postoperative photograph (side view).

Microscopic examination shows chronic inflammation and hypertrophy of all the skin elements. The sebaceous glands are enlarged and numerous, and the thickened dermis is laced with blood vessels.

Numerous methods of treatment such as dermabrasion, shaving with a knife, complete excision, and grafting with free skin grafts or flaps have been described.[2, 3] However, in 1950 a simple electrosurgical treatment proved so successful that the technique forms the bulk of this report.

The operation is generally accomplished in one stage with local anesthesia and requires no more than 45 minutes operating time and 1 day's hospitalization. Any standard electrosurgical unit with both cutting and coagulating currents and a variety of tips may be used. Since this is truly a sculpturing procedure, the surgeon should be comfortably seated with easy access and a good view of the patient's entire face. Using a wire loop and cutting current, the nose is sculptured down to a normal contour (Fig. 20-2, *A*), using care that the underlying cartilages are not exposed. Sufficient skin elements are left to result in spontaneous epithelialization without undue scarring. This may

be carried down to just above the free edge of the nares and over the columella. As the nares are approached, it is safer to insert a finger into the nares so that the rim is not burned, producing contracture and asymmetry. After the upper nose is sculptured, the surrounding skin, which is usually involved by acne-rosacea, is lightly coagulated to produce a blending without a line of demarcation.

The cutting current seals the capillaries and, as larger vessels are encountered, they are picked up with a fine-tooth forceps (Fig. 20-2, *B*) and coagulated (Fig. 20-2, *C* to *E*).

The area is then treated as a second-degree burn. It is covered by Xeroform gauze (Fig. 20-2, *F*) and mild pressure to prevent oozing. This is removed on the third or fourth day, and daily warm saline compresses followed by Xeroform gauze dressings are used. The burn slough separates in 7 to 10 days and healing is usually complete in 2 to 3 weeks.

At first the new epithelium is pink and shiny, but gradually, as normal pores return, the texture and color return to normal (Fig. 20-3). After healing is complete, should one have erred on the side of conservatism and left an irregular mound of tissue, it is a simple office procedure to correct this by further electrocoagulation. The same can be said in the unlikely event of a recurrence some years later.

This method is preferred because:

1. It is a simple and relatively short operation.
2. Minimal hospitalization is required.
3. Local anesthesia suffices.
4. Results are good and improve with the skill of the operator.
5. Complications are few and are either preventable or easily treated.
6. Recurrence is rare.

REFERENCES

1. Wende, G. W., and Bentz, C. A.: Rhinophyma: a pathological analysis of five separate tumors occurring in the same patient, J. Cutan. Dis. **22:**447, 1904.
2. Anderson, R., and Dykes, E. R.: Surgical treatment of rhinophyma, Plast. Reconstr. Surg. **30:**397, 1962.
3. Matton, G., and others: The surgical treatment of rhinophyma, Plast. Reconstr. Surg. **30:**403, 1962.
4. Gurdin, M., and Pangman, W. J.: A simple electrosurgical treatment of rhinophyma, Calif. Med. **73:** 171, 1950.

Aesthetic surgery of the ears

Chapter 21

History of otoplasty for correction of prominent ears

Nicholas G. Georgiade, M.D., F.A.C.S.
Joseph Still, M.D.

Current methods for correction of prominent ear deformities originated with Ely,[1] who in 1881 described a procedure of excision of a full-thickness segment of anterior and posterior skin and intervening cartilage to reposition the ear closer to the scalp. Other early procedures by Keen,[2] Monks,[3] Haug,[4] and Joseph,[5] were also directed toward reducing this cephaloauricular angle. Luckett[6] recognized that the deformity was not the exaggerated angle of the ear to the scalp but an anatomic problem intrinsic to the ear with congenital absence of the antihelical ridge. He proposed that the cartilage to be removed be crescent shaped and located in the region for construction of an antihelix. Following excision, Luckett recommended that the perichondrium be reapproximated by everting the cartilage margins with Lembert sutures, forming an anterior ridge in the region of the antihelix (Fig. 21-1, *A*).

The second major advance in establishing a satisfactory method for correction of prominent ears concerned the physiology of the auricular cartilage. Failure of the early methods to maintain the ears in proper position was the result of the spring action of the cartilage exerting force against the suture line. Morestin[7] in 1903 recommended that the cuts made for excising the cartilage extend well into the superior and inferior poles. Gersuny[8] at the same time recommended that the cartilage of the concha be sharply cross-hatched on the posterior surface. Davis and Kitlowski[9] in 1937 described a procedure in which they emphasized

that the cartilage removed must extend the entire length of the proposed antihelix to break the elasticity.

Two major objections arose to the Davis-Kitlowski procedure. The antihelical ridge formed by everting the cartilage margins produced a sharp abnormal ridge on the anterior auricular surface along the newly created antihelix. Also, because of the amount of skin excised in the sulcus posteriorly, the ears were usually repositioned too close to the scalp (Fig. 21-1, *B*). Various methods were proposed to alter these problems.

Barsky[10] made parallel incisions in the cartilage on either side of the previously described ellipse and then reduced the thickness of the cartilage by shaving.

New and Erich[11] modified the Davis-Kitlowski procedure by placement of through-and-through mattress sutures that were not tied until the post-auricular incision was closed. Both ears could thus be simultaneously repositioned by tying the mattress sutures over cotton pledgets (Fig. 21-1, *C*).

Young[12] stated that if the cartilage was correctly excised, the antihelix could be formed without internal or external sutures. He recommended that the segment excised be in the superior crus, rather than in the inferior, and that only excess skin be removed posteriorly.

Other modifications included those by Mac-Collum,[13] Coe,[14] Seeley,[15] and Seltzer,[16] which varied in the excision of the cartilage.

Becker[17] in 1949 described his method of excising the cartilage for the antihelix in the inferior pole and breaking the elasticity of the cartilage in the posterior surface of the superior pole by making several longitudinal and transverse partial-thickness incisions. A smooth antihelix was then formed by rolling the cartilage posteriorly with internal mattress catgut sutures. Only excess skin was excised (Fig. 21-2, *A*).

Converse[18] completely exposed the posterior perichondrium by undermining the skin and made full-thickness incisions along the helix, in the antihelix, and along the conchal rim. The intervening cartilage was then thinned using an abrader and tubed by suturing the margins posteriorly (Fig. 21-2, *B*).

Mustardé[19] constructed a smooth antihelix by using permanent mattress sutures of 3-0 silk placed through the posterior cartilage to evert the antihelix. No incisions were placed in the cartilage (Fig. 21-3, *A*).

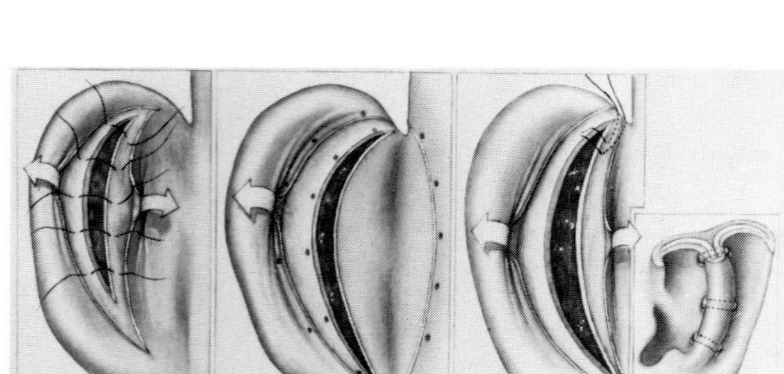

Fig. 21-1. Earlier methods for the correction of prominent ear deformities. These procedures involved excision of varying amounts of cartilage in a linear direction, followed by maintenance of the newly constructed ear with buried sutures. These techniques are generally not used today because of the difficulties in obtaining a natural appearing ear. (See text for further details.)

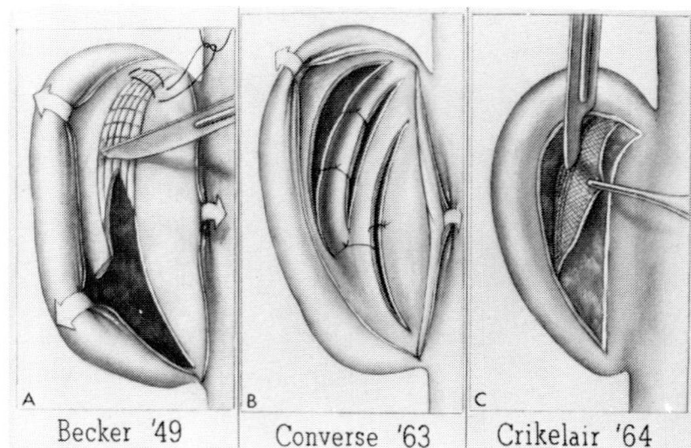

Fig. 21-2. Technique of scoring the cartilage posteriorly (**A**) or thinning the posterior cartilage in order to obtain a more natural roll followed by suturing to obtain the desired roll back of the ear (**B**). Actual dissection of the ear cartilage and scoring the anterior portion of the ear cartilage will allow the cartilage to "roll back" without tension; however, an extensive meticulous dissection is needed to obtain the desired result (**C**).

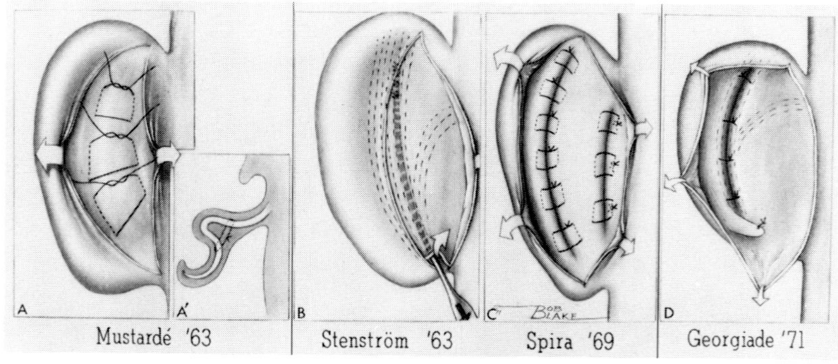

Fig. 21-3. The most commonly used techniques today utilize simpler surgical procedures based on fundamental knowledge of the reaction of cartilage to scoring on the anterior surface. The composite use of posterior placed sutures such as shown in **A** and **A'**, combined with anterior scoring of the cartilage, appears to have been a major improvement in obtaining a uniformly satisfactory appearing ear by use of these simplified techniques (**B** and **C**). With the presence of a prominent concha additional sutures can be inserted with or without the excision of conchal cartilage at the base of the ear (**C**). In order to further improve on the final appearance of the ear and prevent an unnatural appearing protruding lobule and/or superior helix or both, the tail of the helix is dissected free and sutured separately with a mattress suture as is the superior helix in the relative position and distance from the scalp (**D**).

Stenström[20] described his method for rolling the cartilage to produce a smooth antihelix by scratching the anterior auricular cartilage along the newly created antihelix. From laboratory investigation of cartilage, it was found that if cartilage is cut on one surface, it will bend in the opposite direction. His method involved dissection around the tail of the helix and undermining the skin from the anterior surface of the ear. The anterior cartilage was then scratched with a special instrument, thus breaking the spring. The proper fold was maintained by suturing the skin edges, with no sutures being placed in cartilage (Fig. 21-3, *B*).

A different approach to the anterior surface of the auricular cartilage was proposed by Crikelair and Cosman.[21] This approach was through an incision made along the posterior helix. The elasticity of the cartilage was then broken by multiple partial-thickness striations (Fig. 21-2, *C*).

In 1969 Spira and his co-workers[22] stressed the variation in degree of prominence of the ears between different patients and even variation in ears of the same patient. Those with absence of the antihelix were corrected using Mustardé's method. If, in addition, the concha was at an abnormal angle to the scalp, Spira recommended that it be set back using horizontal mattress sutures between the concha and scalp periosteum (Fig. 21-3, *C*).

A combination of these procedures designed to obtain an aesthetically acceptable ear appears to be the procedure of choice to most plastic surgeons today.

REFERENCES

1. Ely, E. T.: An operation for prominence of the auricles, Arch. Otol. **10**:97, 1881.
2. Keen, W. W.: New method of operating for relief of deformity of prominent ears, Ann. Surg. **12**:49, 1890.
3. Monks, G. H.: Operation for correcting the deformity due to prominent ears, Boston Med. Surg. J. **124**:84, 1891.
4. Haug, R.: Eine einfache neue plastische Methode zur Rucklagerung hochgradig abstehender ohrmuschelu, Deutsch. Med. Wschr. **20**:776, 1894.
5. Joseph, J.: Nasenplastik und sonstige Gesichtsplastik, Leipzig, 1928, Curt Kabitzsch; Eselsohren, Verh. Berlin Med. Ges. **27**:206, 1896.
6. Luckett, W. H.: A new operation for prominent ears based on the anatomical deformity, Surg. Gynec. Obstet. **10**:635, 1910.
7. Morestin, H.: De la reposition et du plissement cosmetiques du pavillon de l'orielle, Rev. Orthop. **4**:289, 1903.
8. Gersuny, R.: Ueber einige kosmetische Operationen, Wien. Med. Wschr. **53**:2253, 1903.
9. Davis, J. S., and Kitlowski, E. A.: Abnormal prominence of the ears: a method of readjustment, Surgery **2**:835, 1937.
10. Barsky, A. J.: Plastic surgery, Philadelphia, 1938, W. B. Saunders Co.
11. New, G. B., and Erich, J. B.: Protruding ears: a method of plastic correction, Am. J. Surg. **48**:385, 1940.
12. Young, F.: The correction of abnormally prominent ears, Surg. Gynec. Obstet. **78**:541, 1944.
13. MacCollum, D. W.: The lop ear, J.A.M.A. **110**:1427, 1938.
14. Coe, H. E.: Correction of lop ears, Northwest Med. **41**:126, 1942.
15. Seeley, R. C.: Correction of the congenital protruding ear, Am. J. Surg. **72**:12, 1946.
16. Seltzer, A. P.: The importance of correcting outstanding ears, Ann. Otol. **58**:1012, 1947.
17. Becker, O. J.: Surgical correction of the abnormally protruding ear, Arch. Otolaryng. **50**:541, 1949.
18. Converse, J. M., and Wood-Smith, D.: Technical details in the surgical correction of the lop ear deformity, Plast. Reconstr. Surg. **31**:118, 1963.
19. Mustardé, J. C.: The correction of prominent ears using simple mattress sutures, Brit. J. Plast. Surg. **16**:170, 1963.
20. Stenström, S. J.: A natural technique for correction of congenitally prominent ears, Plast. Reconstr. Surg. **32**:509, 1963.
21. Crikelair, G. F., and Cosman, B.: Another solution for the problem of the prominent ear, Ann. Surg. **160**:314, 1964.
22. Spira, M., and others: Correction of the principal deformities causing protruding ears, Plast. Reconstr. Surg. **44**:150, 1969.

Chapter 22

The normal ear: what is it?

Robert F. Ryan, M.D.
William J. Pollock, M.D.
Edward L. Bitseff, M.D.

"In selecting the best method of correction, one's first major hurdle is to 'See What We Look At'. Possibly the greatest lesson to learn in human surgical art is to see not only one item, but this item in its anatomic environment."[1] In these opening sentences Broadbent and Mathews have framed the major problem facing reconstructive surgeons. To aid us in seeing the normal ear and its relationship to the face, we will follow three approaches: (1) anatomic dissections of normal ears, (2) consideration of abnormal ears, and (3) the artist's concept of ears. Landmarks of the normal ear were in one of the first anatomy lessons for all physicians (Fig. 22-1, *A*).[2] The framework of the external ear closely conforms to the shape of the underlying single sheet of elastic cartilage except for the large, dependent lobule (Fig. 22-1, *B* and *C*). This cartilage tends to be cupped or trough shaped and the circular opening of the external meatus is formed by a fibrous ligament that extends anteriorly from the tragus to the crus of the helix. Hollinshead[3] states that there are six intrinsic and three extrinsic muscles of the external ear, all supplied by the facial nerve. Their only function is to wiggle the ear, a technique most useful in amusing children. Sensation of the skin of the external ear is from several sources. The auriculotemporal branch of the fifth nerve supplies the tragus, anterior limb of the helix and crus, and part of the tympanic membrane. The greater auricular nerve from the second and third cervical nerves supplies the rest of the external ear except for the concha, which is supplied by the auricular branches of the seventh, ninth, and tenth cranial nerves but with considerable variation in different patients.

The arterial supply to the external ear is from the external carotid artery via the posterior auricular and superficial temporal arteries. The external auditory meatus, the canal, and the external surface of the tympanic membrane receive their arterial supply from the deep auricular artery, which is a branch of the internal maxillary artery.

One portion of the external ear skeleton mentioned in many surgical procedures is the helical tail. As seen in Fig. 22-1, *C*, the tail of the helix is tucked down parallel and close to the concha. In Fig. 22-2, *A*, the helical tails of both the right and left ear cartilages are pulled out by hemostats. The helical tail has considerable roll to it and may be quite rigid. Also seen in Fig. 22-2, *A*, is the spine of the helix. This is where the major ligament for attaching the external ear to the head takes origin, and from the same area part of the dense ligaments run to the tragus to form the anterior border of the auditory meatus. These ligaments are shown between the thumb and index fingers in Fig. 22-2, *B*.

From a surgical viewpoint, the creation of the antihelical roll is a major objective in many types of otoplasty. In Fig. 22-2, *B*, the cartilage of the right ear shown in Fig. 22-1, *C* has been divided vertically into halves and the lateral half has been rotated 90 degrees to show the cut surface of the cartilage. The obtuse angle of the antihelical roll

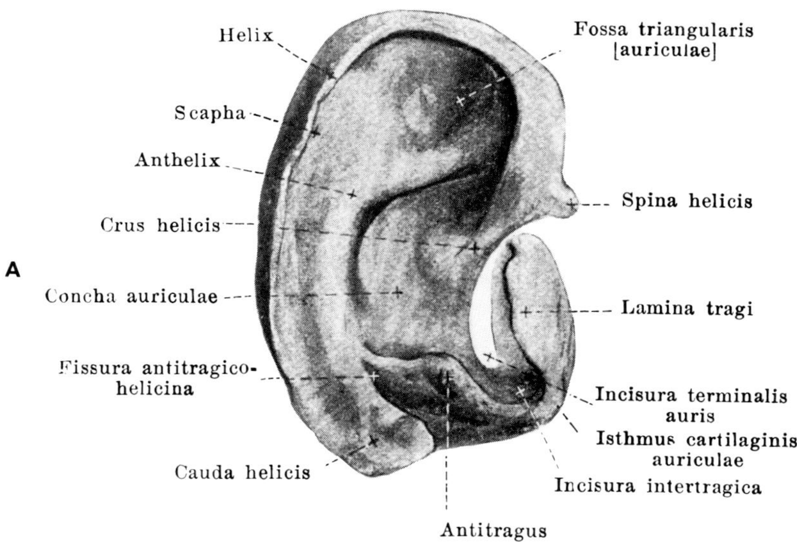

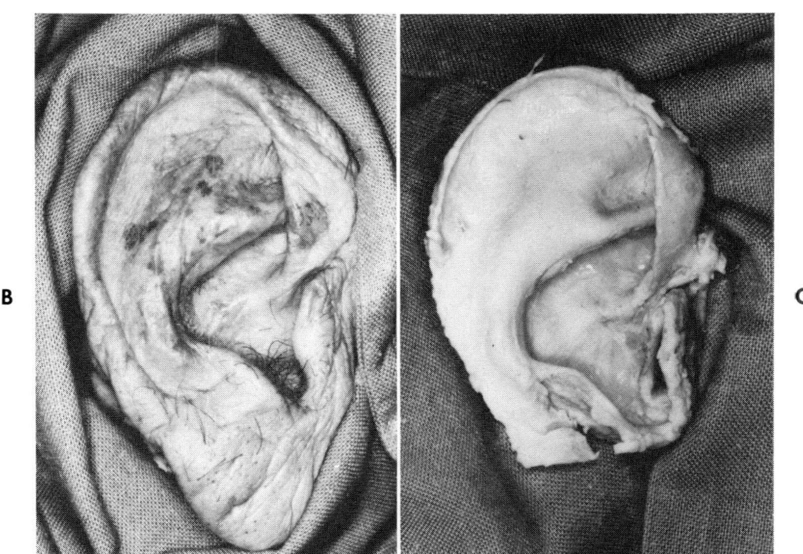

Fig. 22-1. A, Right ear cartilage, isolated and viewed from without. **B,** Photograph of the right ear of a 57-year-old male cadaver. **C,** The skin and subcutaneous tissue have been removed from the ear to show the cartilage. Note how closely the cartilage conforms to the external pattern of the ear. (**A** from Spalteholz, W. In Barker, L. F., editor: Hand atlas of human anatomy, vol. III, Philadelphia, 1923, J. B. Lippincott Co.)

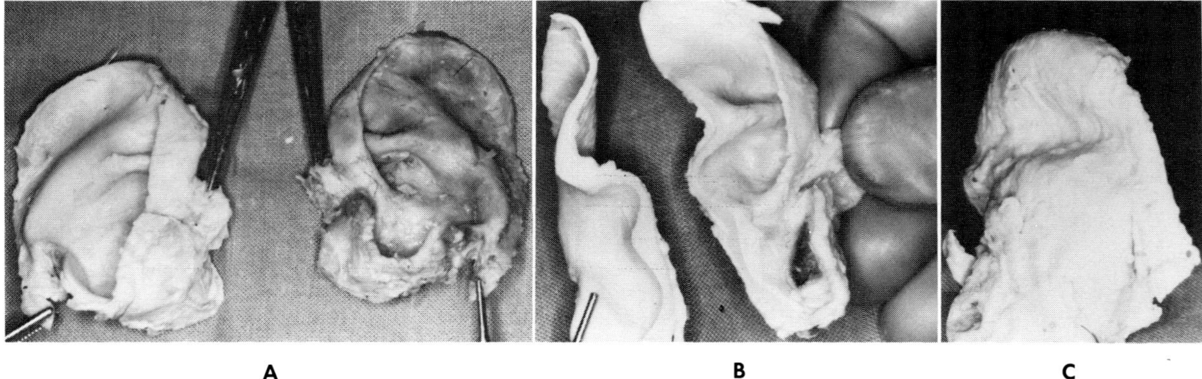

A   B   C

Fig. 22-2. A, Cartilaginous skeleton of both right and left ears from a 57-year-old male cadaver. The hemostats in the lower corners are retracting the helical tail away from the concha. The other two instruments indicate the spine of the helix. **B,** Right ear cartilage after being cut vertically in half and the lateral portion rotated 90 degrees. The curve of the antihelix is apparent. The ligaments extending from the helical spine are held between the thumb and index fingers. One ligament attaches the ear to the skull and the other forms the anterior wall of the auditory meatus where it runs from the tragus to the helical spine. **C,** Posterior surface of the right ear cartilage showing the forward curve of the antihelix. The ear canal is at the lower left.

is apparent (Fig. 22-2, *B*). This roll is readily seen on looking at the posterior surface of the right ear cartilage (Fig. 22-2, *C*).

While it is not the purpose of this chapter to review the embryology and complete development of the external ear, any surgeon doing reconstructive work on the external ear should read the superb article by Blair Rogers.[4] It is often easier to recognize the normal if one studies the abnormal. Rogers has divided the deformed ear into four types: (1) microtia, (2) lop ear, (3) cup ear, and (4) protruding ear.

Microtia is described as a malformed and underdeveloped auricle or a malformed lobule with the rest of the ear being almost totally absent (Fig. 22-3, *A*). According to Rogers, "There are gradations between microtia and 'lopping' just as there are gradations between lopping and cupping." The ears shown in Fig. 22-2, *B* and *C*, show the characteristics of both mild microtia and marked lopping. The ears of two other patients (Fig. 22-3, *D* and *E*) fit Rogers' description of lopping and also fit Potter's description[5] of the renal agenesis with deformed ears. Rogers describes the cup ear as "an essentially malformed protruding ear which has characteristics of both lop ear and protruding ears . . . overdevelopment of its deep, 'cup-shaped' concave concha." The ear shown in Fig. 22-3, *F* and *G,* shows this cupping.

If the ear has excessive conchal cartilage, in-

creased angle of protrusion from the skull, or unfolding of the antihelix but with a normal vertical height, Rogers calls it protruding.

It is of great interest that all of the patients whose ears are shown in Fig. 22-3 represent examples of various ear-renal syndrome with some type of urologic problem.[6, 7] Ureteropelvic obstruction has been the most common anomaly in our series and was present in the patient shown in Fig. 22-3, *F* and *G.*

The third approach to the appreciation of the normal ear is through the eyes of the artist. Primitive art works often show the ear as a mass or general form without much detail (Fig. 22-4, *A*). In other cultures the lobule of the ear is stretched as a sign of beauty (Fig. 22-4, *B* and *C*). Whether as a sign of beauty or just artistic liberty, the pre-Columbian artists of Mayan civilization did show protruding ears (Fig. 22-4, *D*). Even Michelangelo's David has ears that seem more prominent than the standard accepted by many surgeons today, that is, ¾-inch protrusion from the scalp to the ear. In the famous Uffizi Gallery in Florence, one notices the small, low-set ears in the "Portrait of Baldassarre" by Pietro Perugino.

In his book, *Atlas of Human Anatomy for the Artist*, S. R. Peck[8] describes the helix as being shaped like a question mark. Inside the helix is the antihelix, which has two legs. "The upper leg is round and full in form, while the lower leg is little more

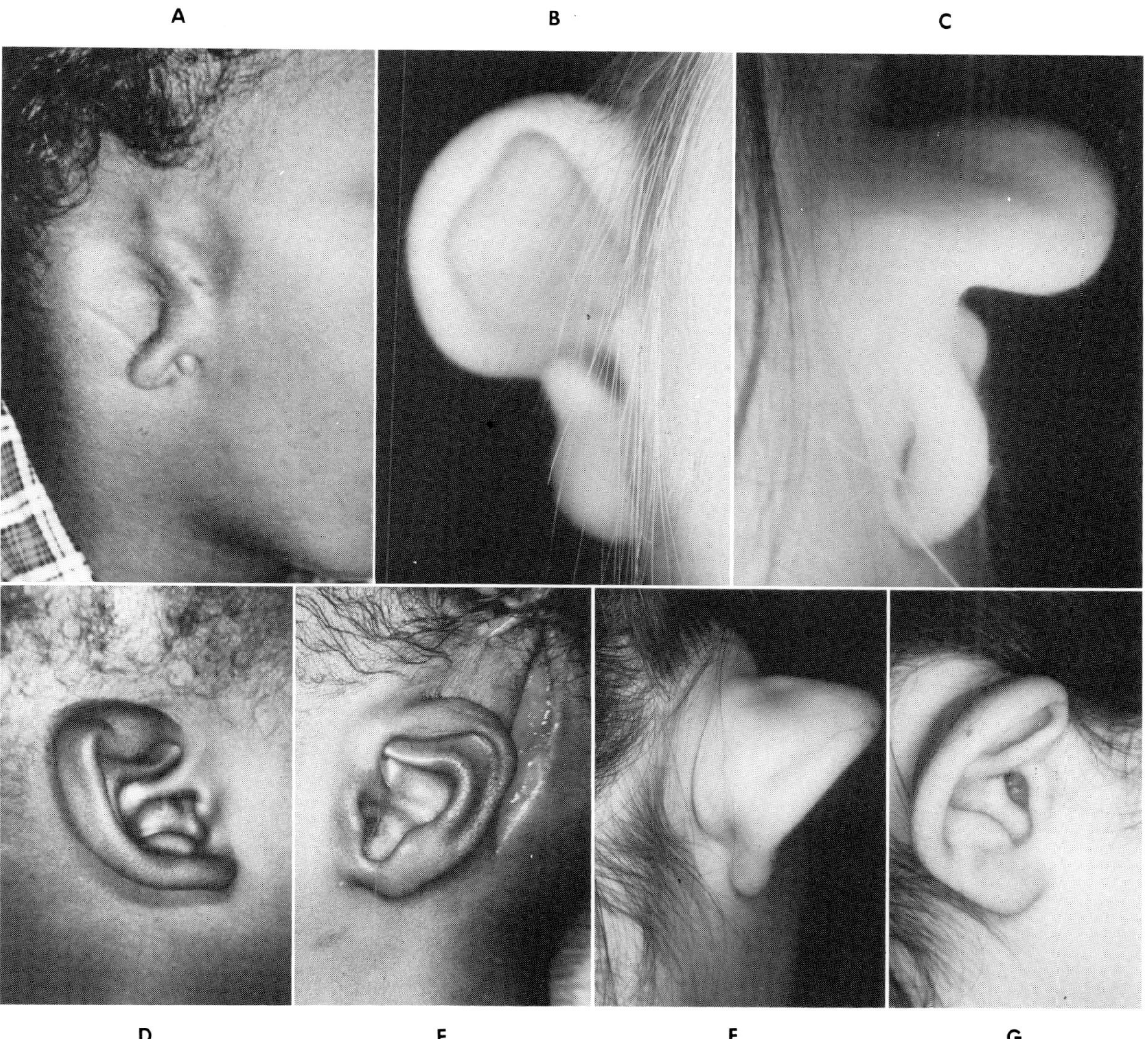

Fig. 22-3. **A,** A 4-year-old male with microtia of the right ear and right renal agenesis. **B** and **C,** Bilobed ear of a 5-year-old female who also had ureteral obstruction on one side and a nonfunctioning kidney on the opposite side. This patient's ear shows characteristics of both microtia and lopping. **D,** Right ear of a newborn male with lopping as defined by Rogers. This patient also had bilateral renal agenesis. **E,** Left ear at autopsy of an infant with bilateral renal agenesis and lopping of both ears. **F** and **G,** Right ear of a 5-year-old female with marked cupping of the ear. The left ear was normal. This patient had marked ureteropelvic obstruction, which was corrected with salvage of the right kidney.

Fig. 22-4. A, Pre-Columbian warrior of the Toltec period. Note the stylized mass at the site of the ears. **B,** Wood sculpturing from the Union of South Africa showing the elongated lobe of the ear. **C,** East African wood sculpturing showing a large loop made in the ear lobe as a sign of beauty. **D,** Pre-Columbian statue said to be of the Mayan period (650-900 A.D.) that shows protruding ears.

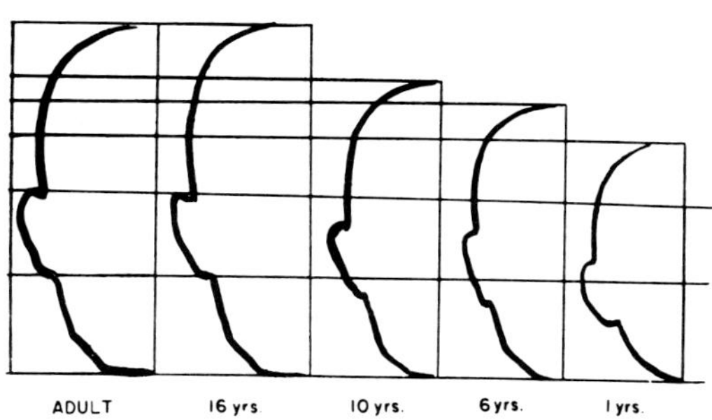

Fig. 22-5. Relative changes in the position of the ears at different ages. (From Broadbent, T. R., and Mathews, V. L.: Artistic relationships in surface anatomy of the face: application to reconstructive surgery, Plast. Reconstr. Surg: **20:**1, 1957.)

than a sharp edge. A pit is formed between the legs, called the triangular fossa. The antihelix then swings downward beside the helix. Its forward extension below acts as a 'curtain rod' for the soft, hanging lobe."

The best description of the relationship of the external ear to other facial features is that given by Broadbent and Mathews.[1] Their following succinct quotation covers the entire subject:

The long axis of the ear normally rests parallel with the nose and this long axis is a line from the highest point on the superior helix to the anterior border of the lobule. The anterior points of attachment of the helix and lobule are in line with the ramus of the mandible. The lobule is on a horizontal line level with the nasal tip. The superior point of attachment of the ear to the head is on a line level with the eye. The helix rises away from the head, at the eye level, to the level of the eyebrow. The ear canal is on a horizontal line halfway between the eye and the base of the nasal tip, and is at the halfway mark from

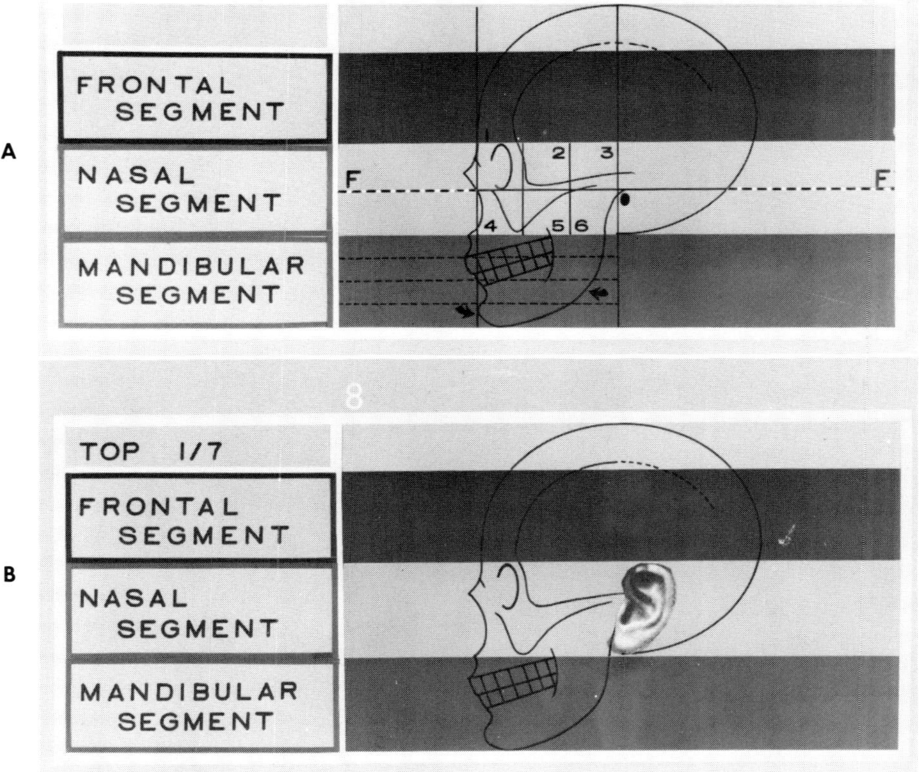

Fig. 22-6. A, Sketch No. 7 from the simplified system for drawing the lateral view of the skull from the VIII Dalinde International Symposium on Esthetic Surgery. The external auditory canal is marked by a dark oval below the Frankfort plane and posterior to the midline vertical plane of the skull. **B,** Completed sketch of the lateral view of the skull with an external ear in place. The anterior border of the helix and anterior portion of the earlobe are on the same line as the posterior border of the ascending ramus of the mandible. The ear is of approximately the same length as the nasal segment.

the back of the skull to the lateral facial plane. The external ear is located posterior to this point. The attachment of the anterior superior helix rim to the skull is one ear length posterior to the lateral canthus of the eye. When both ears are missing, this distance is determined by measuring the distance from the nasal tip to the glabella or the length of the flexor surface of the long finger. These distances are equal to the length of the ear.

These same authors also show graphically the relative size and location of the ear at different ages (Fig. 22-5).

Perhaps one of the best ways to appreciate the relationship of the ear to the rest of the face is by drawing it. Following the instructions for drawing the lateral projection of a skull according to the Dalinde Course[9] gives the picture seen in Fig. 22-6, *A*. If we add an ear to the completed drawing, the length of which is equal to the nasal segment, and place the lobe and anterior border of the helix on a line with the posterior border of the ramus of the mandible, then we have the relationship shown in Fig. 22-6, *B,* the normal ear.

REFERENCES

1. Broadbent, T. R., and Mathews, V. L.: Artistic relationships in surface anatomy of the face: application to reconstructive surgery, Plast. Reconstr. Surg. **20:**1, 1957.
2. Spalteholz, W. In Barker, L. F., editor: Hand atlas of human anatomy, vol. III, Philadelphia, 1923, J. B. Lippincott Co.
3. Hollinshead, W. H.: Anatomy for surgeons, vol. 1, New York, 1954, Harper & Row, Publishers.
4. Rogers, B. O.: Microtic lop, cup and protruding ears: four directly inheritable deformities, Plast. Reconstr. Surg. **41:**208, 1968.
5. Potter, E. L.: Bilateral renal agenesis, J. Pediat. **29:** 68, 1946.
6. Vincent, R. W., Ryan, R. F., and Longenecker, C. G.: Malformation of ear associated with urogenital anomalies, Plast. Reconstr. Surg. **28:**214, 1961.
7. Longenecker, C. G., Ryan, R. F., and Vincent, R. W.: Malformation of the ear as a clue to urogenital anomalies, Plast. Reconstr. Surg. **35:**303, 1965.
8. Peck, S. R.: Atlas of human anatomy for the artist, New York, 1951, Oxford University Press.
9. Presented at the VIII Dalinde International Symposium on Esthetic Surgery, Mexico City, March, 1970.

Chapter 23

Otoplasty: direct surgical approach

Eugene H. Courtiss, M.D.
Richard C. Webster, M.D.
Malvin F. White, M.D., D.M.D.

Before performing any aesthetic surgical procedure, it is vital to analyze the component parts producing the deformity. This preparation permits planning of the techniques required to produce the goal—a harmonious balance of features. Although perfection is the objective, it can never be achieved. Through continuous critical analysis however, improved results can be obtained. These thoughts obviously apply to the nose, breast, eyes, as well as ears.

OBJECTIVES

When treating protruding ears, the aim is to produce gentle folds and smooth contours that are present irrespective of the angle from which the ear is viewed. This objective—a natural ear—is difficult to quantitate. Nevertheless, the "normal" protrusion of an ear has been calculated by several observers[1-5] and has been found to measure 1.7 to 2.0 cm. from the scalp to the helix. Generally the protruding ear exceeds this parameter by 0.5 to 1.5 cm. Care must be taken, however, not to replace one deformity with another. Leaving the patient with a telephone deformity, a plastered-back look, sharp edges where none normally exist, a hidden helix, or drooping of the top of the ear are all to be avoided. Nevertheless, even if some of these defects are present, patient and parent satisfaction after otoplasty is quite high, providing the ear no longer protrudes.[6, 7]

The surgeon should also guard against the tendency to overcorrect. This results in an ear as unnatural looking as a protruding ear, and in

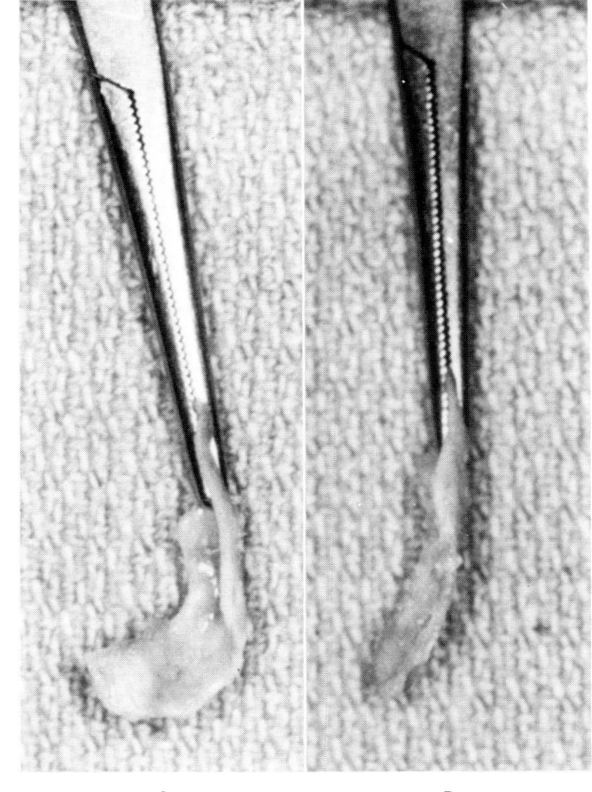

A   B

Fig. 23-1. A, Fresh ala cartilage before morselization. Note bending of cartilage. **B,** Fresh ala cartilage after morselization. Note cartilage is almost straight.

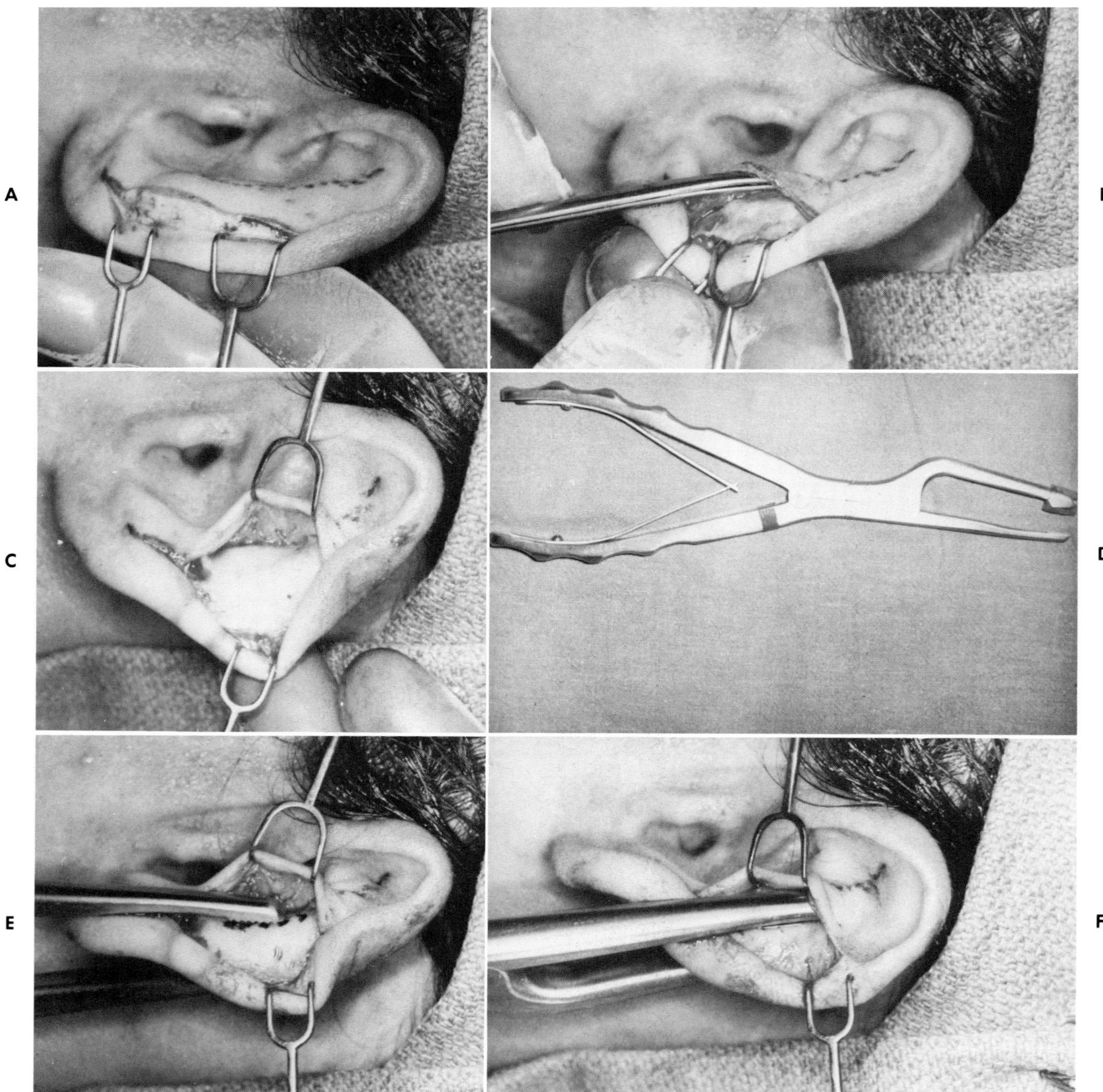

Fig. 23-2. A, Incision and exposure of cartilage. (For demonstration purposes, the antihelix has been dotted with methylene blue.) **B,** Blunt scissors dissection elevating subperichondrial flap as far as helical rim. **C,** Flap elevation from helical rim above to tail of helix below. **D,** Auricular morselizer (available from Weck and Company) padded with suction tubing. **E,** Morselizer in place, guarded blade on posterior surface. Dotted line is present at height of concha. **E,** Morselizer blades closed and advanced. The cartilaginous spring has been broken. **G,** After morselization. Note dotted methylene blue line has moved into scapha as conchal depth is lessened. This is the maximum the concha can be rolled into the scapha. **H,** After suturing. Three or four horizontal mattress sutures hold antihelix and lobule into position. **I,** After closure.

addition, the overcorrected ear is difficult to bring out. Although we do not purport to have the panacea for treating protruding ears, our most predictable and satisfying results have followed the approach to be described.

That proper treatment depends on an accurate analysis of the deformity is a surgical axiom, and otoplasty is no exception.[2, 8] Although there are variations, the generally seen protruding ear is composed of three distinct and definite parts, which may be present either alone or in combination: First is the failure of development of an antihelical fold. Second is an excessively deep concha. Finally, the tail of the helix may protrude excessively, pushing the lobule laterally. In addition, the antitragus, tragus, and Darwin's tubercle all lend to the general shape of the ear and should be managed appropriately, if deformed. Each of these components must be corrected to give a natural ear.

BACKGROUND BRIEFS

In attempting to correct these deformities, many methods have been utilized. Dieffenbach sug-

gested the removal of retroauricular skin in 1845. This was extended by Morestin[9] in 1903 to include cartilaginous excision. Generally, however, unsatisfactory results continued until 1910 when Luckett[10] suggested incising the cartilage. For the next 50 or so years modified Luckett techniques[11-13] prevailed. However, in 25% of cases sharp edges resulted[14]; this stimulated consideration of other methods for controlling cartilaginous contour. Currently surgeons[15, 16] incise, excise, carve, slice, thin by abrasion,[17, 18] scratch,[19, 20] suture,[21, 22] fish-scale, tube, and morselize the delicate cartilage of the ear. It is approached from the front,[16, 16a] the back, and both sides.[17] Actually, the organ is attacked in almost as many ways as there are surgeons doing the operation. Current trends—breaking the spring of the cartilage—are based on the observation[23-25] that dynamic forces are normally present in cartilage. When these forces are broken by superficial incision or perforation (morselization), the cartilage will gently bend *away* from that side.

The key to excellence rests in tailoring the surgery to the individual patient. Unpredictable and often disappointing results frequently follow when

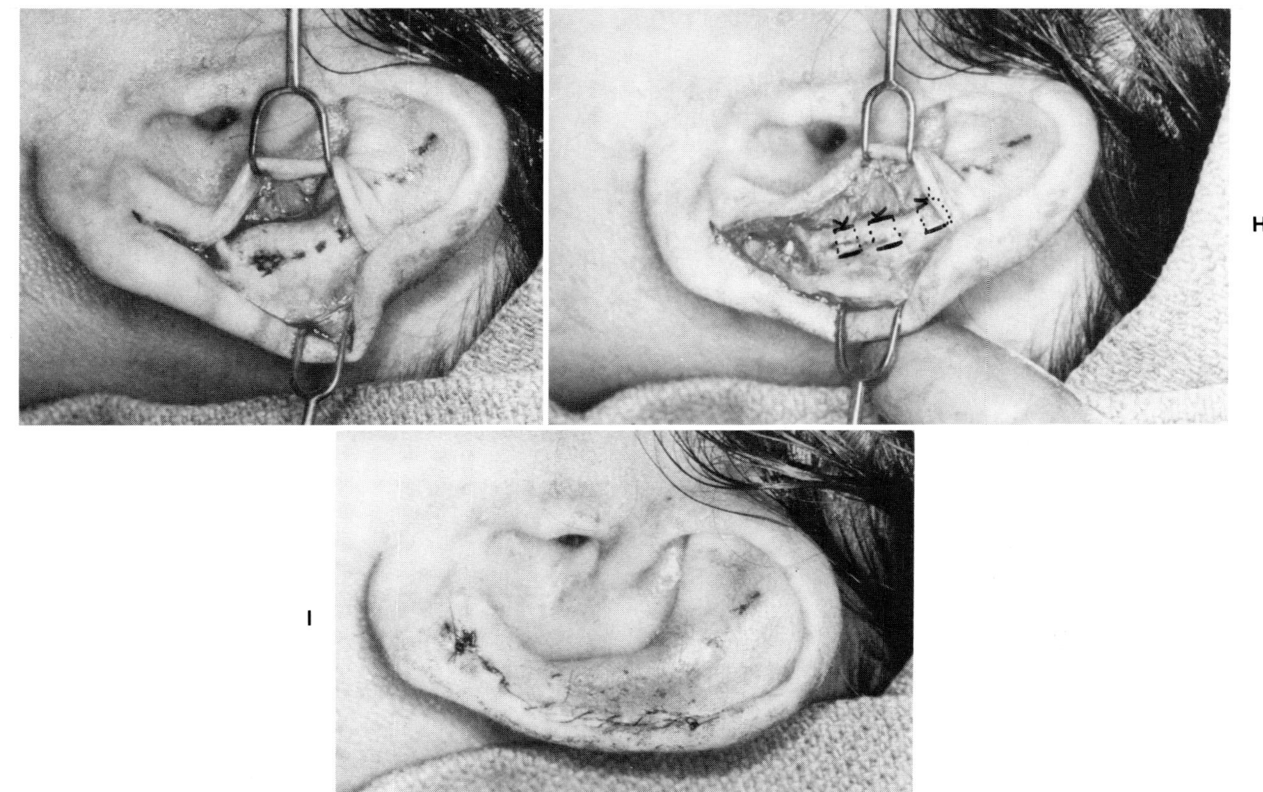

Fig. 23-2, cont'd. For legend see opposite page.

the reverse is true and the patient is simply plugged into a predetermined operative procedure. The operation must be selected for the patient and not the patient for the operation. There is no surgical procedure that is followed by uniformly excellent results; consequently, the best method is the least complicated, and the least complicated is followed by the fewest complications.

Finally, we feel that for too many years our techniques have been obsessed with the antihelix and too little attention has been given to the concha. Generally, poor results follow when the creation of an antihelix is performed in such a way as to simultaneously correct an excessively deep concha by rolling the concha into the scapha. Nevertheless, regardless of the technique used, if the ear does not look satisfactory in the operating room, it will not improve with time. Immediate adjustments must be made before the wound is closed.

In this approach, we begin by adjusting the antihelix using the morselization-suture technique. There is simultaneous correction of the protruding lobule, if present. Initially no effort is made to alter the conchal depth or angle. A smooth rolling antihelix without projection lateral to the helical rim is the first objective. After this is obtained, conchal depth and angle are treated by conchal setback. If the patient presents only a poorly developed antihelix, then only morselization is done; if he has only an excessively deep concha, then only the setback is done. Roughly a third of patients demonstrate poorly developed antihelix alone, a third a deep concha alone, and a third a combination of the two.

ANTIHELIX

The antihelical fold is best created by breaking the spring of the cartilage. This can be done with a knife or toothed instrument,[4, 19, 26]; however, with these instruments it is technically difficult to penetrate the cartilage and not go through the full-thickness of cartilage. With the morselizer, uniform penetration is obtained because of the uniform depth of the teeth (Fig. 23-1). Actually "morselization" is a misnomer. The morselizer does not chew up the cartilage into little morsels, instead, it perforates the cartilage, breaks the spring, and allows it to be molded and held in the desired position by sutures. The sutures maintain the desired position, thereby minimizing the chance for recurrence. The result, therefore, is quite similar to scoring the cartilage with a forceps or knife and inserting sutures via separate stab wounds.[4, 19, 26]

Prior to using the morselization method we used the posteriorly placed buried mattress technique recently popularized by Mustardé.[1, 21, 27] In our hands, however, protrusion recurred in 15% of ears, an unacceptably high failure rate.

TECHNIQUE OF MORSELIZATION

An incision is made on the anterior surface of the ear (just under the helical rim) starting just below the antitragus and extending halfway or three quarters up the ear (Fig. 23-2, *A*). The dissection follows the line of the proposed antihelical fold and continues beneath the perichondrium, which is deviated by spreading the scissors (Fig. 23-2, *B* and *C*). The subperichondrial pocket (1.5 cm. in width) extends to just below the helical rim. A *padded auricular* morselizer is inserted

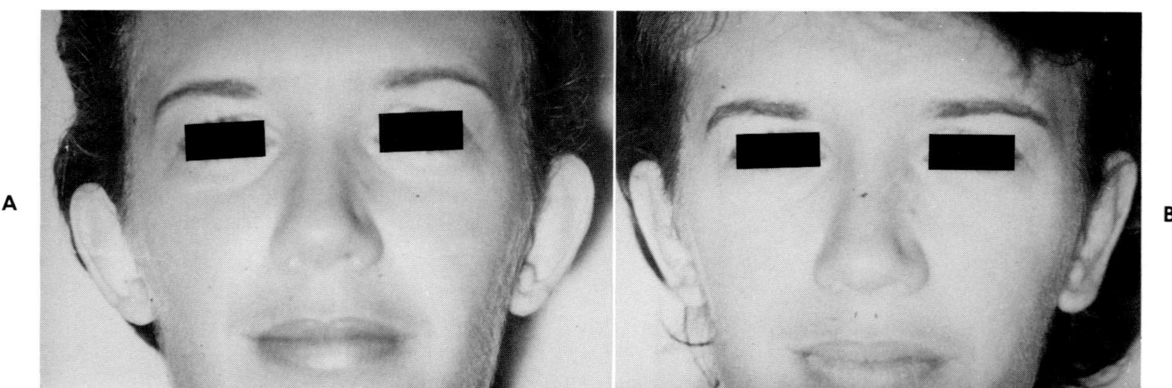

Fig. 23-3. A, Preoperative view. **B,** Postoperative view after bilateral otoplasty using morselization alone.

(padded side against the skin of the posterior side) (Fig. 23-2, *D* and *E*). The instrument is closed with firm pressure, causing the teeth to penetrate the cartilage (Fig. 23-2, *F*). It is easiest to start above and morselize the cartilage as the instrument is withdrawn. A more natural appearing antihelix can be obtained if the morselization does not extend to the helical rim, leaving 0.5 to 1.0 cm. of normal cartilage between the helical rim and the antihelical roll. The morselization is continued until the spring of the cartilage is weakened, although a minor degree of residual "memory" may be present. This is effectively controlled by anteriorly placed, horizontal, buried mattress sutures (Fig. 23-2, *G* and *H*). These two to four sutures fix the ear in the desired position and are the key to the method. The morselization merely breaks the spring; the sutures fix the position of the ear. Any nonabsorbable suture material is satisfactory. At this point the antihelix should be well formed and no

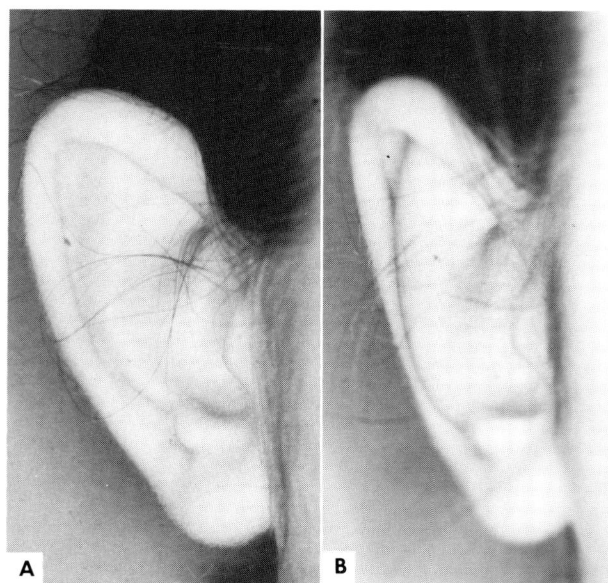

Fig. 23-4. A, Preoperative view. **B,** Postoperative view, morselization with suture of tail of helix.

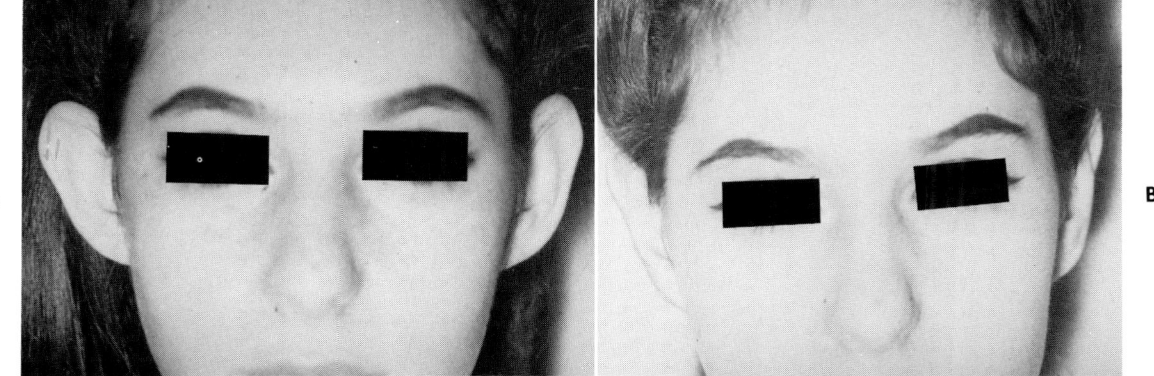

Fig. 23-5. A, Preoperative view. **B,** Postoperative view, morselization alone.

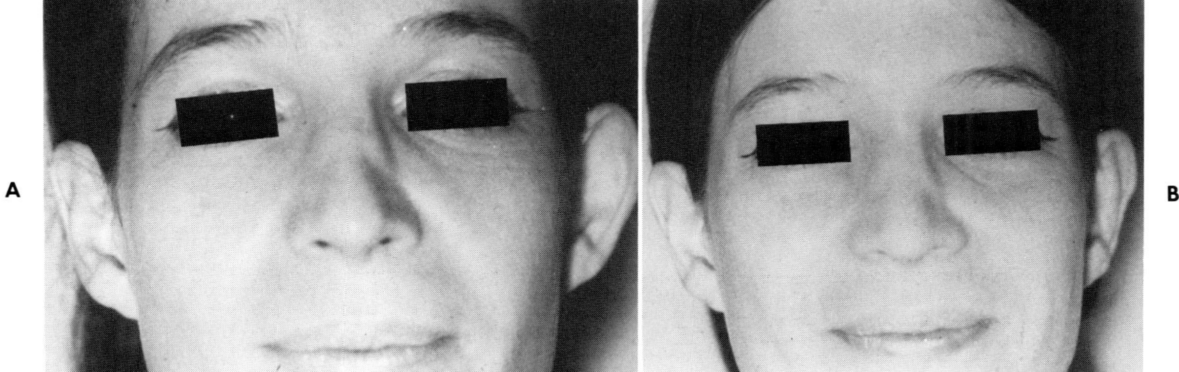

Fig. 23-6. A, Preoperative view. **B,** Postoperative view, left otoplasty alone, morselization alone.

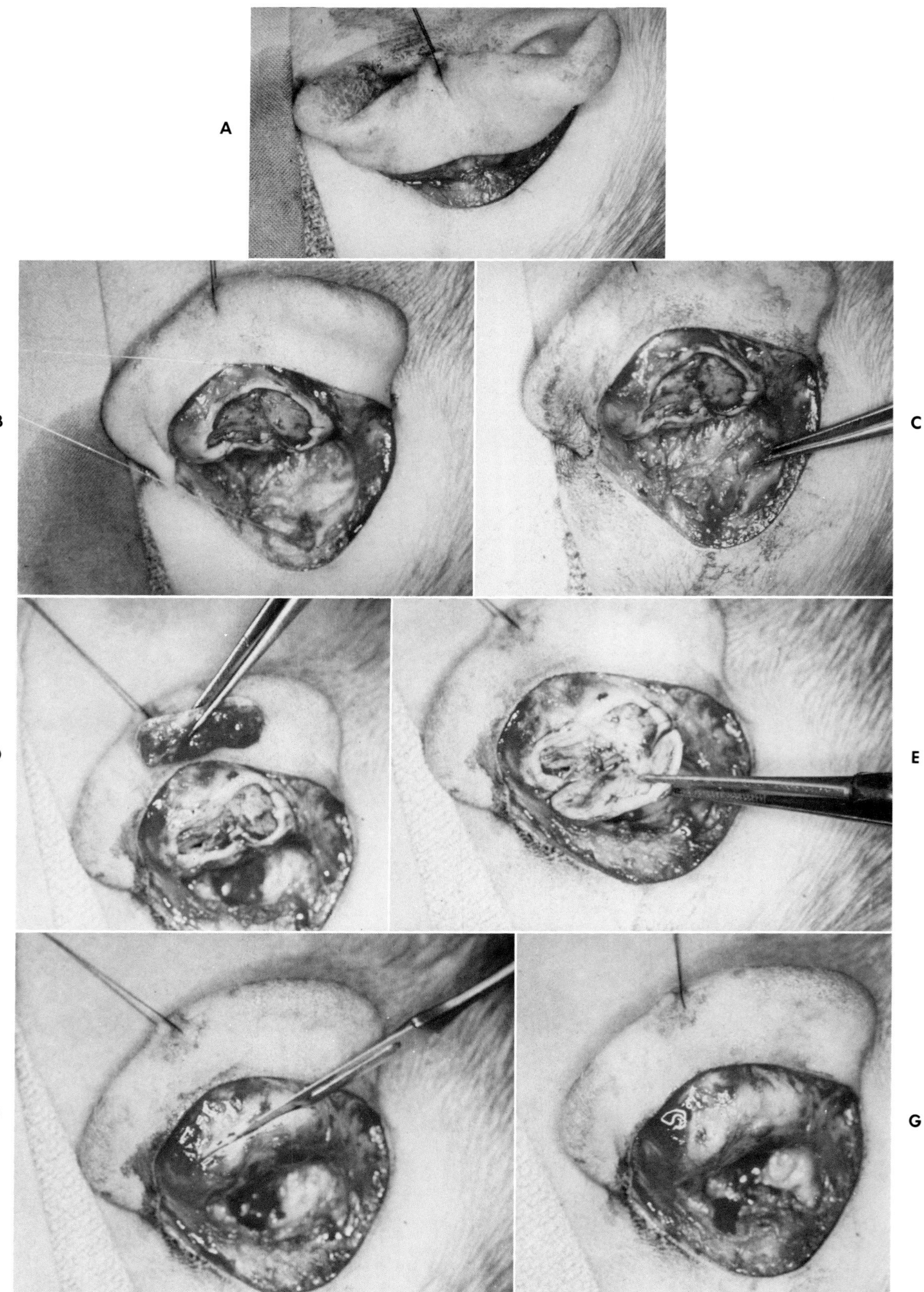

Fig. 23-7. A, Skin excised. **B,** Incision into conchal wall. **C,** Cartilaginous disk almost excised. **D,** Disk removed from most medial wall of concha. **E,** Forceps on posterior auricular muscle. This should be excised as necessary to gain space, thereby allowing the ear to fall into scalp. **F,** Suture placed bringing concha into scalp. One should be careful not to slide ear forward and thereby partially obliterate external auditory canal. **G,** Suture tied and ready for skin closure.

change will occur later. After antitragus and tragus are carefully examined and appropriately trimmed, the flap is simply replaced and sutured with material of the surgeon's choice (Fig. 23-2, *I*). The results of morselization are demonstrated in Figs. 23-3 to 23-6.

CONCHAL SETBACK

If an excessively deep concha is of a *minor* degree, it may be managed by rolling the concha into the scapha. This can be done with any of the techniques that create an antihelical fold. However, if a *moderate* or *severe* conchal protrusion is present, this produces a large scapha and/or protrusion of the antihelix lateral to the helix, both of which are undesirable secondary deformities.[6]

A considerably more versatile method is the use of the conchal setback. Others[28-30] have suggested simple suturing of the concha to the mastoid periosteum. We have extended the method by excising cartilaginous disks and have markedly improved our results. The technique brings the entire ear closer to the head without disturbing or distorting the helix, antihelix, or lateral conchal wall. By removing cartilage from the depth of the ear, the entire auricle is set in closer to the scalp. Incidently, minor antihelical deformations are corrected by setback alone. This is caused by rolling the ear around an anterior pivot point, as the setback is done.

Technique. An ellipse of skin (2 to 4 cm. in width) is excised from the posterior surface of the

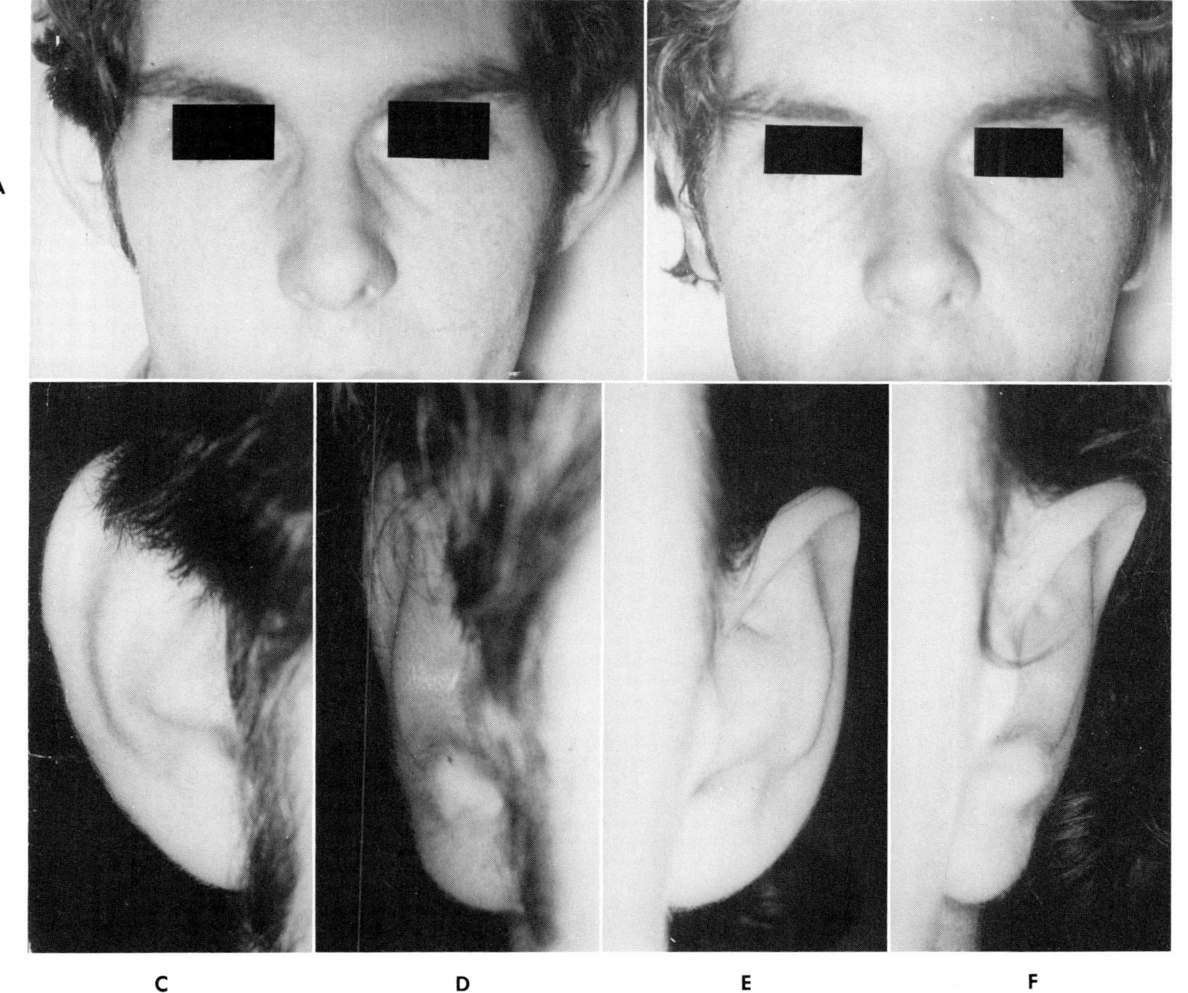

Fig. 23-8. A, Preoperative view. **B,** Postoperative view, setback alone. **C,** Preoperative view of right ear, same patient. **D,** Postoperative view, setback alone. **E,** Preoperative view of left ear. **F,** Postoperative view, setback alone.

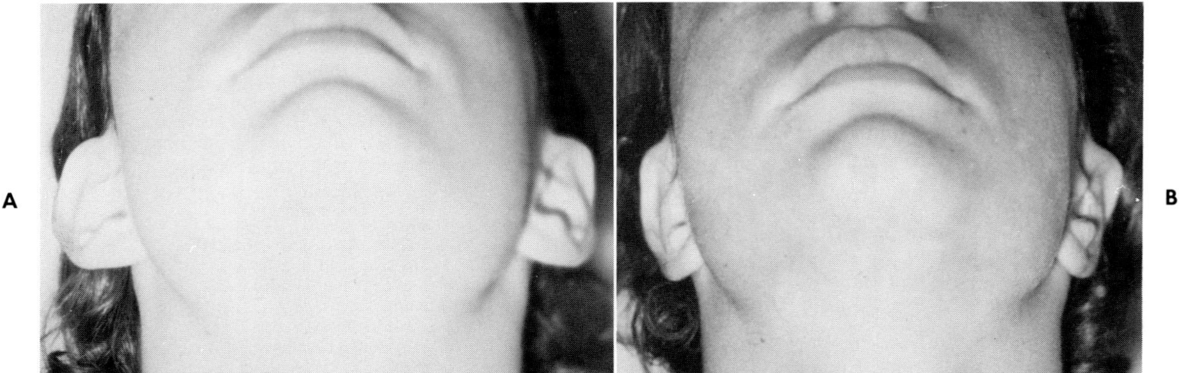

Fig. 23-9. A, Preoperative view. **B,** Postoperative view, setback with suture of tail of helix, no morselization.

ear (Fig. 23-7, *A*). The ellipse, centered on the sulcus, is taken equally from the ear side and from the mastoid side; this exposes the posterior conchal wall. Disks of conchal cartilage are then removed from the most medial portion of the concha until the ear easily rests the desired distance from the scalp (Fig. 23-7, *B* to *I*). The cartilaginous excision can be angled to set in the upper or lower part of the ear to a greater or lesser degree. As the cartilage is excised, damage to the skin lateral to the concha should be avoided, if possible. If a "drainage hole" is placed, it heals quite kindly. Occasionally the posterior auricular muscle is thick and restricts the distance the ear can be set back. When this occurs, the muscle should be excised (Fig. 23-7, *E*). After hemostasis is secured (by electrocautery), the entire ear is brought against the mastoid periosteum and held in place with one or more catgut or nylon sutures (Fig. 23-7, *F*). The stitches run between the mastoid periosteum and the remaining concha—they simply fix the ear in the desired position (Fig. 23-7, *G*). The incision is closed in a routine fashion. One should note that the setback does not depend upon the removal of skin to maintain the position of the ear and, therefore, the retroauricular sulcus is preserved. Hematoma formation is minimized by very careful contour dressing. The most effective and easiest dressings are pieces of cotton soaked in mineral oil and contoured into the desired convolutions of the ear. If a hematoma does develop, it can be evacuated through a little stab wound in the bottom of the concha. The dressing remains in place for 1 week, at which time it is removed and no additional dressing or nightband is required.

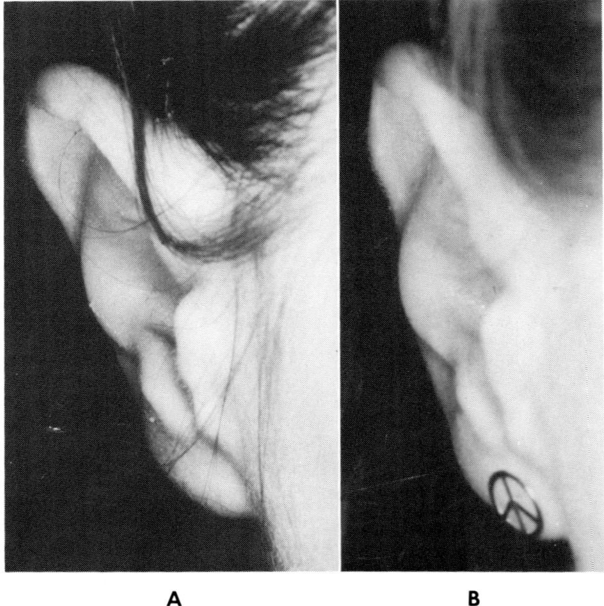

Fig. 23-10. A, Preoperative view. **B,** Postoperative view, setback alone, 4 months.

The expected results after conchal setback are seen in Figs. 23-8 to 23-10.

PROTRUDING LOBULE

The protruding lobule, occasionally a most frustrating problem, is best approached by suturing the tail of the helix to the concha (Fig. 23-4). This can be accomplished from the anterior or posterior side, depending on whether antihelix formation and/or conchal depth require correction. Occasionally, excision of the tail[17] and/or diagonal excision of skin from the back of the lobule is required.[31] Closure may have to include an adhesion-like suture to the cheek.[8]

DISCUSSION

It is a common fallacy to believe that otoplasty for protruding ears is an easy operation. Others[24] have critically analyzed their results and have found up to 30% to be fair to poor; usually, this is the result of technical errors. However, technical failure of such an unacceptably high frequency makes one question the basic soundness of the method.

SUMMARY

The specific components comprising the protruding ear deformity must be analyzed and treated directly. The antihelix is created by morselization, the excessively deep concha by cartilaginous disk excision and setback, and the protruding lobule by suture. These approaches can be combined in the same operative session to give predictable results. In addition, attention should be paid to such contour defects as may be present in the antitragus, tragus, or Darwin's tubercle.

REFERENCES

1. Adamson, J. E., Horton, C. E., and Crawford, H. H.: The growth pattern of the external ear, Plast. Reconstr. Surg. **36:**466, 1965.
2. Rubin, L. R., and others: An anatomic approach to obtrusive ear, Plast. Reconstr. Surg. **29:**360, 1962.
3. Spira, M., and others: Analysis and treatment of the protruding ear, Transactions of IV International Congress, Rome, 1967, p. 1090.
4. Stenström, S. J.: Cosmetic deformities of the ears. In Grabb, W. C., and Smith, J. W.: Plastic surgery, Boston, 1968, Little, Brown, and Co., p. 513.
5. Tanzer, R. C.: The correction of prominent ears, Plast. Reconstr. Surg. **30:**236, 1962.
6. McDowell, A. J.: Goals in otoplasty for protruding ears, Plast. Reconstr. Surg. **41:**417, 1968.
7. Ju, D. M. D.: The psychological effect of protruding ears, Plast. Reconstr. Surg. **31:**424, 1963.
8. Converse, J. M., and Wood-Smith, D.: Technical details—the surgical correction of the lop ear deformity, Plast. Reconstr. Surg. **31:**118, 1963.
9. Morestin, M. H.: De la reposition et du plissement cosmetiques du pavillon de l'oreille, Rev. Orthop. **14:**289, 1903.
10. Luckett, W. H.: A new operation for pominent ears based on the anatomy of the deformity, Surg. Gynec. Obstet. **10:**635, 1910; reprinted in Plast. Reconstr. Surg. **43:**83, 1969.
11. Jayes, P. H., and Dale, R. H.: The treatment of prominent ears, Brit. J. Plast. Surg. **41:**193, 1952.
12. McEvitt, W. G.: The problem of the protruding ear, Plast. Reconstr. Surg. **2:**481, 1947.
13. Tamarian, J. A., and Mirehouse, O.: Reconstruction of the antihelix and scapha in selected cases of protruding ears, Arch. Otol. **60:**597, 1959.
14. Strombeck, J. O.: Results of survey for protruding ears, Acta Chir. Scand. **122:**138, 1961.
15. Chongchet, V.: A method of antihelical reconstruction, Brit. J. Plast. Surg. **16:**268, 1963.
16. Ju, D. M. D., Li, C., and Crikelair, G. F.: The surgical correction of protruding ears, Plast. Reconstr. Surg. **32:**283, 1963.
16a. Crikelair, G. F., and Cosman, B.: Another solution for the problem of the prominent ear, Ann. Surg. **160:**314, 1964.
17. Courtiss, E. H., Webster, R. C., and White, M. F.: Otoplasty for protruding ears: a simplified controlled approach, presented to American Society for Aesthetic Plastic Surgery, Boston, May, 1971.
18. Stark, R. B., and Saunders, D. E.: Natural appearance restored to the unduly prominent ear, Brit. J. Plast. Surg. **15:**385, 1962.
19. Stenström, S. J.: A "natural" technique for the correction of congenitally prominent ears, Plast. Reconstr. Surg. **32:**509, 1963.
20. Barnes, W. E., and Morris, F. A., Jr.: Otoplasty: an improved technique, Southern Med. J. **59:**681, 1966.
21. Mustardé, J. C.: The correction of prominent ears with buried mattress sutures: current trends in plastic surgery, London, 1964, Butterworth & Co. (Publishers) Ltd., p. 233.
22. Mustardé, J. C.: The treatment of prominent ears by buried mattress sutures: a ten-year survey, Plast. Reconstr. Surg. **30:**382, 1967.
23. Fry, H. J. H.: Cartilage and cartilage grafts, Plast. Reconstr. Surg. **40:**426, 1967.
24. Gibson, T., and Davis, W. B.: The distortion of autogenous cartilage grafts: its causes and prevention, Brit. J. Plast. Surg. **10:**257, 1958.
25. Cloutier, A. M.: Correction of outstanding ears, Plast. Reconstr. Surg. **23:**412, 1961.
26. Kaye, B. L.: A simplified method for correction of the prominent ear, Plast. Reconstr. Surg. **40:**44, 1967.
27. Davenport, G., and Bernard, F. D.: Experience with the mattress suture technique in the correction of prominent ears, Plast. Reconstr. Surg. **36:**91, 1965.
28. Furnas, D. W.: Correction of prominent ears by concha-mastoid sutures, Plast. Reconstr. Surg. **42:**189, 1968.
29. Neuner, O.: A simple method for the correction of prominent ears, Plast. Reconstr. Surg. **47:**111, 1971.
30. Spira, M., and others: Correction of the principal deformities causing protruding ears, Plast. Reconstr. Surg. **44:**150, 1969.
31. Baumgartner, P. H.: A technical hint for the correction of prominent ears, based on the method of converse, Plast. Reconstr. Surg. **37:**66, 1966.

Chapter 24

Another solution for the problem of the prominent ear*

George F. Crikelair, M.D.
Bard Cosman, M.D.

Anomalies of the external ear are frequent. Those defects that give the ear a protruding appearance are especially common. The social and emotional impact of this deformity is widely appreciated.

Attempts at operative correction at one time included measures such as excision of postauricular skin alone or direct suture of the ear to the mastoid area. The unpleasant or temporary results of these procedures led to their abandonment. Luckett's analysis of the defect involved in the protruding ear represented a second stage in the development of otoplasty operations.[1] Procedures based on his technique, however, often produce obvious operative stigmata.[2]

Methods more recently introduced have attempted to achieve correction of the outstanding ear deformity without the appearance of operation.[2-12] The multiplicity of techniques testifies to the difficulty of achieving this end. In the last 20 years an otoplasty procedure coming close to this goal has been evolved at the Columbia-Presbyterian Medical Center. Several of its features have received but slight previous emphasis. This paper details these items of technique and presents a critical review of the results achieved in a recent 5-year period.

TECHNIQUE

A postauricular incision is outlined parallel to the free border of the helix and about 1 cm. from

it (Figs. 24-1 and 24-2). The incision extends from the cephaloauricular angle superiorly to within the fleshy lobule inferiorly. Using this line as the outer border, an ellipse of skin can be removed. The skin excision is an advantage in exposure but is not essential for this nor is it necessary to keep the ear in position postoperatively.

Needles are passed posteriorly from the anterior surface of the ear just beneath the helix overhang, tipped with methylene blue, and withdrawn. After the initial skin incision the wound edges are slightly undermined and the methylene blue markings are identified in the cartilage. The auricular cartilage is cut through along this line from the cephaloauricular angle to where the tail of the helix begins (Fig. 24-3). This separates the rim of the ear from the major ear cartilage.

The auricular cartilage is then freed subperichondrially on its anterior surface (Fig. 24-4). It is to be noted that the posterior perichondrium is not disturbed. The anterior dissection is carried forward in the upper part of the ear until the inferior crus of the antihelix is exposed. In the middle and lower portion of the ear the edge of the cavum concha is similarly exposed. If the cavum is significant in the ear protrusion, its exposure is made more nearly complete. Despite this freeing, the cartilage still maintains a protruding position (Fig. 24-4, C).

The anterior cartilage surface is lightly striated in at least three or four directions (Fig. 24-5). The striations do not extend through the full thickness of the cartilage. As this is done the cartilage

*Reprinted from Annals of Surgery **160**:314-324, 1964.

136

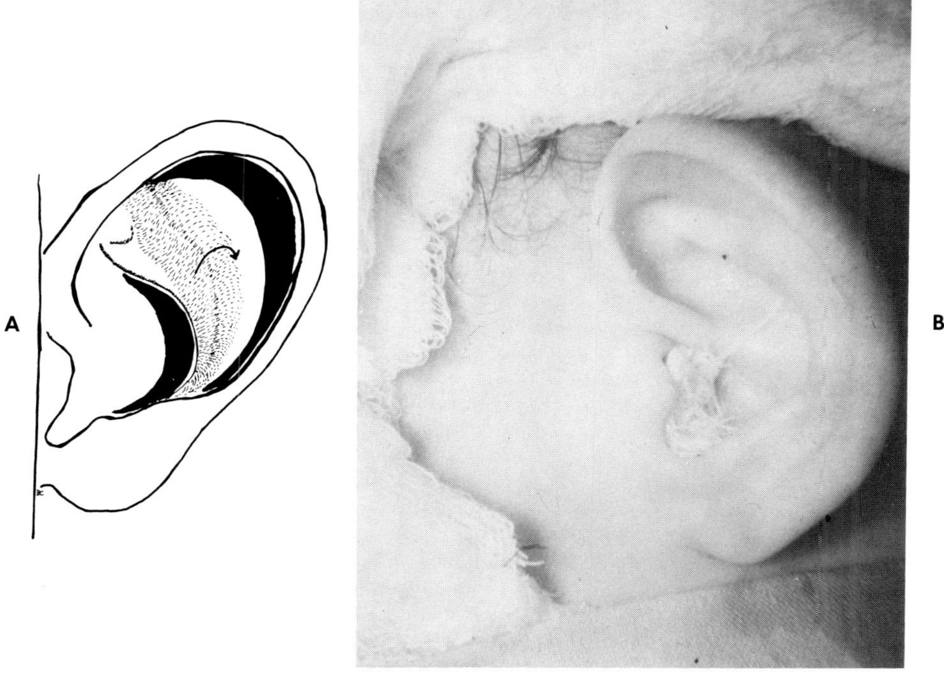

Fig. 24-1. A, Prominent ear with combined concha and antihelix deformity. The darkened ellipses indicate desirable reductions in size; the arrow indicates the antihelix with superior crus to be constructed. **B,** Prominent ear at start of operation. (See Fig. 24-10.)

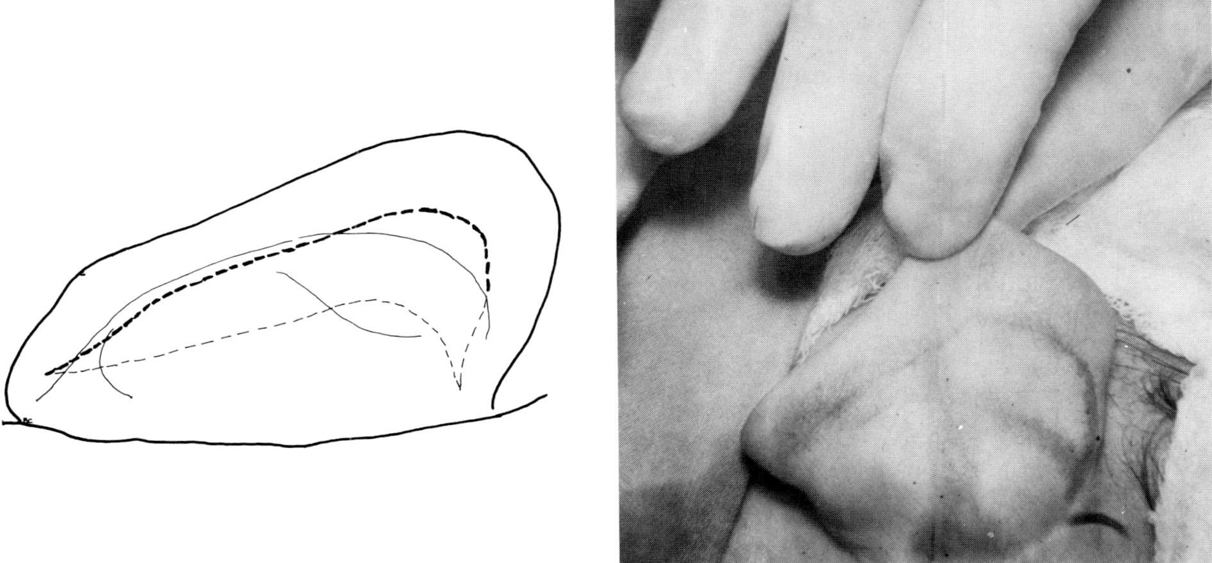

Fig. 24-2. Postauricular incision indicated; possible elliptical skin excision also outlined.

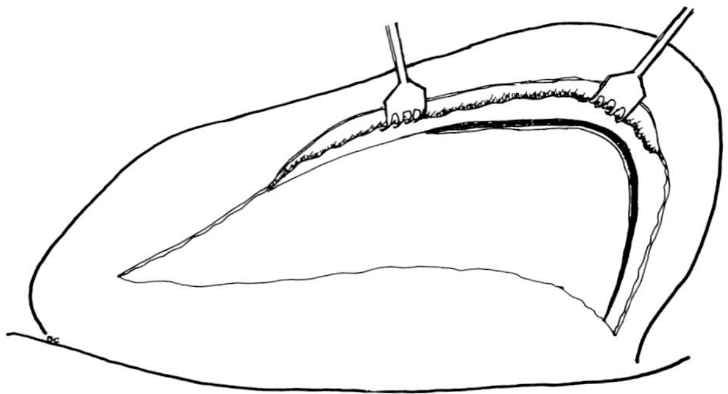

Fig. 24-3. Auricular cartilage cut through to anterior surface.

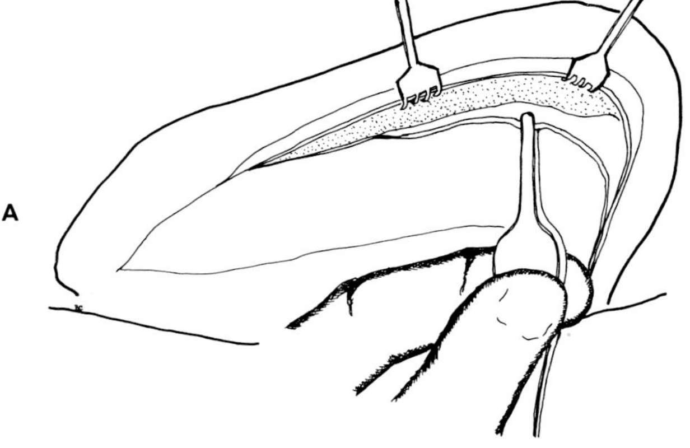

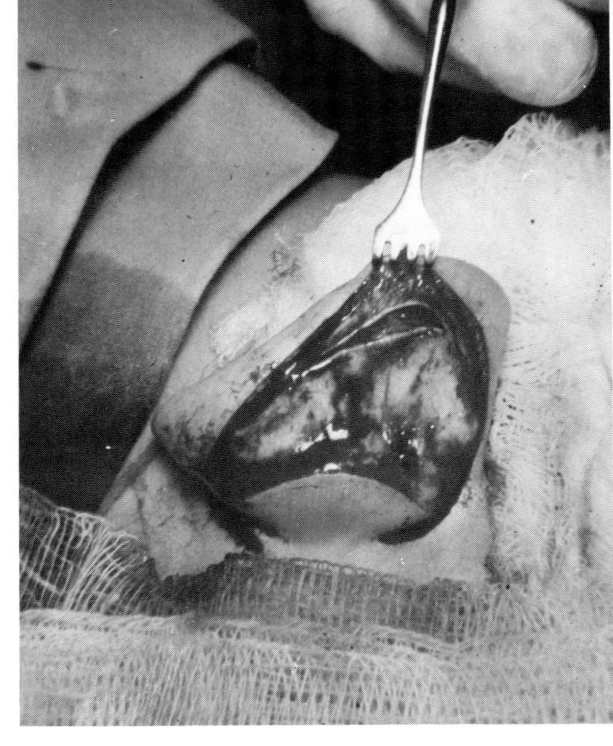

Fig. 24-4. A and **B,** Subperichondrial auricular cartilage dissection on anterior surface. **C,** Fully exposed cartilage maintains protruding position.

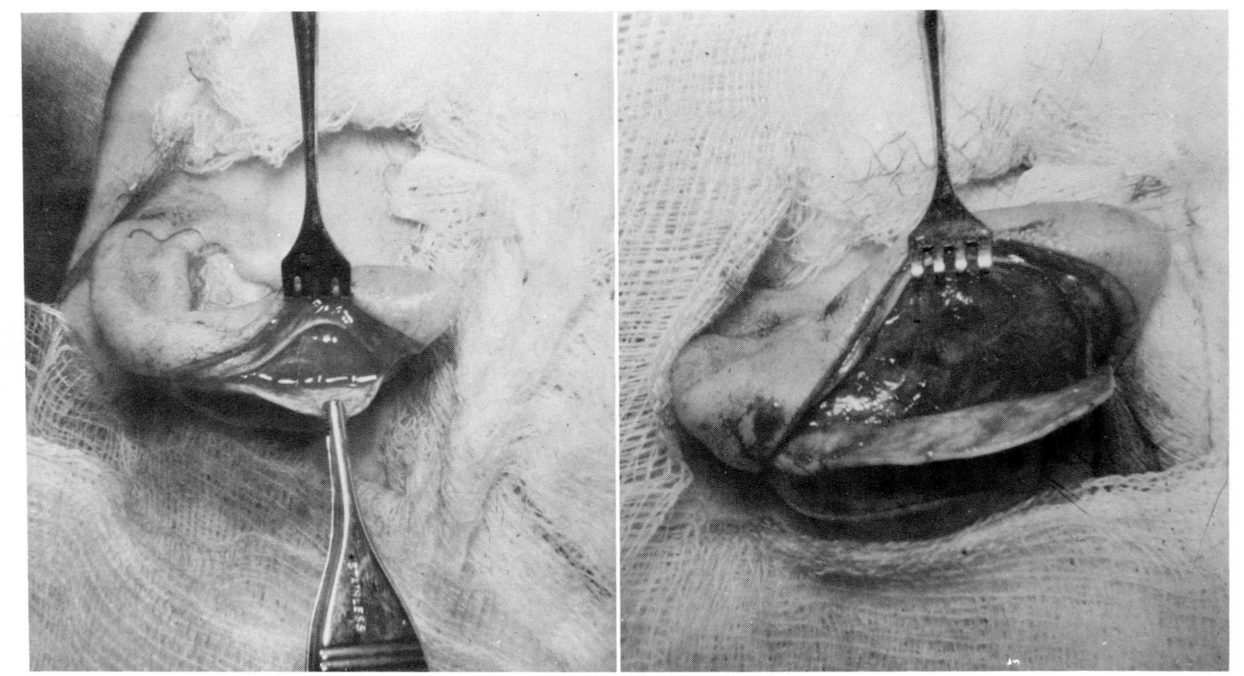

will be observed to bend back on itself, forming the smooth roll of the antihelix with its superior crus, both in proper position (Fig. 24-6). The location of the antihelix bend is independent of the location of the striations and appears to be inherent to the cartilage. The acuteness of the roll can be increased by increasing the number and decreasing the spacing of the striations.

The invariability of this phenomenon must be experienced to be believed. Intrinsic forces within the auricular cartilage similar to those described for rib cartilage appear to be involved.[13] The observation was made several years ago that anterior dissection of the auricular cartilage is associated with some backward bending and that striation could accentuate this.[3, 4] The intentional use of anterior dissection and striation as the basis of an otoplasty technique has been the subject of recent report.[14]

The now redundant auricular cartilage edge can be trimmed. The amount to be excised can be determined by replacing the helix over the auricular cartilage and observing the excess from behind. The width removed affects the width of the new

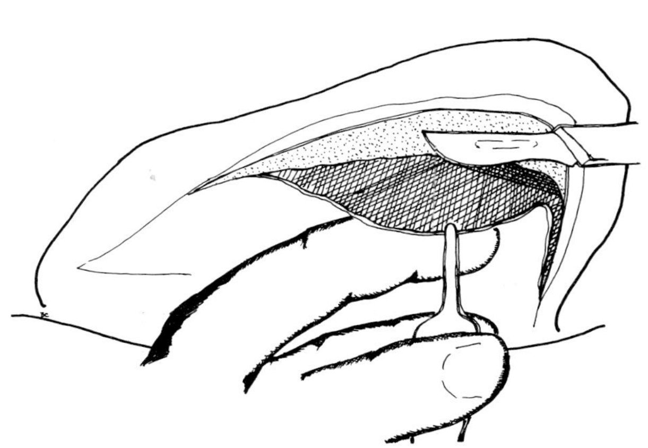

Fig. 24-5. Striation of anterior surface of cartilage.

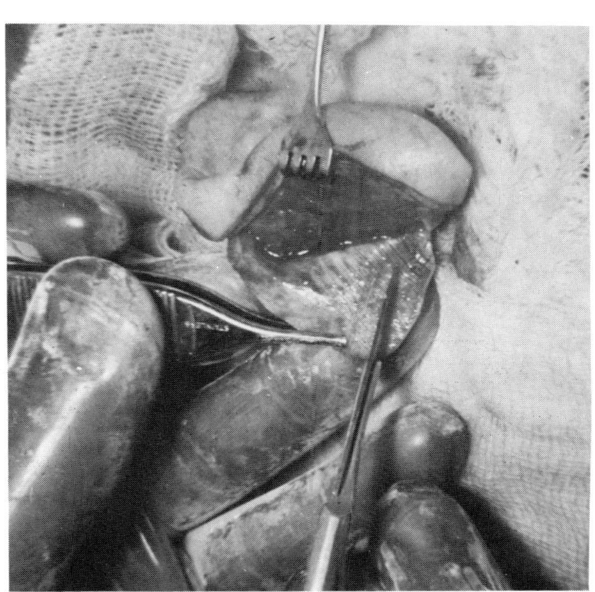

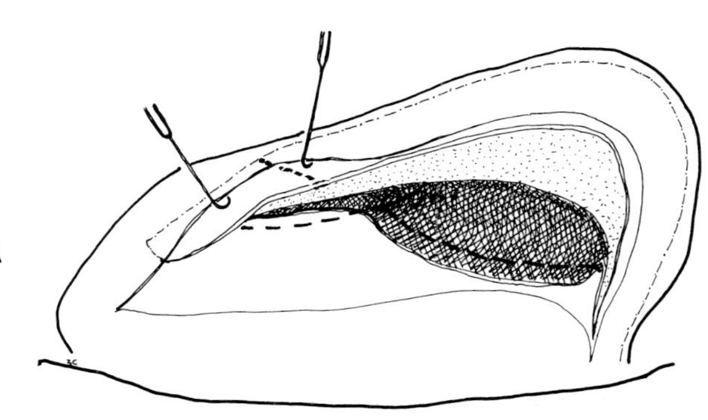

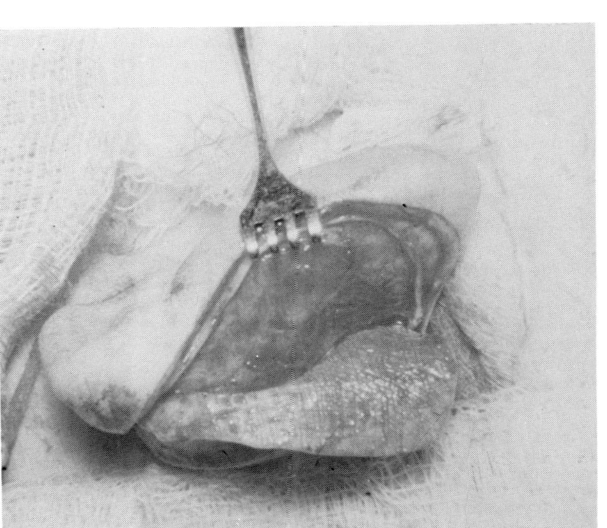

Fig. 24-6. A, Trimming of auricular cartilage and concha edge to be carried out as indicated by heavy dash lines. Tail of helix excision is similarly indicated. B, Smooth antihelical roll is formed by the cartilage after anterior striation. Compare with Fig. 24-4, C.

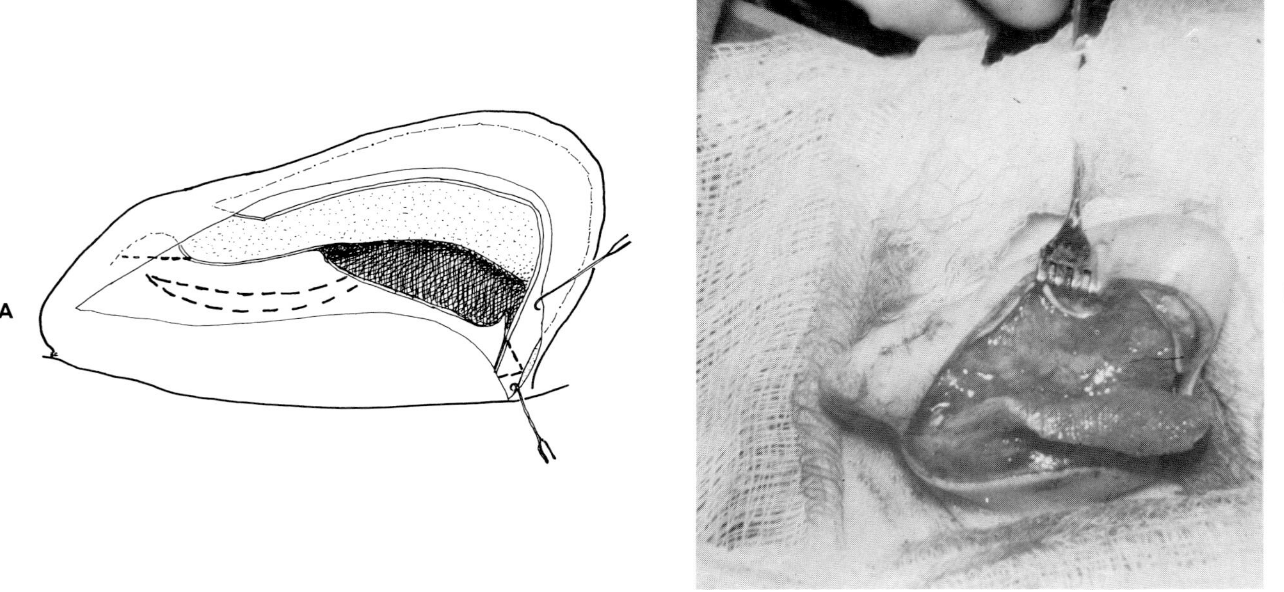

Fig. 24-7. A, Tail of helix excised. Tentative excisions of superior helix wedge, concha ellipse, and antitragus prominence are indicated by heavy dash line. **B,** Appearance after trimming of auricular cartilage edge and excision of a helix wedge.

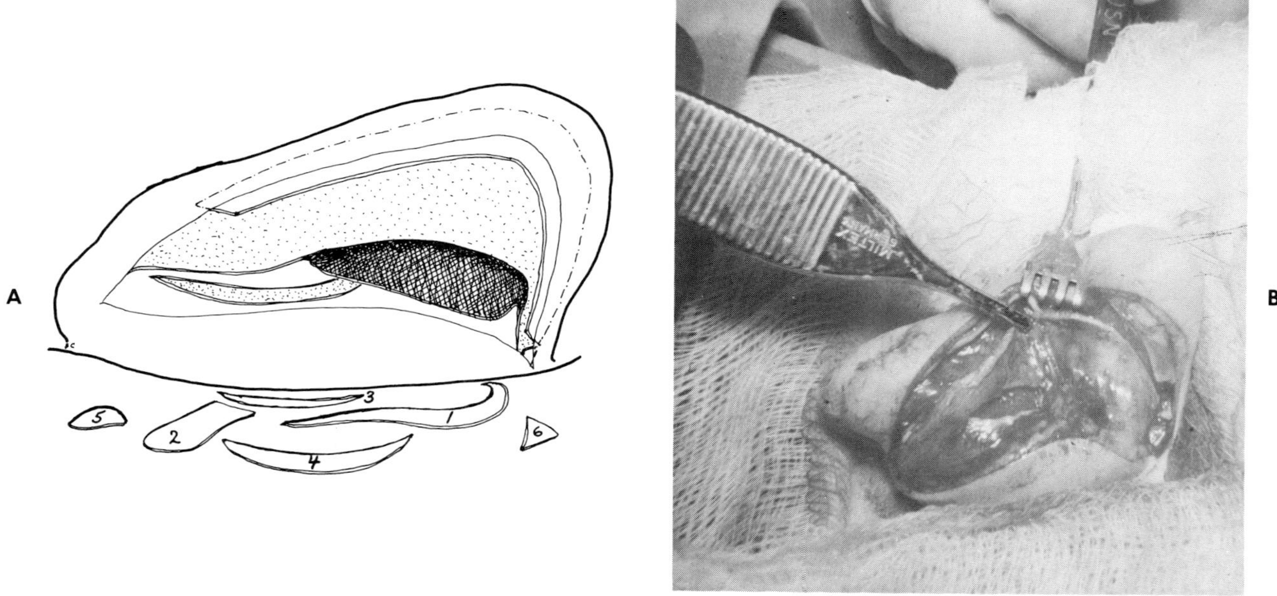

Fig. 24-8. A, Pieces of cartilage excised in maximum reduction: *1,* auricular cartilage edge; *2,* tail of helix; *3,* edge of concha; *4,* ellipse of concha; *5,* antitragus prominence; *6,* helix wedge. The first three are the usual excisions; the remainder are excisions dictated by the individual case. **B,** Location of elliptical excision in concha.

scaphoid fossa. The amount trimmed from the superior pole affects the vertical length of the ear (Fig. 24-6).

The edge of the cavum concha is trimmed (Fig. 24-6, *A*). Striating the edge of the concha does not cause it to bend, as does the auricular cartilage superior to it. The cartilage in this area is thick enough, however, so that the trimmed edge is not sharp. If the conchal width and height are not great this trimming may be all that is needed. If greater reduction is desired an ellipse of cartilage can be excised from the concha. The ellipse lies just beneath the concha edge and the overhanging rim of the inferior crus of the antihelix (Figs. 24-7 and 24-8).

The tail of the helix is next dissected free and excised. This reduces protrusion of the lobule. If the antitragus cartilage contributes to lobule protrusion or if it is prominent in its own right, it is dissected free and trimmed appropriately (Figs. 24-6 to 24-8).

The anterior skin flap is then replaced on the cartilage and molded into its new contours. The anterior aspect is examined. Residual prominence of the conchal edge can be excised. Following this the superior crus of the helix may be cut through at the cephaloauricular angle superiorly and/or a triangle of cartilage excised there. This makes the helix a bucket handle that can both turn back with the new antihelix and lie against it, conform-

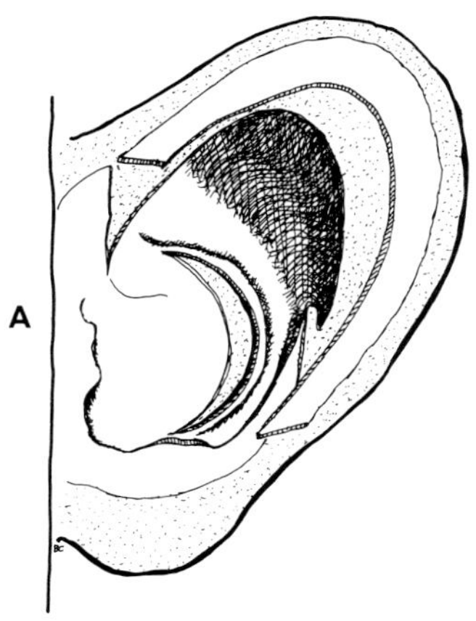

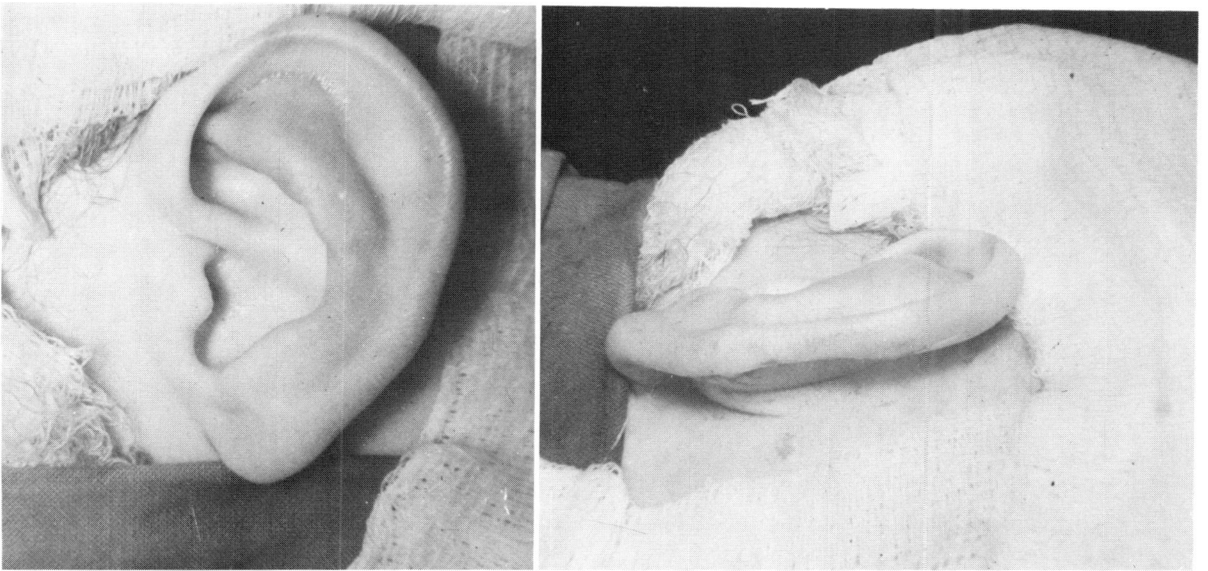

Fig. 24-9. A, Anterior cutaway view showing effect of cartilage excisions with formation of free helix bucket-handle. **B** and **C,** Immediate postoperative result in the ear shown in Fig. 24-4, *B*. Cartilage striation and excision have been carried out as depicted. Postauricular incision has been closed with 5-0 plain catgut. Note how the ear maintains the new position.

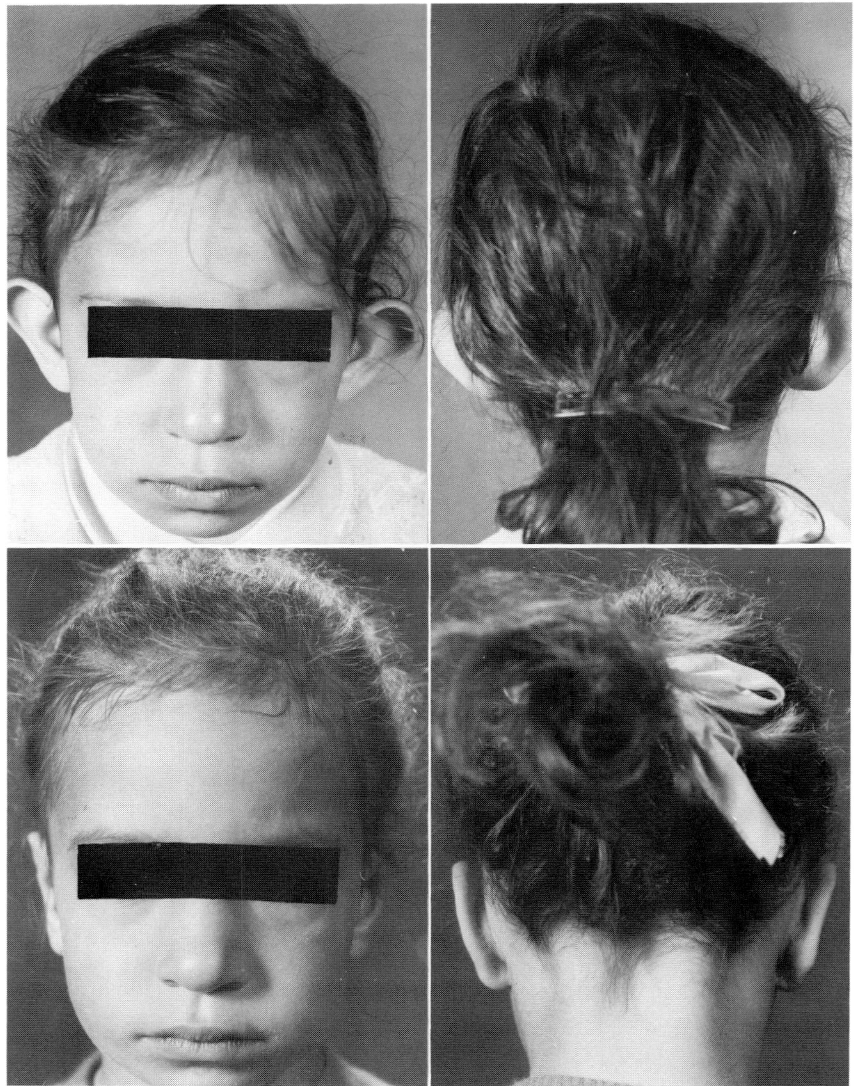

Fig. 24-10. Pre- and postoperative views of patient whose operation was depicted in Figs. 24-1 to 24-9. Good correction of both concha and antihelix deformity with little evidence of operation. Right ear excellent, left ear good.

ing to the latter's reduction in size (Figs. 24-7 to 24-9).

The cartilage maintains the new position without any suturing (Fig. 24-9). The postauricular incision is closed. At this point, modification of the lobule end of the incision by variably shaped excisions can conveniently reduce a large lobe. A carefully molded dressing completes the procedure (Fig. 24-10). The method has applicability in some cases for cup ear (Fig. 24-11). An anterior approach utilizing this basic technique has been presented.[15]

OPERATIVE SERIES

In the period 1957 to 1961, 116 patients had otoplasties performed by the Plastic Surgery Service, Presbyterian Hospital. Of these patients sixty-five were males and fifty-one females. The average age of the males was $9\frac{2}{5}$ years; that of the females was $14\frac{1}{2}$ years. This age differential has previously been noted.[16] The youngest patient operated upon was $3\frac{1}{2}$ years old. Ninety-seven of these patients had their procedures done by the method described. In forty-five of these ninety-seven patients involving eighty-nine ears, pre- and post-

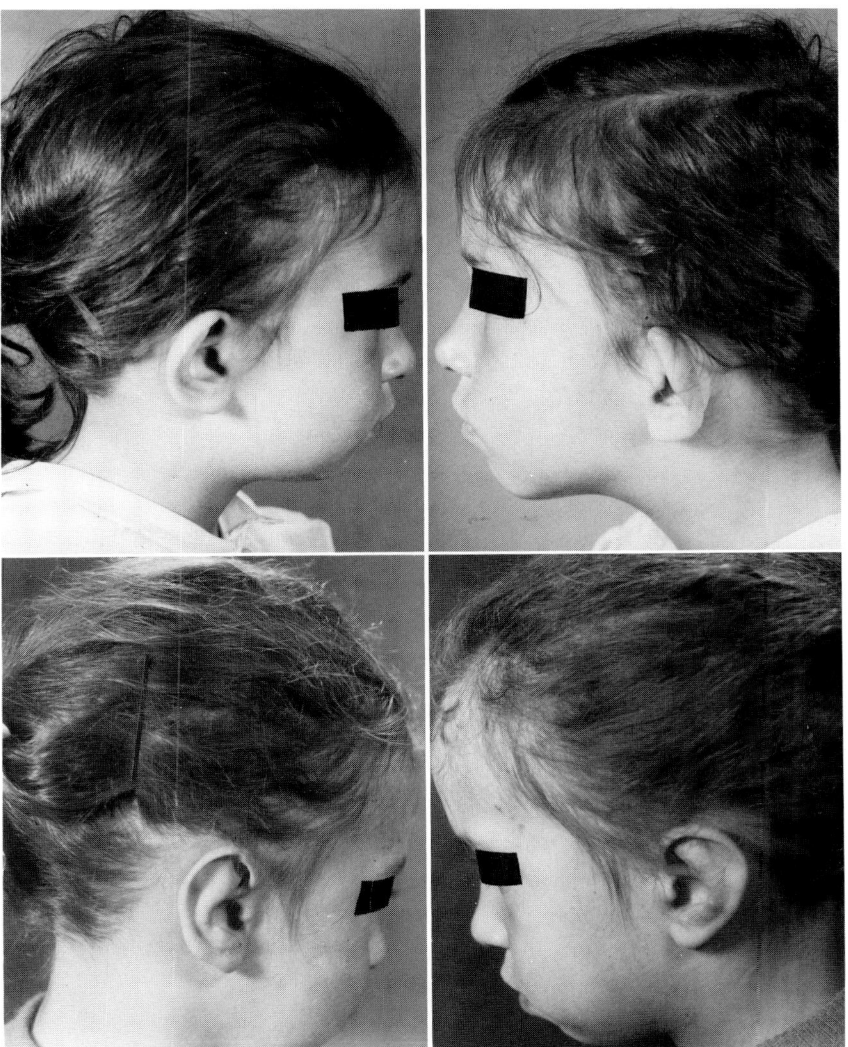

Fig. 24-10, cont'd. For legend see opposite page.

operative photographs and adequate follow-ups were available. The average length of follow-up in this group was 18 months, with twenty-two patients followed 12 months or longer. Table 5 shows the nature of the defects causing the ear protrusion in this series. Table 6 lists the relatively few complications encountered.

CRITERIA OF RESULTS

Within the broad limits allowed by the normal variation between the ears of different individuals as well as between the ears of a single individual,

Table 5. Deformities causing protrusion—eighty-nine ears

Antihelix, predominately failure of formation	21
Concha, predominately excessive depth and width	6
Antihelix and concha deformity combined	32
Antihelix, concha, and lobule deformity	20
Cup deformity	10

Table 6. Complications—eighty-nine ears

Hematoma	1
Suture line opening with spontaneous healing	1
Crusting on antihelix	2
Hypertrophic scar	1

criteria can be established to evaluate operative results. Factors of general appearance and of specific detail are involved. Within the first category it is axiomatic that the appearance of protrusion must be corrected. The ear should lie parallel to the mastoid area but without loss of the cephalo-auricular sulcus.

In the second category five areas must be considered. The helix is one. Its curve should be smooth without buckling or overhang. The superior crus of the antihelix should be softly rounded and with prominence equal to or slightly less than that of the helix. The sharp fold produced by some otoplasty techniques is an operative defect specifically to be avoided. The cavum concha should be neither excessively wide nor excessively deep. The antitragus should be in line with the antihelix curve and not unduly prominent. The lobule should be in line with the helix, not outstanding, and not oversized for the rest of the ear.

A result satisfying these criteria was termed excellent. Minor deficiency in up to three of these areas was compatible with a good result. If more minor deficiences were present or if one major defect was noted in the helix, antihelix, or concha, the result was termed fair. More major defects led to classification as poor. Failure to correct the overall impression of protrusion was in itself ground for terming the result poor. Based on these criteria the results in eighty-nine ears are listed in Table 7.

Table 7. Results—eighty-nine ears

	Excellent	*Good*	*Fair*	*Poor*
Number	20	39	16	14
Percent	22.5	43.8	18	15.7

DISCUSSION OF DEFECTS

With the exception of one instance of slight protrusion developing in the second postoperative month, the faults seen late were those found immediately after operation. Late adverse changes did not occur in any significant degree. The distribution of defects is shown in Table 8.

Significant residual protrusion was caused in some cases by insufficient striation of the antihelix but in more instances by failure to reduce a wide deep concha (Fig. 24-12, *A*). A revision operation was carried out in three cases. The improvement achieved indicated that the defects were the result of an initial technical error.

The most common faults involving the helix were slight overhang of the upper third and some buckling in that area or in the middle third (Fig. 24-12, *B* and *C*). The overhang seemed to be caused by leaving too wide a rim of helix. The buckle or notch occurred when the ear size reduction was very great, when the cut through the superior hinge of the bucket handle was performed in the

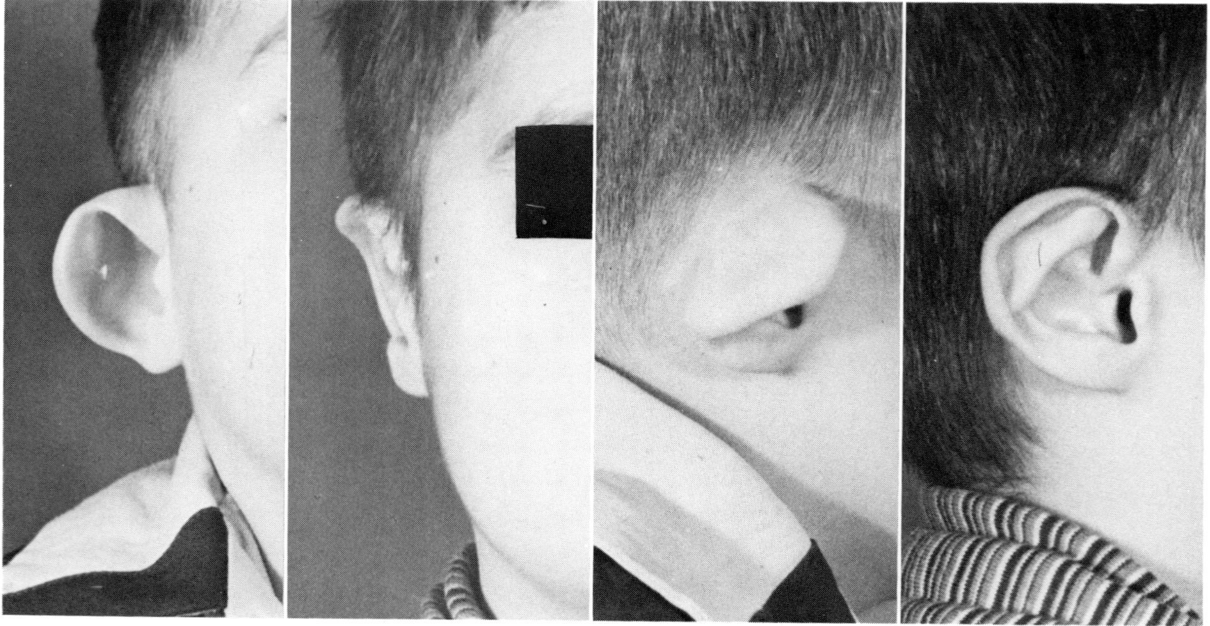

Fig. 24-11. Pre- and postoperative photographs showing result of the described technique in a case of cup ear.

Table 8. Postoperative defects—eighty-nine ears

Significant residual protrusion	*Helix*		*Antihelix*		*Concha*		*Antitragus*		*Lobule*	
10	Overhang:	15	Too prominent:	9	Too deep:	10	Too prominent:	2	Too prominent:	12
	Buckle:	16	Too reduced:	2	Too wide:	5	Cartilage spur:	1	Too large:	2
			Too sharp:	1	Too reduced:	1				
			Cartilage spur:	1						
			Poorly formed:	1						

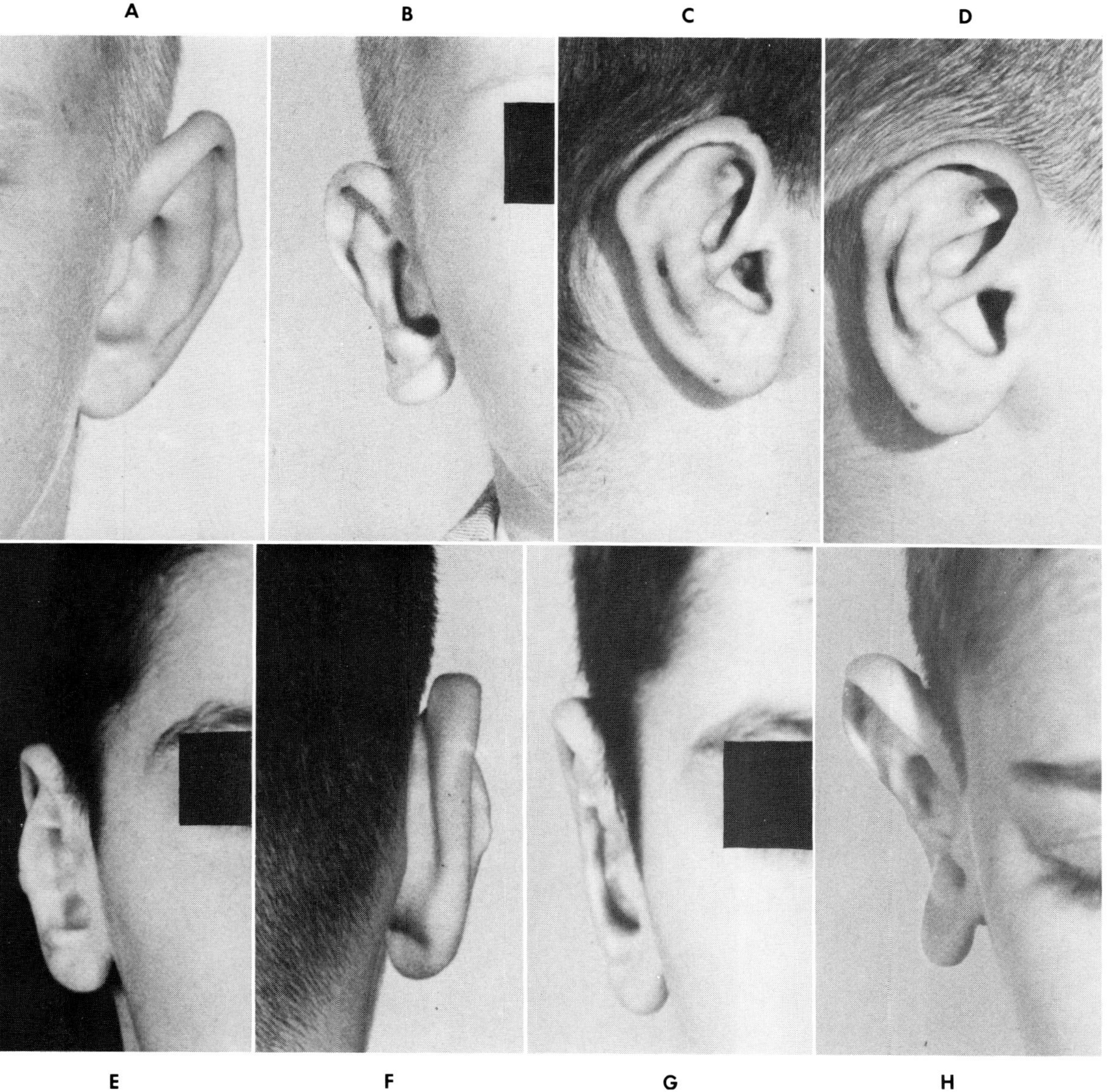

Fig. 24-12. A, Residual protrusion caused by insufficient reduction of a wide concha. **B,** Slight helix overhang together with failure to reduce lobule, producing "telephone ear" deformity. **C and D,** Helix buckle 1 month postoperatively with improvement in 6 months. **E and F,** Prominent antihelix associated with pronounced antihelix bend and failure to reduce concha adequately. **G,** Ear seen in E and F after revision of cartilage. **H,** Telephone ear deformity associated with too great a reduction at concha edge, helix overhang, and failure to reduce lobule.

visible part of the helix rather than at the cephaloauricular base of the superior helix crus, or when no division was made at all. Alone among the defects discussed here, the passage of time tended to ameliorate helical buckling (Fig. 24-12, *C* and *D*).

Undue prominence of the antihelix was the most frequent fault in that area. Some cartilages show so pronounced a tendency to curl when anteriorly striated that too great a backward bend is formed. Failure to reduce the concha also gives rise to apparent prominence of the antihelical roll (Fig. 24-12, *E* to *G*). The ease of forming this curl is indicated in the infrequent occurrence of a flat or poorly formed antihelix.

Too great a reduction of the middle portion of the antihelix results in the telephone ear deformity, in which relative superior helix and inferior lobule prominence is associated with a depressed midportion of the ear. Aesthetic sense must guard against this error (Fig. 24-12, *H*). Too sharp an antihelix was noted in one instance only, and it seems likely that in this case the cartilage was cut through rather than striated. Cartilage beading does not occur on the striated surface and in those cases revised the area has been found to be smooth.

Failure to correct the depth or width of the concha was a frequent error until the need for excision of a concha ellipse was accepted. Careful initial dressing obviates the skin folds that might be anticipated with this maneuver. Placing the elliptical excision within the conchal edge rather than in the cephaloauricular sulcus preserves the sulcus better. Too great an excision of the concha can also lead to telephone ear deformity.

Antitragal prominence was rare. However, it was observed that the antitragal cartilage sometimes played a part in lobule prominence persisting after the tail of the helix was excised. Removal of a portion of the antitragus when needed, lobule soft tissue excision, and careful removal of the helix tail should prevent the defect of lobule prominence.

SUMMARY

The goal of operation for prominent ears is correction of the deformity without signs of the operation. A technique is presented that in large measure achieves this goal. The use of striation on the anterior surface of the cartilage to achieve a softly rounded antihelix without the use of sutures and alteration of the concha by elliptical excision beneath the conchal edge appear to be individual technical features deserving emphasis. The entire procedure permits great flexibility in the manipulation of the total size of the ear as well as the size and shape of its component parts.

The use of this method in eighty-nine ears has been reviewed and errors discussed. It is felt that the defects noted represent mistakes in the application of the technique rather than faults inherent to it. In the great majority of cases a satisfactory result is obtained that successfully avoids the stigmata of operation.

REFERENCES

1. Luckett, W. H.: A new operation for prominent ear based on the anatomy of the deformity, Surg. Gynec. Obstet. **10:**635, 1910.
2. Strombeck, J. O.: Results of surgery for protruding ears: a follow-up study of two series, one treated with a modification of Luckett's operation and one with a "new" technique, Acta Chir. Scand. **122:**138, 1961.
3. Cloutier, A. M.: Correction of outstanding ears, Plast. Reconstr. Surg. **28:**412, 1961.
4. Farrior, R. T.: A method of otoplasty; normal contours of the antihelix and scaphoid fossa, Arch. Otolaryng. **69:**400, 1959.
5. Holmes, E. M.: A new procedure for correcting outstanding ears, Arch. Otolaryng. **69:**409, 1959.
6. Mustardé, J. C.: The correction of prominent ears using simple mattress sutures, Brit. J. Plast. Surg. **16:**170, 1963.
7. Stark, R. B., and Saunders, D. E.: Natural appearance restored to the unduly prominent ear, Brit J. Plast. Surg. **15:**385, 1963.
8. Straith, R. E.: Correction of outstanding ears, Plast. Reconstr. Surg. **24:**277, 1959.
9. Tamerin, J. A., and Mirehouse, O.: Reconstruction of the antihelix and scapha in selected cases of protruding ears, Arch. Otolaryng. **70:**597, 1959.
10. Tanzer, R. C.: The correction of prominent ears, Plast. Reconstr. Surg. **30:**236, 1962.
11. Converse, J. M., and others: A technique for surgical correction of lop ears, Plast. Reconstr. Surg. **15:**411, 1955.
12. Converse, J. M., and Wood-Smith, D.: Technical details in the surgical correction of the lop ear deformity, Plast. Reconstr. Surg. **31:**118, 1963.
13. Gibson, T., and Davis, W. B.: The distortion of autogenous cartilage grafts, its cause and prevention, Brit. J. Plast. Surg. **10:**257, 1959.
14. Chongchet, V.: A method of antihelix reconstruction, Brit. J. Plast. Surg. **16:**268, 1963.
15. Ju, D. M. C., Li, C., and Crikelair, G. F.: The surgical correction of protruding ears, Plast. Reconstr. Surg. **32:**283, 1963.
16. Ju, D. M. C.: The psychological effect of protruding ears, Plast. Reconstr. Surg. **31:**434, 1963.

Chapter 25

Protruding ears and their correction

Nicholas G. Georgiade, M.D., F.A.C.S.

A short operative procedure to transform an unsightly protruding ear to one that is cosmetically acceptable is the aim in reconstructive surgery of the ear. The adverse psychologic effects to the patient even at the age of 4 or 5 years is considerable, and the ability to transform the auricular deformity to one of acceptable dimensions will change this youngster's entire outlook on life.

Much has been written over the years in describing various techniques and their modifications in an attempt to attain a normal appearing ear.[1-8] These procedures have been noted in Chapter 21.

It has become apparent to me over the past 19 years that most techniques fail to adequately outline their various goals. A satisfactory correction of the protruding ears should include the following points:

1. The antihelical fold should be created so as to have a soft curvature or roll and not project more than the helix.
2. The lobule of the ear in its final position should not project more laterally than the helix.
3. The superior portion of the helix of the ear should not protrude and generally should be about 5 mm. closer to the head than the middle portion of the ear.
4. There should be a definite posterior auricular sulcus.

The technique to be described here encompasses the best of many plastic surgeons' techniques and is a composite of their suggestions[6-11] as well as my modifications.

Following suitable anesthesia (xylocaine 1% with 1:100,000 epinephrine) these steps are executed in the following order.

1. The protruding ear is pressed back to the head and a judgment is made as to the size of ellipse of posterior skin to be excised so that the ear will be in the desired position closer to the head. This ellipse is then outlined in brilliant green, with care being taken to maintain the area of excision slightly above (lateral) to the auriculocephalic sulcus (Fig. 25-1, *A* and *B*).

2. The remaining posterior auricular skin is undermined, exposing the tail of the helix. This cartilage is dissected free from its fibrous attachment to the concha, exposing a small area of anterior conchal subcutaneous tissue (Fig. 25-1, *C*).

3. A Joseph elevator is inserted in this conchal, helical tail space onto the anterior surface of the antihelix cartilage. A path is created subcutaneously by "wiggling" the Joseph elevator along what is judged to be the new antihelix, inferior crus, and scapha areas (Fig. 25-1, *C*).

4. A special Stenström auricular rasp is inserted in the same tunnel, or one half of a Brown-Adson pickup may be used if the Stenström rasp is not available. Through this tunnel, previously created by the Joseph elevator, the anterior surface of the cartilage is scratched over the entire area previously described, with a wider area of scratching in the scaphal area. This procedure will allow the cartilage to unfold like a fan, effectively taking the "spring" out of the cartilage and creating a normal roll[10] (Fig. 25-1, *D*).

5. Short 25-gauge needles are inserted along the curvature of the newly created antihelix and scapha area and onto the posterior surface. Approximately four 4-0 white nylon vertical mattress sutures are inserted into the cartilage, further creating and positioning the antihelix and helix.

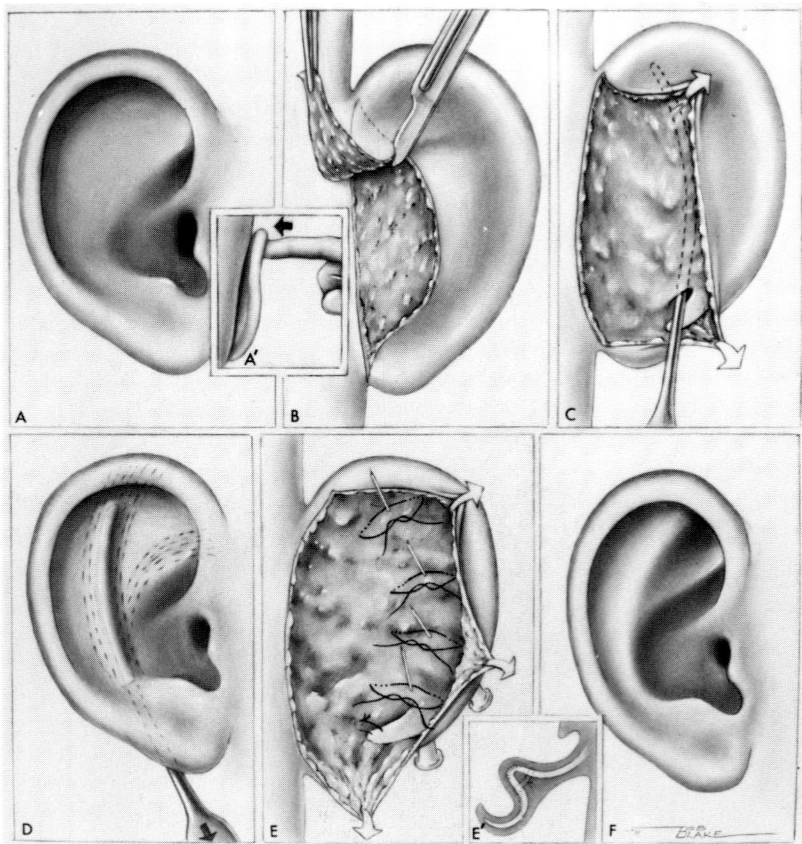

Fig. 25-1. **A,** An estimate of the distance of the ear to be positioned is carried out by finger pressure, noting the amount of redundant posterior auricular skin. **B,** The previously estimated size of redundant skin is excised. **C,** Following undermining of the remaining posterior auricular skin, the tail of the helix is dissected free and separated from the conchal cartilage. The Joseph type periosteal elevator is inserted into the medial space onto the anterior surface of the helical cartilage, undermining the skin to the superior crus in a **Y** type of undermining. **D,** A curved rasp is used to scratch the anterior surface of the ear as shown. **E,** Nylon (4-0 white) sutures are inserted in the manner as shown. Particular attention is placed in repositioning the tail of the helix to position the lobe in close apposition to the posterior auricular area. A second critical suture is placed in the superior helix area to position the superior crus in a satisfactory position and cephaloauricular relationship. **F,** A soft natural roll of the helix is created in the manner outlined, which will closely simulate an unoperated natural appearing ear.

Fig. 25-2. **A,** Preoperative ear is shown with the size of the skin excision outlined in brilliant green dye. **B,** The inferior tail of the crus is shown. Note the space between the tail and main portion of the cartilage (concha). **C,** Sharp scissors is shown dissecting the fibrous tissue in the conchal-helical space. **D,** The Joseph elevator is shown on the anterior cartilage surface of the helix. Dissection is "fanned" out as shown over the superior and inferior crura. **E,** A sharp rasp type instrument is inserted from the posterior approach onto the anterior cartilaginous surface. **F,** The rasp is now shown on the anterior surface and the cartilage is scratched over the areas desired to result in a natural roll of the antihelix. **G,** Small 27-gauge needles are placed along the desired antihelix fold. **H,** The 4-0 white nylon sutures are placed in a manner to create a natural antihelix fold, with care being taken to overlap the tail of the helix over the conchal cartilage.

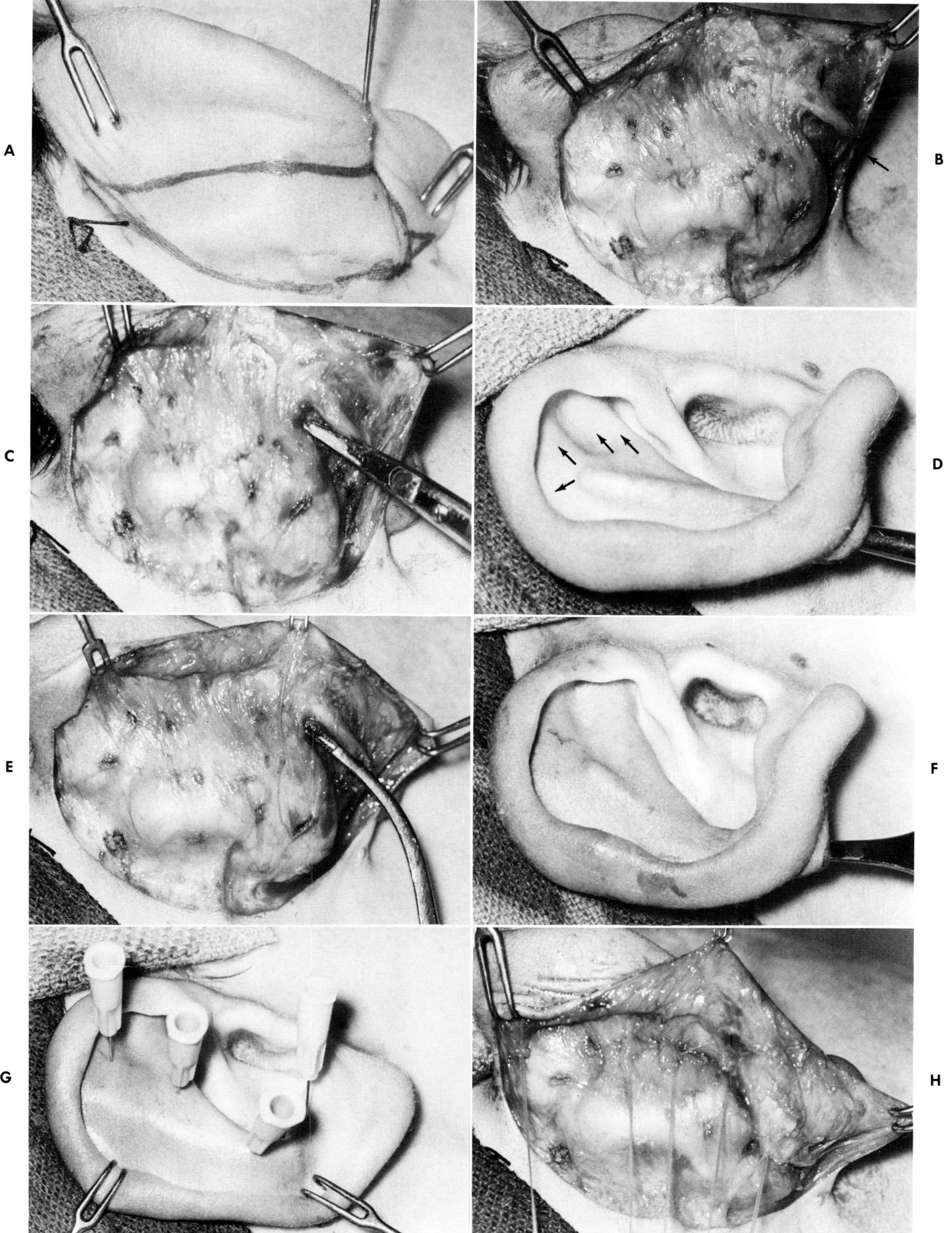

Fig. 25-2. For legend see opposite page.

A separate horizontal mattress type suture is used to bring the tail of the helix, which was previously separated from the fibrocartilages of the antitragus (concha), inferiorly over the conchal cartilage in order to turn the antitragus and lobule medially to prevent it from protruding. A nylon mattress type suture is also placed in the superior helix to adjust the position of the superior ear in relation to the head (Fig. 25-1, *E*).[11]

6. At this time, prior to tying the antihelix sutures, if there is a prominent concha, three mattress sutures are taken through the mastoid periosteum and conchal cartilage, bringing the conchal cartilage close to the process.[9]

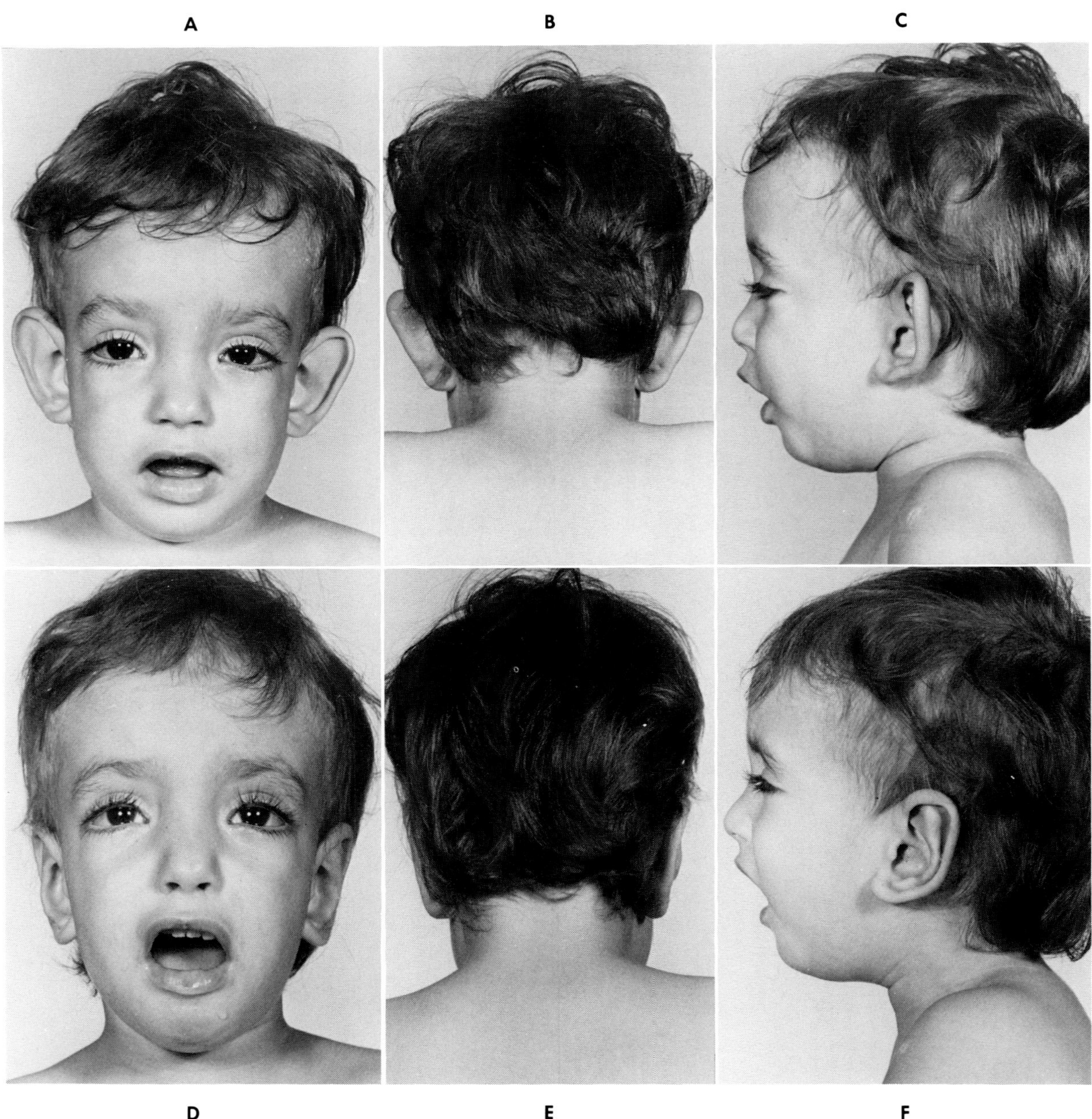

Fig. 25-3. A, A 3-year-old boy is shown with large fully developed and quite prominent ears. **B,** Posterior view of the same child. **C,** A lateral view of the ear is shown. Note the marked lack of an antihelix fold. **D** and **E,** Postoperative appearance of the same boy. Note the natural relative positioning of the auricles and lobules. **F,** Lateral view of the repositioned ear. Note the nice, natural appearing antihelix.

7. Following creation of the antihelix and tail of the helix by suturing the 4-0 white nylon sutures, a number of interrupted 5-0 white dacron sutures are inserted subcutaneously along the line of closure.

8. A running 3-0 chromic catgut suture is then used to approximate the skin edges in the auricular sulcus. This suture is not removed but usually will separate within 12 to 14 days and is removed during the normal cleaning process of the posterior auricular area (Fig. 25-1, *F*).

9. Moistened cotton mechanic's waste is carefully used to fill in the contours over the antihelix, scapha, helix, and conchal areas, following which

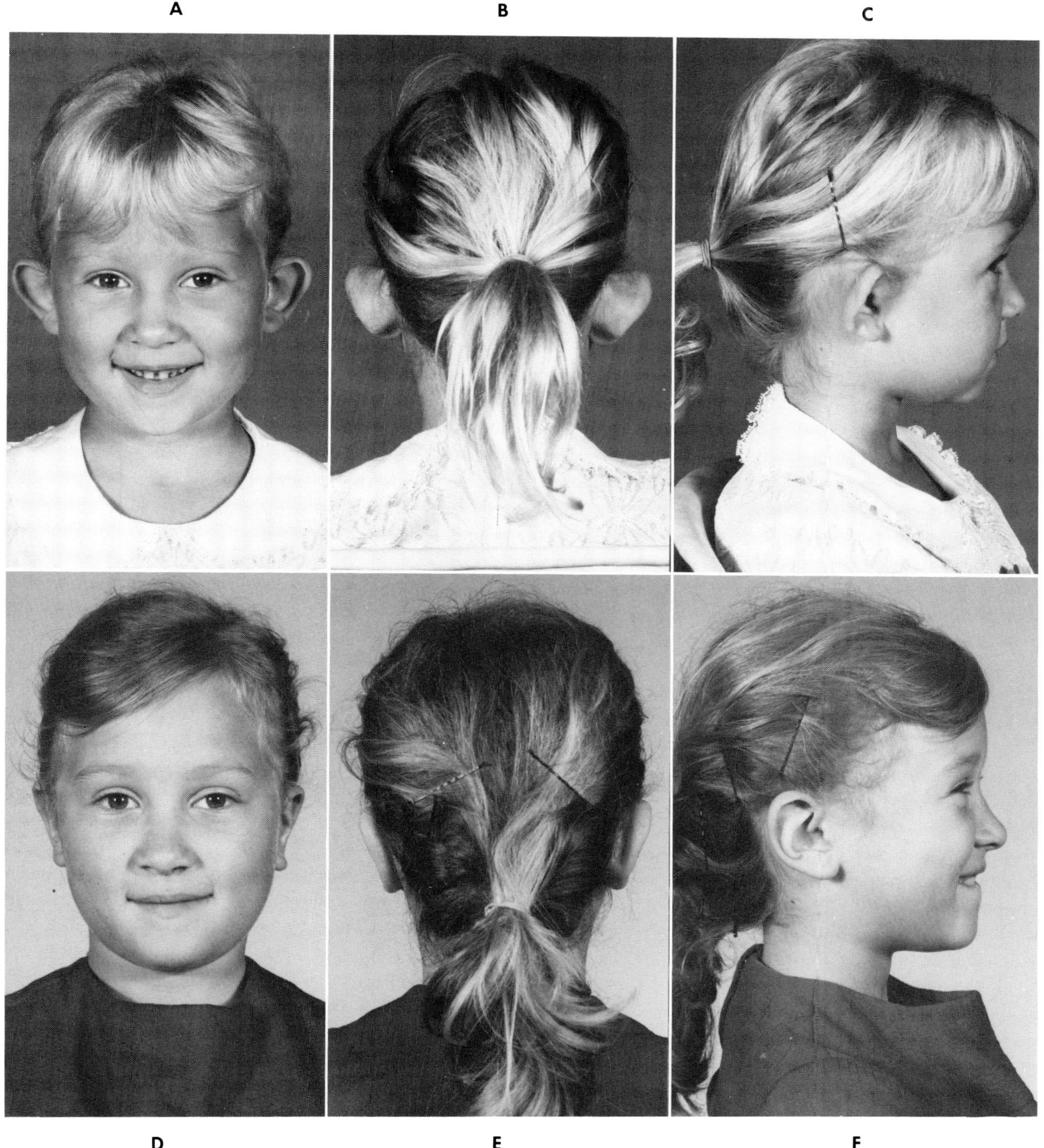

Fig. 25-4. A to **C,** Preoperative appearance of a 5-year-old girl with prominent ears with lack of antihelix. **D** to **F,** Postoperative appearance almost 2 years later.

a Kerlix gauze head roll is applied. Fig. 25-2 illustrates the clinical application that was shown schematically in Fig. 25-1.

The ears are reevaluated the following day and any superficial anterior auricular blebs that may be present are aspirated with a small needle and the ears are redressed again for 6 additional days.

Following this a protective Kerlix gauze is applied over the ears every night for 2 additional weeks to minimize the chances of trauma to the ears while asleep.

Three patients varying in ages from 3 to 8 years are shown with various types of prominent ears (Figs. 25-3 to 25-5.)

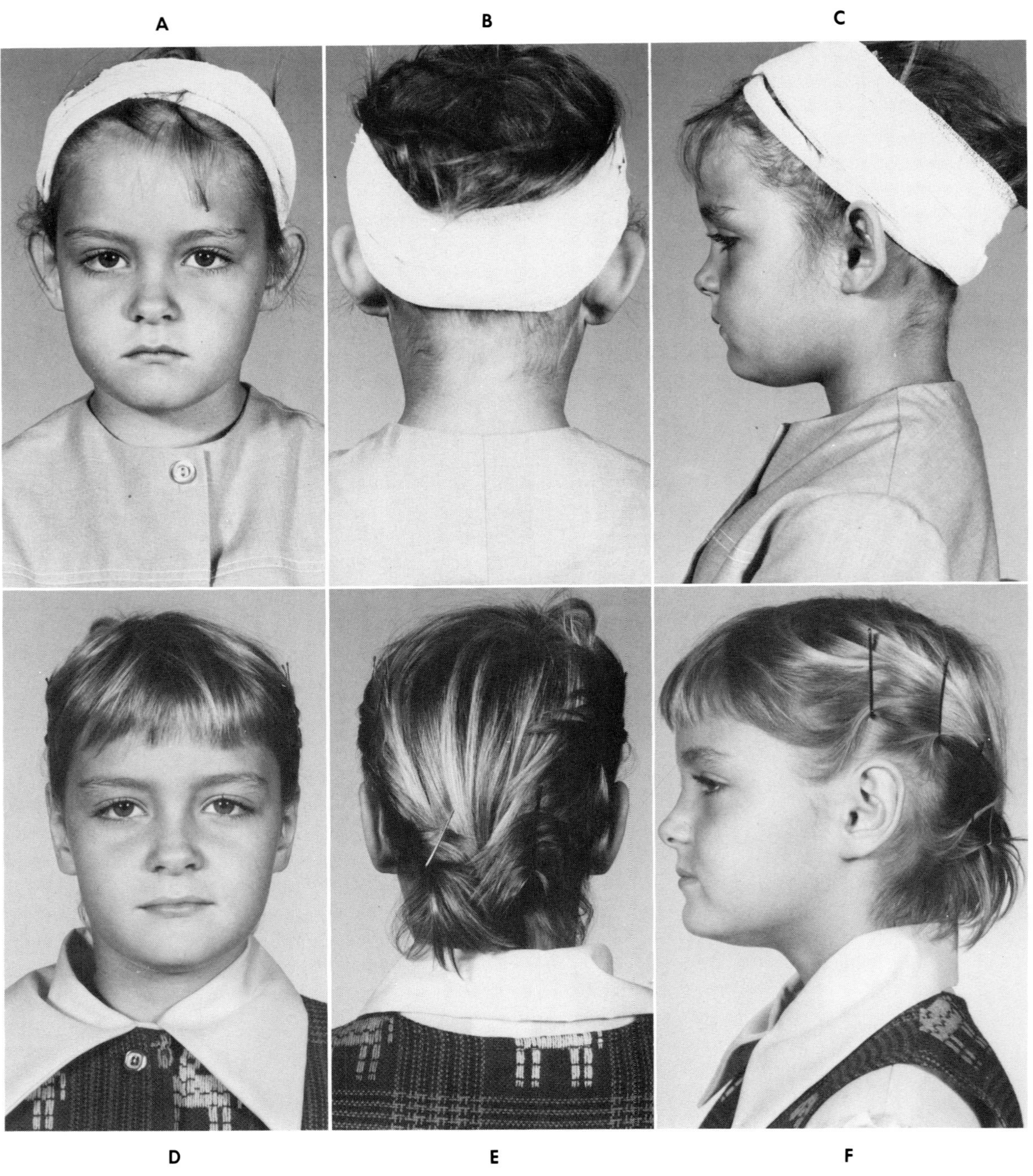

Fig. 25-5. A to **C,** Preoperative appearance of an 8-year-old girl with asymmetry of the auricles. **D** to **F,** Postoperative appearance 2½ years later.

SUMMARY

A relatively simple, short, accurate technique has been described for attaining a natural appearing ear. The procedure is almost always carried out under local anesthesia and some sedation. Eighty-four ears have been corrected utilizing the described technique without any untoward effects or complications.

REFERENCES

1. Davis, J. S., and Kitlowski, E. A.: Abnormal prominence of the ears: a method of readjustment, Surgery **2**:835-848, 1937.
2. Luckett, W. H.: A new operation for prominent ears based on the anatomical deformity, Surg. Gynec. Obstet. **10**:635-637, 1910.
3. New, G. B., and Erich, J. B.: Protruding ears: a method of plastic correction, Am. J. Surg. **48:** 385-390, 1940.
4. Young, F.: The correction of abnormally prominent ears, Surg. Gynec. Obstet. **78**:541-550, 1944.
5. Converse, J. M.: Acquired deformities of the auricle, Plast. Reconstr. Surg. **28**:1107-1121, 1964.
6. McDowell, A. M.: Goals in otoplasty for protruding ears, Plast. Reconstr. Surg. **41**:17-27, 1968.
7. Stenström, S. J.: A "natural" technique for correction of congenital prominent ears, Plast. Reconstr. Surg. **32**:509-517, 1963.
8. Mustardé, J. D.: The correction of prominent ears using simple mattress sutures, Brit. J. Plast. Surg. **16**:170-176, 1963.
9. Spira, M., and others: Correction of the principle deformities causing protruding ears, Plast. Reconstr. Surg. **44**:150-154, 1969.
10. Gibson, T.: The distortion of autogenous cartilage grafts: its cause and prevention, Brit. J. Plast. Surg. **10**:257, 1957.
11. Goulian, D., and Conway, H.: Prevention of persistent deformity of the tragus and lobule by modification of the Luckett technique of otoplasty, Plast. Reconstr. Surg. **26**:399-404, 1960.

Chapter 26

A simplified auriculoplasty

Bernard L. Kaye, M.D., D.M.D., F.A.C.S.

Many different methods have been used to correct prominent ears in which the deformity is a result of lack of development of the antihelical fold. Luckett[1] is credited with first showing the necessity for surgically recreating this fold (1910). Since then numerous techniques have been advocated to form the antihelical fold, most of them involving weakening of the posterior surface of the cartilage by incising, resecting, abrading, or slitting.[2-6] Virtually all of these techniques required a long posterior auricular incision, usually with removal of some posterior skin. During the past decade several operations have been described that create the antihelical fold by striating the anterior surface of the cartilage.[7-10] Mustardé presented a simple method of auriculoplasty in 1963,[10] in which he formed the antihelical fold by the insertion of mattress sutures alone, making no attempt to weaken the cartilage either anteriorly or posteriorly.

Because of occasional failures with Mustardé's method, and because of a single but uncomfortable experience with keloiding of standard posterior auricular incisions, I tried to devise a method for creating an antihelical fold that would eliminate the need for extensive posterior auricular incisions, dissection of flaps, or excision of skin. This method was also designed to weaken the anterior surface of the antihelical fold by striating with simple, nonspecialized instruments and to allow the placement of mattress sutures without large incisions.[11]

OPERATIVE TECHNIQUE

Local anesthetic solution (xylocaine 1% with epinephrine 1:100,000) is infiltrated anteriorly over the region of the antihelix, as well as medial, lateral, and posterior to the antihelix, to aid in dissection and in passing of mattress sutures. It is easier on the patient if the most medial infiltrations, anteriorly and posteriorly, are made first. A 1.5-cm. transverse incision is made on the *posterior* surface of the ear, at or just superior to the level of the inferior end of the tail of the helix. This provides exposure for the tail of the helix, which may be mobilized or excised if desired.[12] Through this posterior incision the cartilage is penetrated and an anterior skin tunnel is elevated over the antihelical region and posterior crus with a pair of small, flat-tipped scissors (Fig. 26-1). In cases where the auricular prominence is confined to the upper half of the ear, the antihelix may be approached from above through a transverse stab incision made on the posterior surface of the ear at the superior end of the posterior crus.

After elevating the anterior skin tunnel over the antihelical region, the posterior skin is also elevated with a thrust of the scissors along the posterior surface of the antihelical cartilage, creating a posterior tunnel. This maneuver should produce posterior scarring and additional holding power for the new antihelical fold.

One arm of a slender pair of toothed forceps is introduced through the posterior incision onto the anterior surface of the antihelix and posterior crus. These areas are striated vertically in a slightly curved direction to weaken the antihelix and form a new antihelical fold (Fig. 26-1, *B*). One can use the forceps as they are, although it is even more convenient to make a pair of handy striating instruments by splitting apart a pair of slender toothed forceps with a hammer and screwdriver. (The ear morselizer, a double action "tenderizing" device, has also been used for this step, and

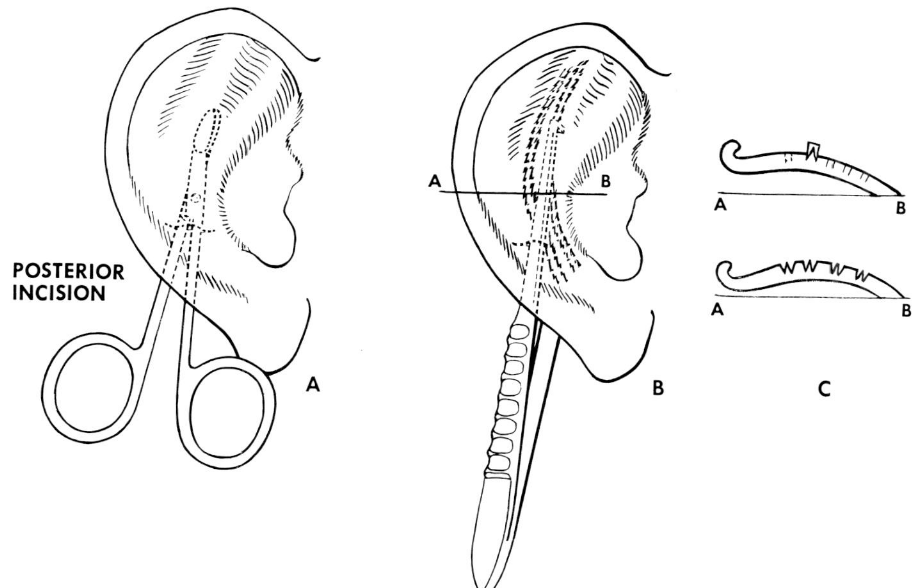

Fig. 26-1. A, 1.5-cm. posterior transverse incision made near inferior end of tail of helix. After penetrating cartilage, anterior skin tunnel is elevated over antihelix and posterior crus with small, flat-tipped scissors. Skin is also elevated over posterior aspect of antihelix and posterior crus (not shown). **B,** Anterior perichondrium and cartilage of antihelix and posterior crus are striated vertically with one arm of a pair of toothed forceps to break spring of antihelical cartilage. Antihelix is then "creased" between thumb and index finger to give it more definition (see text). **C,** Diagrammatic cross section of ear, showing how anterior striation releases anterior tension; anterior slits open, allowing cartilage to fold back posteriorly, recreating antihelical fold. (From Kaye, B. L.: A simplified method for correcting the prominent ear, Plast. Reconstr. Surg. **40:**44-48, 1967.)

I find that it works effectively and rapidly to weaken the cartilage.) Simple striation with a forceps produces equally good end results.

The right-handed surgeon, striating through an inferior incision, will find it much easier to do this on both ears from the right side of the operating table. If he uses a superior incision, or if he is left handed, he should striate both ears from the left side of the table.[7]

After striation or morselization, the antihelix can be made more definite and prominent by folding and compressing it between the thumb and index finger, much as one creases a fold of paper. This step emphasizes the antihelical fold by stretching the anterior perichondrium and opening the vertical striations (Fig. 26-1, *C*).

Three small stab incisions, 3 to 5 mm. in length, are made in the skin of the anterior surface of the concha, medial to the future antihelix, where the inner portions of the mattress sutures will lie (Fig. 26-2, *A*). Care must be taken to incise *skin only;* cutting perichondrium and cartilage could result in the mattress suture pulling through. Each in-

cision is undermined with fine scissors to provide room to bury the knot of the mattress suture. If desired, the undermining may be carried out to the extent of interconnecting the stab incisions by subcutaneous tunnels. Occasionally it may be convenient to introduce the mattress suture through one stab incision, bring it back through an adjacent stab incision, and return it through the subcutaneous tunnel to the first stab incision where it is tied (Fig. 26-3, *B*, inferior mattress suture). This maneuver can help save the day if the usual mattress suture cuts through the medial cartilage and perichondrium.

Now 5-0 polypropylene (Proline, Ethicon) or polyethylene (Davis & Geck) is used for the mattress sutures instead of the heavier multifilament material previously described. The finer monofilament material has proved to be more satisfactory, and it is felt strongly that only monofilament material should be used for the mattress sutures. Using a medium or large cutting needle, the suture is passed through the superior end of the stab incision, through the cartilage on each side of the antihelical fold, and brought out through the skin of the scapha. During this step, care is taken to pass the suture *subcutaneously* on the posterior aspect of the antihelix (Fig. 26-2, *B*). The needle is then reinserted *through the same needle hole,* passed subcutaneously in a caudal direction on the lateral surface of the scapha, and brought out

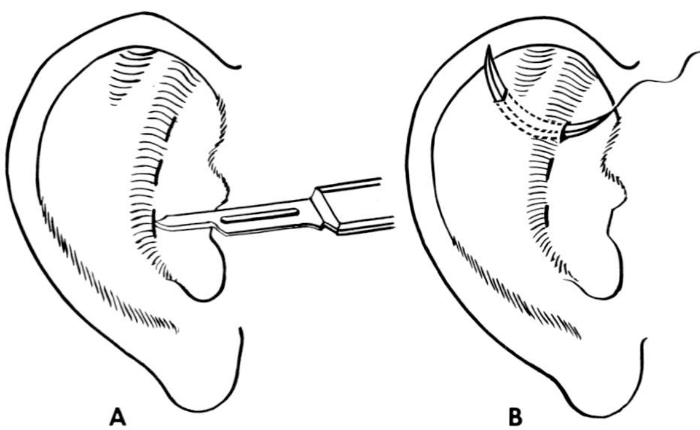

Fig. 26-2. A, Small stab incisions along medial side of antihelix made through *skin only.* They are undermined (not shown) to provide room for burying knot of mattress sutures. **B,** Suture passed through antihelical fold, keeping it *subcutaneous posteriorly.* (From Kaye, B. L.: A simplified method for correcting the prominent ear, Plast. Reconstr. Surg. **40**:44-48, 1967.)

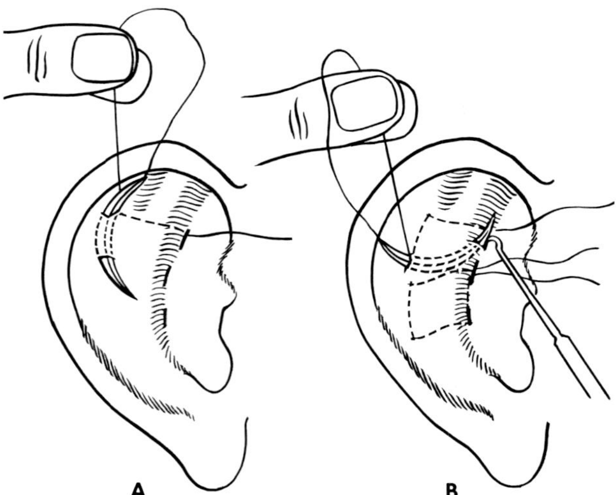

Fig. 26-3. A, Needle reintroduced through same hole and passed subcutaneously on lateral aspect of antihelix. Tension on suture helps convert hole to slit temporarily. **B,** Needle is returned through same hole, passed through both sides of antihelical fold (subcutaneous posteriorly), emerging through its own stab incision (upper suture) or alternatively through lower stab incision and then back to its own incision (lower suture). (From Kaye, B. L.: A simplified method of correcting the prominent ear, Plast. Reconstr. Surg. **40**:44-48, 1967.)

through the lateral surface again, 1 to 1.5 cm. below its previous point of emergence (Fig. 26-3, *A*). The needle is reintroduced through the *same hole* from which it just emerged, and it is passed medially through both sides of the fold of cartilage, again subcutaneously on the posterior aspect. (It is easy to pass a suture needle back through its own skin hole by simply applying tension to the suture and momentarily converting the hole to a slit.) The suture is brought back out through the inferior end of the small stab incision through which it was introduced, or, if desired, through an adjacent stab incision and then back to the original stab incision by burrowing beneath the intervening conchal skin (Fig. 26-3, *B*).

The mattress sutures are tied after all of them have been placed. The amount of folding can be determined by visual judgment or by direct measurement. One half of a surgeon's knot can be used to hold the ear back temporarily to evaluate the degree of folding. This knot can be tightened or loosened to adjust the set-back before putting in the second throw. If desired, the patient can even be given a mirror at this stage and permitted to evaluate the amount of setback himself.[13] After the knots of the mattress sutures are tied, they are cut short and buried beneath the conchal skin flaps that have been created by undermining the edges of the incisions (Fig. 26-4, *A*). Each stab incision is closed with one or two fine sutures (Fig. 26-4, *B*), and the posterior incision is closed. The patient wears a standard ear dressing for 2 to 3

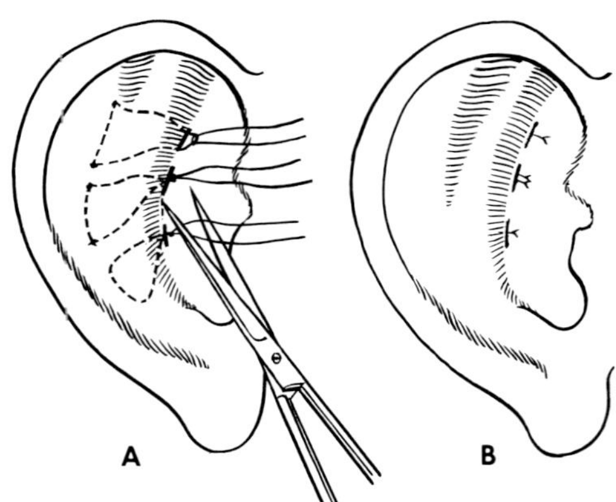

Fig. 26-4. A, The three mattress sutures are tightened appropriately, tied, cut short, and buried under the flaps of the stab incisions. **B,** Stab incisions closed with one or two fine sutures. Posterior incision also closed (not shown). (From Kaye, B. L.: A simplified method for correcting the prominent ear, Plast. Reconstr. Surg. **40**:44-48, 1967.)

days (longer in young children) and a night headband for about 4 weeks.

THE EXCESSIVELY DEEP CONCHA

We have not found it necessary to resect conchal cartilage and skin to reduce the excessively deep concha. Instead, a bit of the lateral concha is rolled into the new helix by bringing the vertical striations more medially onto the lateral portion of the concha. This maneuver transfers conchal tissue to the antihelix and scapha. Although it makes the scapha somewhat wider, it is aesthetically acceptable.[6, 7] On the other hand, if one wishes to make a long posterior incision, one can suture the concha to the mastoid process as advocated by Furnas[14] and then proceed with the reconstruction of the antihelix as described previously.

THE PROTRUDING EAR LOBE

Occasionally one may accomplish a very satisfactory cartilaginous setback, only to be left with annoying protrusion of the ear lobes. These can be pulled in nicely by resecting a transverse, elliptical wedge of skin and subcutaneous tissue from the

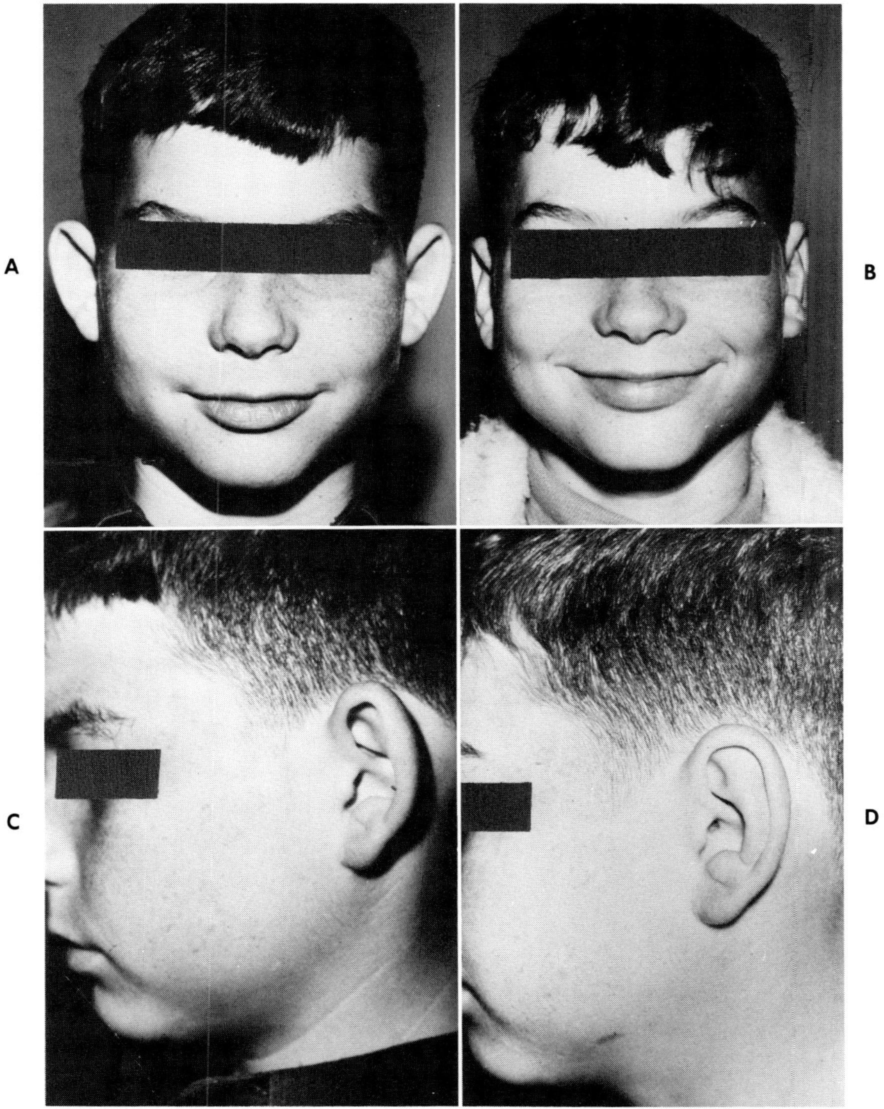

Continued.

Fig. 26-5. Representative cases illustrating result obtained by Kaye's technique. (From Kaye, B. L.: A simplified method for correcting the prominent ear, Plast. Reconstr. Surg. **40:**44-48, 1967.)

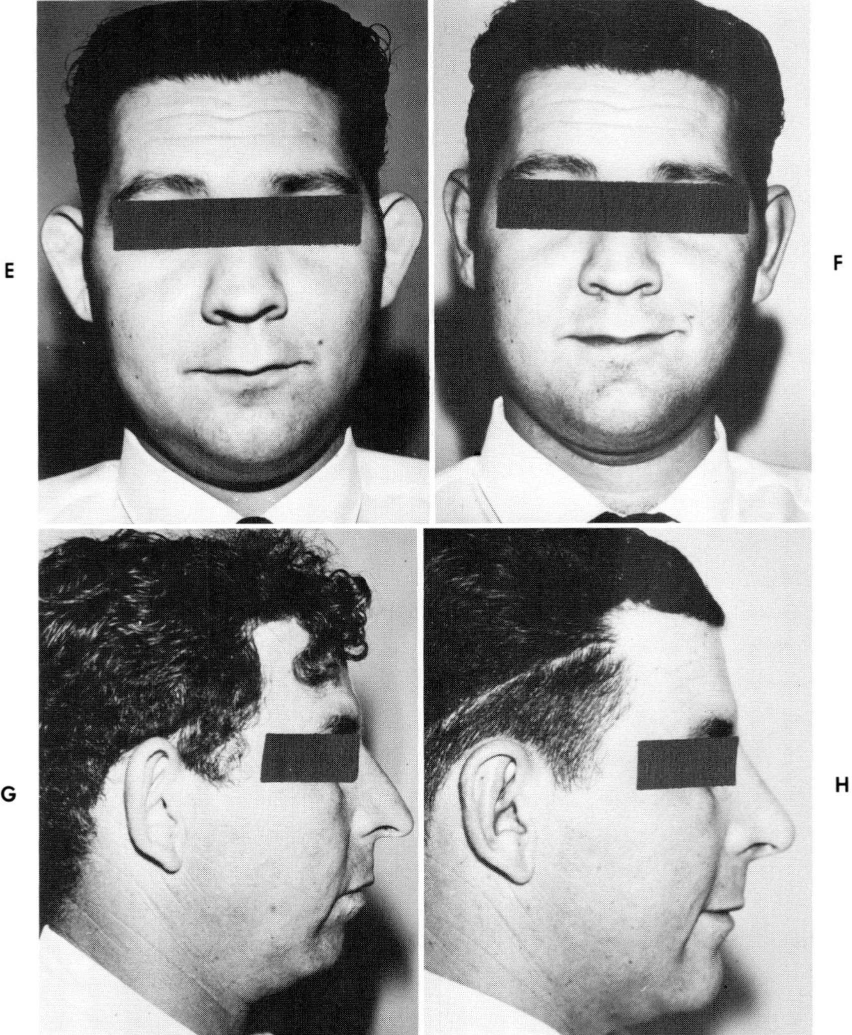

Fig. 26-5, cont'd. For legend see p. 157.

posterior base of each lobe and closing the defect transversely.

DISCUSSION

When the anterior perichondrium over the region of the antihelix is adequately striated, the ear folds back significantly, even before the mattress sutures are tied. This relaxation obtained by anterior striation indicates that the ear is held out in its deformed position by *anterior tension;* elimination of anterior tension tends to correct the deformity[15] (Fig. 26-1, *C*). However, one should not be tempted to perform the operation by striation alone, without the use of mattress sutures.

The posterior and lateral portions of the mattress sutures can be passed quite easily with the simple expedient of ballooning the tissues slightly with anesthetic solution. The needle can be reintroduced through its own hole by exerting tension on the proximal portion of the suture, a maneuver that momentarily converts the hole to a slit.

The ability to return the needle of the mattress suture through the same hole that it made is a definite convenience. It allows the surgeon to "come up for air" whenever he cannot make a complete pass with his needle. If, for example, one cannot make the posterior subcutaneous passage of the uppermost mattress suture from lateral to medial in one jump, the solution is to simply emerge at some point through the posterior skin and come back in through the same hole. One can then direct the needle through its final destination (the uppermost conchal slit incision).

Lately, as a variation of a variation of Wright,[16] we have tried making our slit incisions for the mattress sutures in a transverse direction on the posterior aspect of the ear in order to bury the knot in thicker covering tissue. This was done with the hope of reducing or eliminating the occasional late protrusion of the knot of the mattress suture through the thin conchal skin. This method of placing the mattress suture is trickier than the basic technique, and more time is needed to evaluate its efficacy. We recommend to those employing the simplified technique for the first few times that the standard conchal approach be used for inserting and tying the mattress sutures.

Occasionally one may encounter late protrusion of a mattress suture knot through the conchal skin. At this point several months may have elapsed and it will usually be quite safe to cut off the knot, with or without removal of the monofilament mattress suture, and allow the skin to close over.

Finally, this method of inserting a mattress suture through a stab incision can also be used as a simple office technique to correct occasional failures seen with other methods of recreating the antihelical fold.

Advantages of the simplified auriculoplasty include small incisions, minimal dissection, no skin resection, little need for hemostasis, decreased operating time, decreased risk of postoperative bleeding and infection, flexibility permitting numerous modifications according to the desires of the surgeon and the demands of the individual case, and shortening of time required for wearing dressings. The results obtained by this simplified technique are comparable to those obtained by more complicated methods (Fig. 26-5).

SUMMARY

A simplified technique of auriculoplasty is presented in which an anterior approach is used for the introduction of mattress sutures after weakening the cartilage by striating the anterior surface. The operation involves a minimum of dissection and disturbance of tissue, thereby minimizing or eliminating problems of postoperative discomfort, risk of infection, hemostasis, and hematoma. This technique is not only effective by itself but is also compatible with other ancillary techniques commonly used in auriculoplasty.

REFERENCES

1. Luckett, W. H.: A new operation for prominent ears based on the anatomy of the deformity, Surg. Gynec. Obstet. **10**:635-637, 1910.
2. Barsky, A. M.: Plastic surgery, Philadelphia, 1938, W. B. Saunders Co., p. 178.
3. Converse, J. M., and others: A technique for surgical correction of lop ears, Plast. Reconstr. Surg. **15**:411-418, 1955.
4. Holmes, E. M.: A new procedure for correcting outstanding ears, Arch. Otolaryng. **69**:409, 1959.
5. Stark, R. B., and Saunders, D. E.: Natural appearance restored to the unduly prominent ear, Brit. J. Plast. Surg. **15**:385-397, 1962.
6. Paletta, F. X., Ship, A. G., and Van Norman, R. T.: Double spring release in otoplasty for prominent ears, Am. J. Surg. **106**:506-510, 1963.
7. Stenström, S. J.: "Natural" technique for correction of congenitally prominent ears, Plast. Reconstr. Surg. **32**:509-518, 1963.
8. Ju, D. M. C., Li, C., and Crikelair, G. F.: The surgical correction of protruding ears, Plast. Reconstr. Surg. **32**:283-293, 1963.
9. Crikelair, G. F., and Cosman, B.: Another solution for the problem of the prominent ear, Ann. Surg. **160**:314-324, 1964.
10. Mustardé, J. C.: Correction of prominent ears using simple mattress sutures, Brit. J. Plast. Surg. **16**:170-176, 1963.
11. Kaye, B. L.: A simplified method for correcting the prominent ear, Plast. Reconstr. Surg. **40**:44-48, 1967.
12. Goulian, D., and Conway, H.: Prevention of persistent deformity of the tragus and lobule by modification of the Luckett technique of otoplasty, Plast. Reconstr. Surg. **26**:399-404, 1960.
13. Morgan, B. L.: Personal communication.
14. Furnas, D. W.: Correction of prominent ears by concha-mastoid sutures, Plast. Reconstr. Surg. **42**:189-193, 1968.
15. Fry, H.: Nasal skeletal trauma and the interlocked stresses of the nasal septal cartilages, Brit. J. Plast. Surg. **20**:146-158, 1967.
16. Wright, W. R.: Otoplasty goals and principles, Arch. Otolaryng. **92**:568-572, 1970.

Chapter 27

Macrotia*

Frederick J. McCoy, M.D.

Macrotia, as a clinical entity requiring treatment, is a rare condition. Disproportionately large ears that are symmetrical and not protruding are seldom a problem presented to the plastic surgeon for treatment. A unilateral macrotia, on the other hand, prompts a request for treatment early in childhood.

For purposes of discussion, this condition can be divided into the following types:

1. Bilateral macrotia
2. Bilateral macrotia with protrusion
3. Unilateral macrotia caused by congenital hypertrophy
4. Unilateral macrotia resulting from tumor
   a. Hemangioma
   b. Lymphangioma
5. Lobular hypertrophy

In an effort to establish guidelines for the size of "normal ears," Rubin and his co-workers[1] found that the average vertical height of the adult human ear is 6.5 cm., the width from tragus to Darwin's tubercle 3.5 cm., and the protrusion from the mastoid process to the helical rim 1.8 cm. (Table 9). Of more importance than the actual measurement, of course, is the *relative* size of the ears when compared with that of the head, neck, and other features. Therefore, ears measuring 7.5 cm. or even more might appear perfectly normal in an individual with a proportionately large head and thick neck. (Fig. 27-1). Largeness, per se, does not become a clinical problem unless it is accompanied by protrusion or, when it is unilateral, causes a conspicuous asymmetry (Fig. 27-2).

---

*I wish to express my appreciation to Dr. Allen Hughes for his sketches of the ear.

BILATERAL MACROTIA WITH PROTRUSION

Bilateral macrotia becomes a problem of the protrusion rather than actual size and is managed by any one of the standard operative techniques to reduce the auriculocephalic angle and perhaps also to adjust the concha. It is amazing at times to see how much smaller ears appear to be when a proper antihelical fold has been created and they have been drawn back closer to the head.

UNILATERAL MACROTIA

When one ear is conspicuously larger than the one on the normal side, it frequently is also more protuberant. Again, it is usually possible to solve the problem by a simple pinback procedure. If this is insufficient, however, a careful analysis will show which segment of the auricle needs reduction. This will usually be found to be in the region of the scapha or the lobule or both.

Table 9

Decade	1st	2nd	3rd	4th	5th	6th	7th	8th
Average height (cm.)	5.8	6.3	6.5	6.4	6.5	7.3	7.5	7.5
Average width (cm.)	3.5	3.7	3.4	3.4	3.6	4.0	4.1	4.0
Length lobule (cm.)	1.8	1.8	1.4	1.5	1.5	2.0	1.9	2.0
Cephaloauricular angle (degrees)	28	30	30	31	30	31	31	30
Distance from mastoid to helix (cm.)	1.4	1.9	1.8	1.8	1.8	2.2	2.2	2.4
Depth of concha (cm.)	1.6	1.8	1.5	1.5	1.6	1.8	1.5	2.1

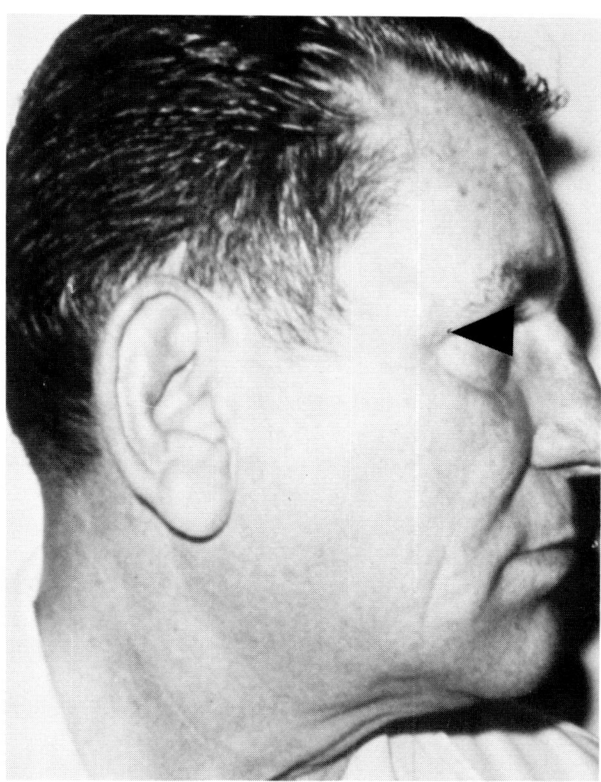

Fig. 27-1. Note that the actual measurements of the ears are greater than normal yet well proportioned relative to the size of the head and neck.

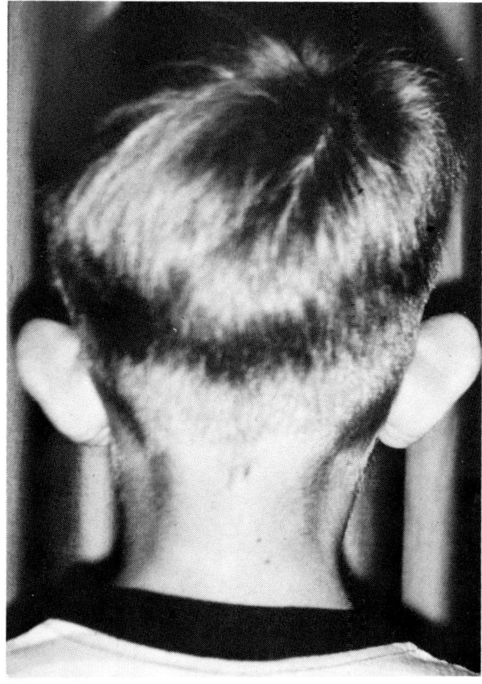

Fig. 27-2. This patient demonstrates a combination of macrotia with asymmetrical protrusion.

Many operative procedures have been suggested for reduction of the upper portion of the ear, and most have been disappointing because of a common tendency to produce a flattening of the natural arc of the helix rim above the point of Darwin's tubercle[2] (Fig. 27-3). A careful study of the best of the proposed methods (Fig. 27-4) shows that some of them remove too much supporting cartilage beneath this point instead of farther forward.

A modification of Kolle's procedure is preferred (Fig. 27-4, *C*), avoiding the triangular cut into the antihelix whenever possible. The initial cut is made through the anterior skin and the underlying cartilage at the lower edge of the helical sulcus, then a 90-degree turn is made to divide the helix well anteriorly. Through this incision the posterior skin is freed above for 4 mm. and below for 12 mm. The mobilized helix is moved to the desired position to achieve the necessary reduction and maintain good conformation. This line is marked to indicate the exact areas of skin and cartilage to be sacrificed. The cartilaginous edges are accurately approximated with 6-0 white silk or nylon knots posteriorly, and the skin edges are closed with 5-0 nylon.

One of the problems common to all ear cartilage surgery is the noticeable ridge at the cut that shows through the thin anterior skin. It may take a year or more, but eventually the irregularity becomes evident. Where a choice can be made, therefore, the cuts should be tucked out of sight in the sulcus beneath the overhanging helix or deep in the concha.

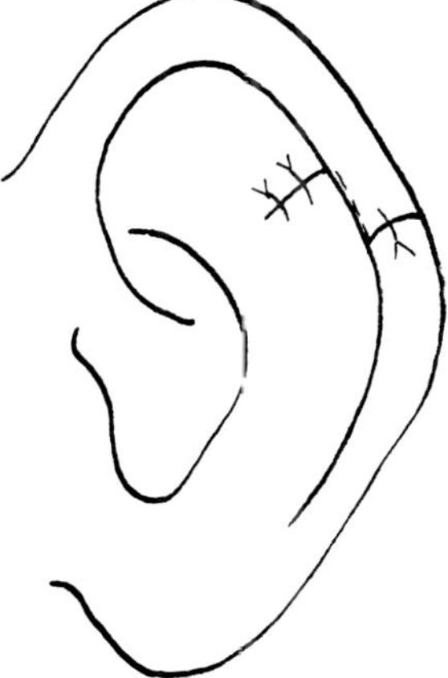

Fig. 27-3. Loss of normal curvature of the helix results from most attempts at reduction of the scapha through wedging procedures.

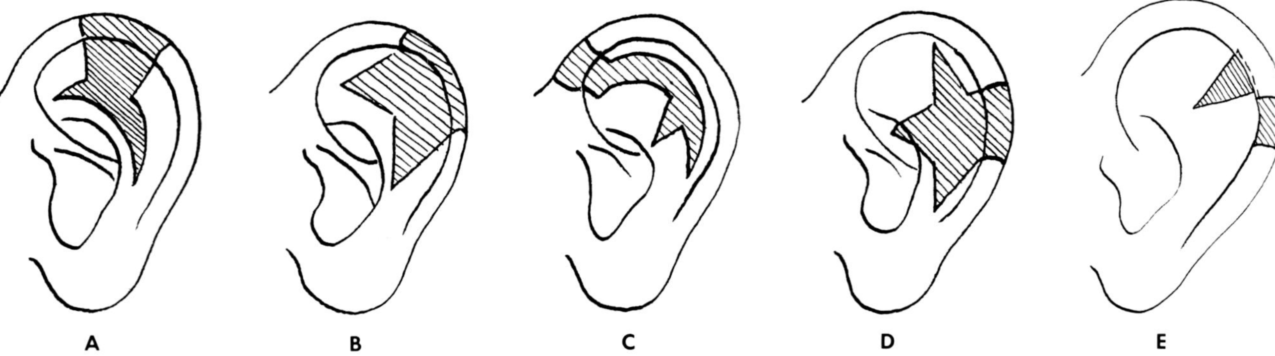

Fig. 27-4. Proposed methods for reduction of the upper portion of the ear: **A,** Cheyne and Burghard; **B,** Day; **C,** Kolle; **D,** Cockeril; **E,** Burian. (**A, B,** and **C** redrawn after Converse, J. M.: Reconstructive plastic surgery, vol. III, Philadelphia, 1964, W. B. Saunders Co., p. 1083; **D** and **E** redrawn after Burian, H.: The plastic surgery atlas, vol. 2, New York, 1968, The Macmillan Co., p. 487.)

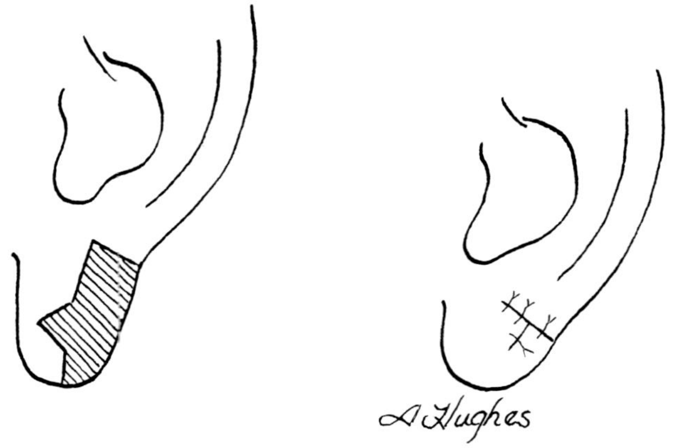

Fig. 27-6. This method of reducing the lobe is effective but, like most wedging procedures, tends to pull the lobe outward. (Redrawn after Burian, H.: The plastic surgery atlas, vol. 2, New York, 1968, The Macmillan Co.)

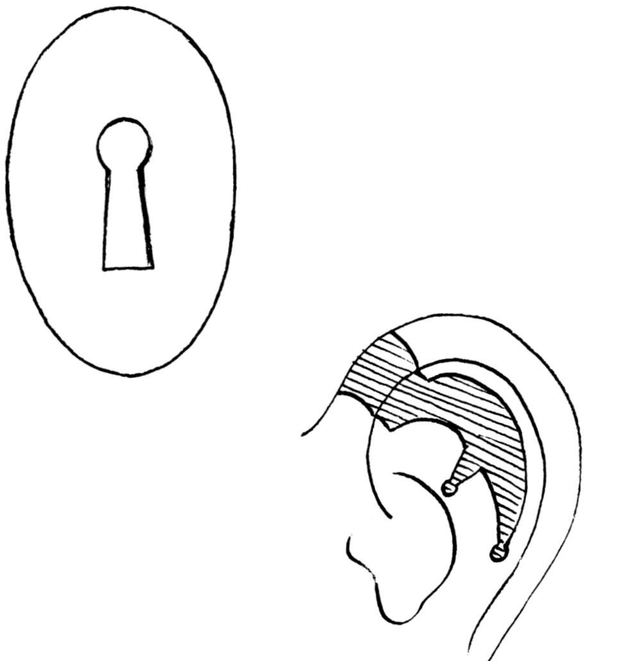

Fig. 27-5. The "keyhole principle" applied to the points of triangle removed from cartilage. This avoids the buckling and sharp projections seen through the skin.

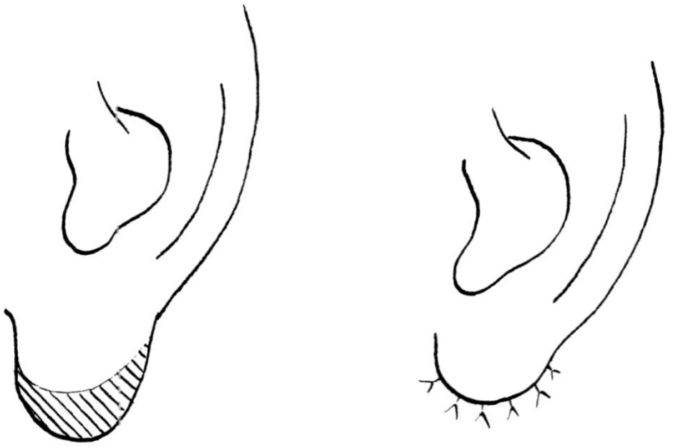

Fig. 27-7. Simple resection of the excess lobule allows for more accurate shaping. The scar is out of direct view and is negligible.

A related problem is created where a triangular wedge of cartilage is removed. When the wedge is closed, a buckling tends to occur at the apex, causing a protruding point that is quite noticeable. This can be avoided by adding a "keyhole" design to the triangle, excising a small circle of cartilage to relieve the compression and warping of the fulcrum or apex (Fig. 27-5).

Should the lobule require reduction, it can be done by the method of Burian (Fig. 27-6) or one of its modifications.[3] Care must be taken to avoid pulling the lobule outward, noticeably increasing the angle between it and the jawline.

A more straightforward approach is a simple tailoring of the border exactly to requirements. The resultant scar is negligible and far less observable than the wedging procedure involving the more exposed anterior surface. This maneuver is a valuable adjunct in rhytidoplasty in the elderly patient, who often has redundant lobes (Fig. 27-7).

Unilateral macrotia resulting from tumor. Hemangioma and lymphangioma may occasionally involve the ear and produce enlargement, not only by the simple addition of tumor bulk but also by hypertrophy of the auricle itself. In most cases the hemangioma is of the involutional type, and regression may be expected to occur sometime prior to age 6 or 7. Even if the resolution fails to occur or does so insufficiently, surgical extirpation is much more easily accomplished after age 10.

Where the mass is primarily subcutaneous with little intracutaneous involvement, it is best approached through an incision extending the full length of the tumor on a line midway between the helix and the auriculocephalic sulcus. This avoids skin flaps too long for the available blood supply. If the skin is involved, it is removed with the mass and a thick split-thickness skin graft is laid over the perichondrium, which should be carefully preserved.

The anterior surface of the auricle should be handled in much the same way except that the incision, whenever possible, should be in the helical sulcus. Bolster dressings held with through-and-through mattress sutures are useful in maintaining proper contouring, though great care must be taken to allow for the additional pressure of postoperative edema, which could produce necrosis at the sutures. Again, if the skin is not salvageable and the perichondrium can be preserved, a full-thickness split graft should be applied. It is often surprising how little bleeding there is in excising

a hemangioma that has been allowed time to mature.

CUPPED EAR

The cupped ear is characterized by a hooding effect of the upper helix folding over and obscuring the flat, underdeveloped antihelix. This produces a contracted, misshapen auricle that seems to protrude from the head at a downward angle. At times these ears are normal in size when unfolded, but some are actually smaller, as a result of underdevelopment in the upper helix.

In milder cases, reconstruction of the antihelix to a normal configuration will correct 90% of the deformity. Then, if the upper border of the superior helix has a reasonably good arch, simple paring of the overhanging fold will complete the repair. The scar along the free helical margin is not noticeable if stitch marks are avoided (Fig. 27-8). Most cases of lop or cupped ears, however, are more difficult. There is a shortening along the line from below Darwin's tubercle to the crus helicis (Fig. 27-10, *A*). Any attempt to unfurl the hooded portion of the ear without relieving this contracture is doomed to failure.

Ragnell suggests exposing, through a **posterior** incision, the entire folded cartilage of the upper ear, which is then expanded by splitting the cartilage into sliding flaps.[4] He suports this latticework arrangement with a stainless steel wire. Stephenson,[5] using a similar approach, achieves excellent correc-

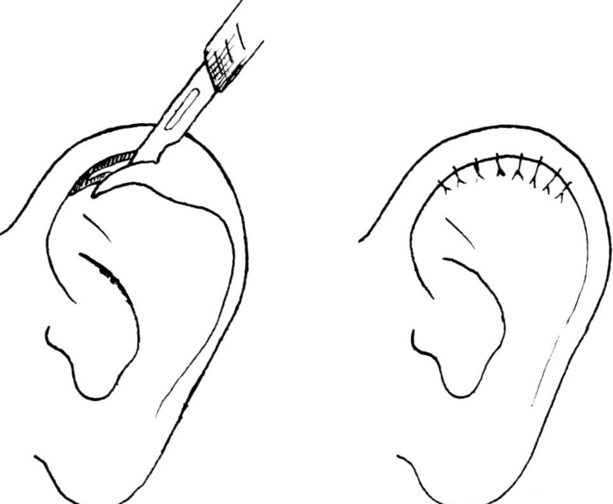

Fig. 27-8. The "hooded" portion is pared by a through-and-through cut. The marginal scar helps to sharpen the edge naturally. Contractures do not occur.

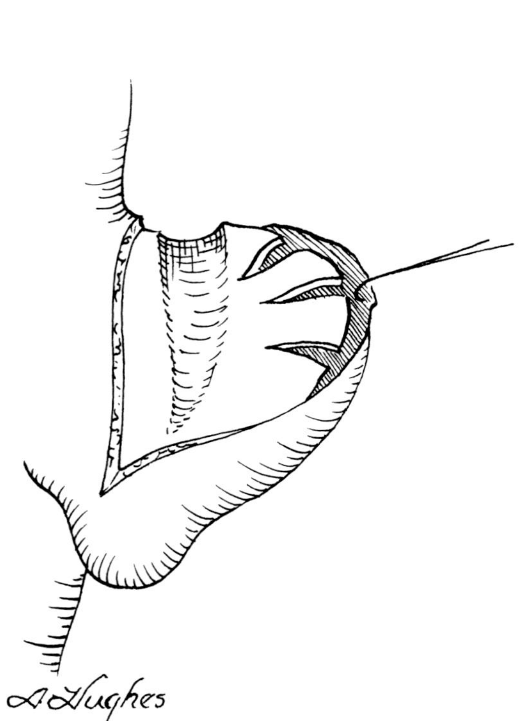

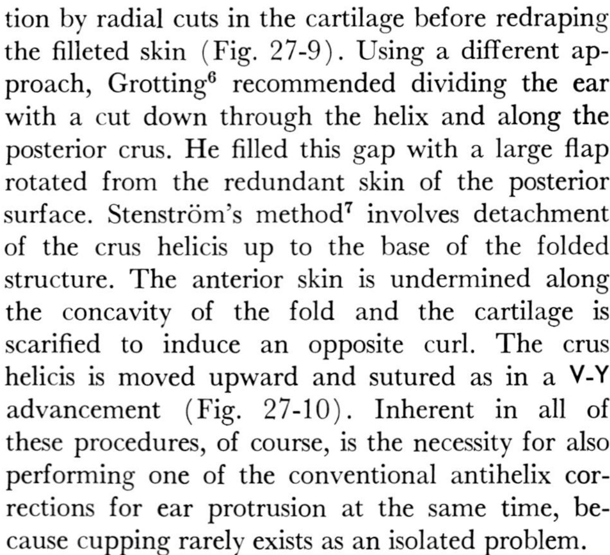

Fig. 27-9. The cartilage has been filleted and radially cut. Thinning has been carried out to afford molding the antihelix fold. The skin is then redraped. (Redrawn after Stephenson, K. L.: Correction of a lop ear type deformity, Plast. Reconstr. Surg. **26:**540, 1960.)

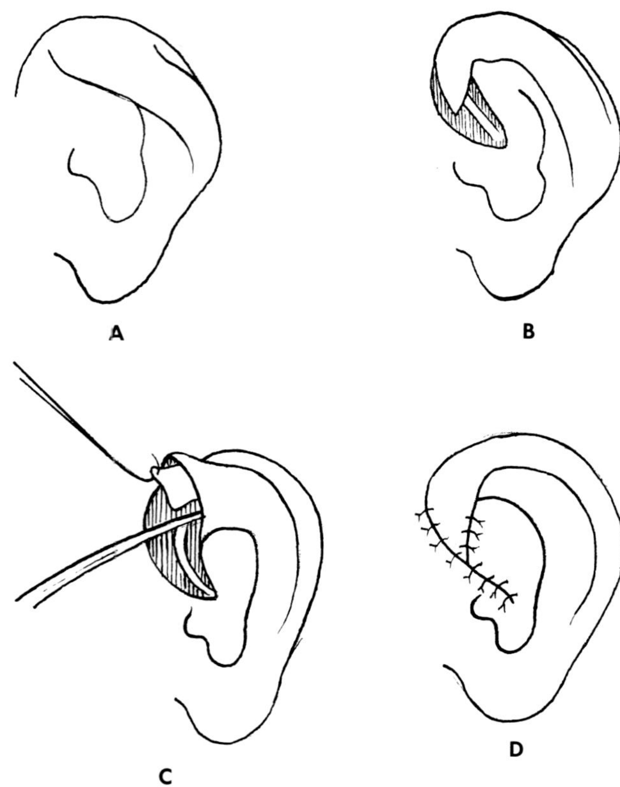

Fig. 27-10. Method of Stenstrom. **A,** Typical deformity of moderately cupped ear. **B,** Freeing of anterior crus of the helix by through-and-through cut. **C,** Extension of the cut in cartilage subcutaneously. **D,** Retraction and closure, utilizing the **V-Y** principle. (Redrawn after Grabb, W. C., and Smith, J. W.: Plastic surgery, Boston, 1968, Little, Brown & Co.)

tion by radial cuts in the cartilage before redraping the filleted skin (Fig. 27-9). Using a different approach, Grotting[6] recommended dividing the ear with a cut down through the helix and along the posterior crus. He filled this gap with a large flap rotated from the redundant skin of the posterior surface. Stenström's method[7] involves detachment of the crus helicis up to the base of the folded structure. The anterior skin is undermined along the concavity of the fold and the cartilage is scarified to induce an opposite curl. The crus helicis is moved upward and sutured as in a **V-Y** advancement (Fig. 27-10). Inherent in all of these procedures, of course, is the necessity for also performing one of the conventional antihelix corrections for ear protrusion at the same time, because cupping rarely exists as an isolated problem.

Despite one's best efforts, perfection in the man-agement of this difficult group of patients is elusive, and even the most experienced plastic surgeon will often be humbled.

REFERENCES

1. Rubin, L. R., and others: An anatomic approach to the obtrusive ear, Plast. Reconstr. Surg. **29:**360, 1962.
2. Converse, J. M.: Reconstructive plastic surgery, vol. 3, Philadelphia, 1964, W. B. Saunders Co.
3. Burian, F.: The plastic surgery atlas, vol. 2, New York, 1968, The Macmillan Company.
4. Ragnell, A.: A new method of shaping deformed ears, Brit. J. Plast. Surg. **4:**202, 1951.
5. Stephenson, K. L.: Correction of a lop ear type deformity, Plast. Reconstr. Surg. **26:**540, 1960.
6. Grotting, J. K.: Otoplasty for congenitally cupped protruding ears using a post auricular flap, Plast. Reconstr. Surg. **22:**164, 1958.
7. Grabb, W. C., and Smith, J. W.: Plastic surgery, Boston, 1968, Little, Brown & Company.

Chapter 28

Surgical management of the auricular lobule

Morrison D. Beers, M.D., F.A.C.S.

ANATOMY

Anatomically the auricular lobule is a component of the auricle. The auricular lobule is a soft tissue mass suspended from the auricular fibrocartilaginous framework at the inferior border and in a dependent position to the auricular cartilage framework. The lobule is composed of integument enclosing fibroareolar and adipose tissues with dense fibrous septal partitions.

The vascular supply to the auricle and lobule is derived from the branches of the superficial temporal and posterior auricular arteries, which are branches of the external carotid artery. Branches of the occipital artery also supply the auricle. The venous drainage of the auricle and lobule is by way of the superficial and posterior auricular veins. The lymphatic drainage of the external ear is into the adjoining auricular lymph node system.[1]

The innervation of the auricle and lobule is predominantly derived from the cutaneous branches of the cervical plexus. The major portion of the auricle and most of the lobule are innervated by the posterior branch of the great auricular nerve originating from the second and third cervical nerves. The auricular branch of the smaller occipital nerve also supplies the auricle and lobule.[2]

Although the lobule is a small component of the auricle, it is conspicuous in its aesthetic relationship to the auricle and the impression of beauty that it creates. Its aesthetic effect extends to the facial appearance as a factor in facial symmetry and balance.[3]

DEVELOPMENT OF THE AURICULAR LOBULE

Although the development of the auricle in the human embryo has been studied in great detail, particular details as to the development of that part of the auricle forming the ear lobule are not available in the literature. Most authors believe that ear lobules are a later development in the evolution of the auricle in the human and that they are derived from the mesenchymal cells of the second branchial arch (hyoid arch).

His (1885) gave a classic account of the development of the auricle in humans from the six auricular tubercles surrounding the first gill cleft. He thought the lobule itself was derived from tubercle 6 and that the taenia lobularis is a remnant of hillock 6.[4]

Gradenigo (1888) presented a study of development of the auricle in a series of mammals including man.[5] According to him, the auricle develops from the fusion of two elevations immediately adjacent to the tubercles, but not from the tubercles themselves. These elevations he termed helix mandibularis and helix hyoidalis. The lobule appears later in man than in mammals as a growth of the lower end of the helix hyoidalis.

According to Schaeffer (1892-1893), the ear lobule is a later acquisition and is found only in anthropoids and man. Streeter (1922) believes the ear lobule forms as a free fold between the taenia lobularis and the lower end of the helix, principally at the expense of or as an elaboration of the taenia lobularis. He thinks that tubercle 6 does not contribute to the formation of the ear lobule. Embryologic and developmental defects occur in this rather complicated area at the juncture of the branchial and postbranchial skin regions and are reflected in clinical manifestations such as clefts, fissures, cysts, sinus tracts, and neoplasms.[6]

DEFECTS OF THE AURICULAR LOBULE

Types of defects of the auricular lobule that are clinically encountered are congenital, developmental, traumatic, and neoplastic. Congenital defects occur as cleft or coloboma of the lobule, absence of lobule, adherence of lobule to neck, accessory sinus tracts, and accessory auricular appendages. Developmental defects such as abnormally small or abnormally large lobules occur. The traumatic defects clinically encountered consist of avulsion, amputation, infection (cellulitis, abscess formation), burns (thermal, electric, chemical), and cicatrices (hypertrophic, contractile, keloidal). The neoplastic lesions of the lobule encountered are epidermoid cysts, pyogenic granulomas, lipomata, verrucae, keratoses, angiomas, and other tumor entities.[7]

SURGICAL PROCEDURES IN THE MANAGEMENT OF THE LOBULE

Many and varied surgical procedures are utilized in the management of abnormalities of the lobule. Historically, they extend from the early period of plastic surgery and essentially consist of excisional, flap, and skin grafting techniques. The surgical procedures reviewed are for the correction of such abnormalities as the adherent ear lobe, the cleft ear lobe, the hypertrophic or large ear lobe, ear lobe reconstruction, cyst excision, tumor resection, scar excision, and the aesthetic repositioning of the protruding ear lobule. The corrections of adherent scar lobules is usually performed by surgical separation at the site of attachment to the neck skin.[8] The procedure of Miller published in his text *Cosmetic Surgery* in 1924 is a basic procedure and is currently applicable for the correction of this condition.[2] Adherent lobes occur frequently and in some instances, may be a racial characteristic, as in the Japanese.[8] The condition may occur as the result of burns or surgical procedures such as rhytidoplasty.

The correction of the ear lobe clefts caused by congenital or traumatic factors follows the usual plastic surgery principles of meticulous closure by a straight line technique, notching, or Z-plasty procedures. Congenital clefts are encountered rarely; however, many acute traumatic clefts occur and immediate repair is indicated.

Numerous surgical procedures are available for the reduction of large ear lobes. Padgett reduced the abnormally large ear lobe by excision of the peripheral posterior margin.[8] In 1970, Santos advocated resection of a horizontally positioned wedge from the anterior portion of the ear lobe, which he introduced as a new technique.[9] Santos reported that this type of procedure prevented the formation of a marginal notch of the lobe. It is noted that Miller reported such a procedure in

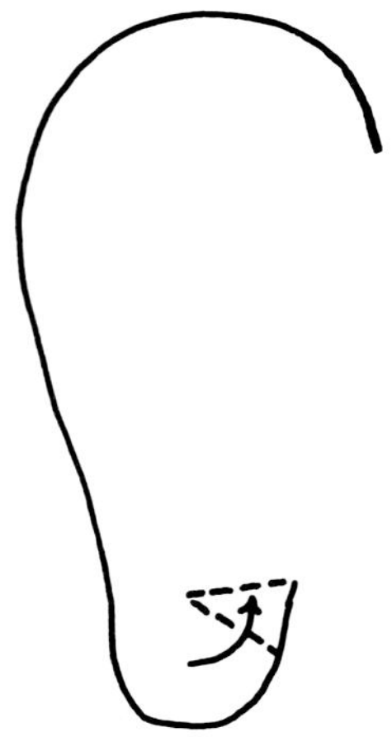

Fig. 28-1. Santos procedure of anterior wedge resection indicated for the reduction of large ear lobule.

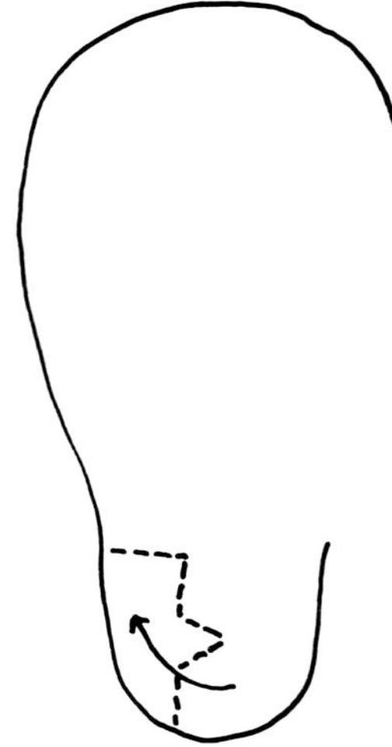

Fig. 28-2. Joseph procedure of posterior rectangular and triangular excision for the reduction of the large ear lobule.

1924.[10] Joseph reduced large ear lobes by the resection of a posteriorly positioned vertical rectangle with a small triangular excision in the midportion of the rectangular resection. When closed, the inferior portion of the lobe was rotated upward and posteriorly to provide a step type closure with a smooth posterior margin.[11] Tanzer and Converse recommend a similar type of excision for the correction of the hypertrophic lobule utilizing the "tongue and groove" principle. Miller advocated a simple vertical elliptical excision of tissue bulk from the posterior surface and closure in a transverse plane. His procedure may also be used for the shortening of a long ear lobule.[10]

RECONSTRUCTION OF THE EAR LOBULE

The ear lobule may be reconstructed by numerous techniques, all of which essentially are flap procedures. The first under consideration is that of Dieffenbach.[12] In this repair, a vertical flap of neck skin inferior to the defect is utilized (Fig. 28-3). The flap is secured to the ear margin at the site of the defect and a first stage delay is performed. In the second stage of this Dieffenbach procedure, the neck flap is detached distally and folded upward on itself to provide the posterior surface of the lobule. A case of traumatic loss of the ear lobule is demonstrated herein utilizing the

Dieffenbach technique (Fig. 28-4). This 22-year-old male, deeply concerned with this defect, had loss of his right ear lobule as the result of an automobile accident. Repair was completed as shown (Fig. 28-5).

The next procedure is that of Gavello, who reported a bilobed type of procedure in 1907 (Fig. 28-6). This one-stage procedure uses a bilobed flap that is folded under the anterior flap to complete the formation of the ear lobule.[7]

The procedure of Nelaton and Ombredanne[13] was reported in 1907. Their technique consisted of the use of a posterior inferior pedicle flap attached to the site of the defect of the lobular base. Ten days later the flap was mobilized, detached distally, and folded onto the posterior portion of the flap, similar to the Dieffenbach procedure.

In 1949 Wynn reported a new procedure for lobule reconstruction.[14] In his technique, a transverse incision is made at the base of the defect and extended posteriorly to the posterior neck full-thickness skin and, by wide undermining, an extensive flap was developed. The mobilized flap was then vertically imbricated and advanced anteriorly to produce a double fold flap to form a completed lobule as a one-stage procedure (Fig. 28-7).

In 1958, Converse reported a one-stage reconstruction of the lobule in which an inferiorly based neck skin pedicle flap was utilized as the anterior

Fig. 28-3. Dieffenbach procedure diagram for the vertical neck pedicle flap reconstruction of the ear lobule.

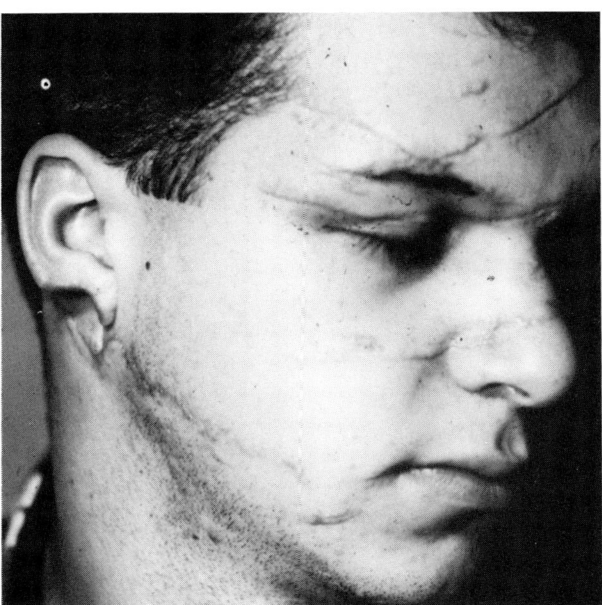

Fig. 28-4. Patient D. M., 22-year-old male with traumatic loss of right ear lobule from an automobile accident. Preoperative photograph.

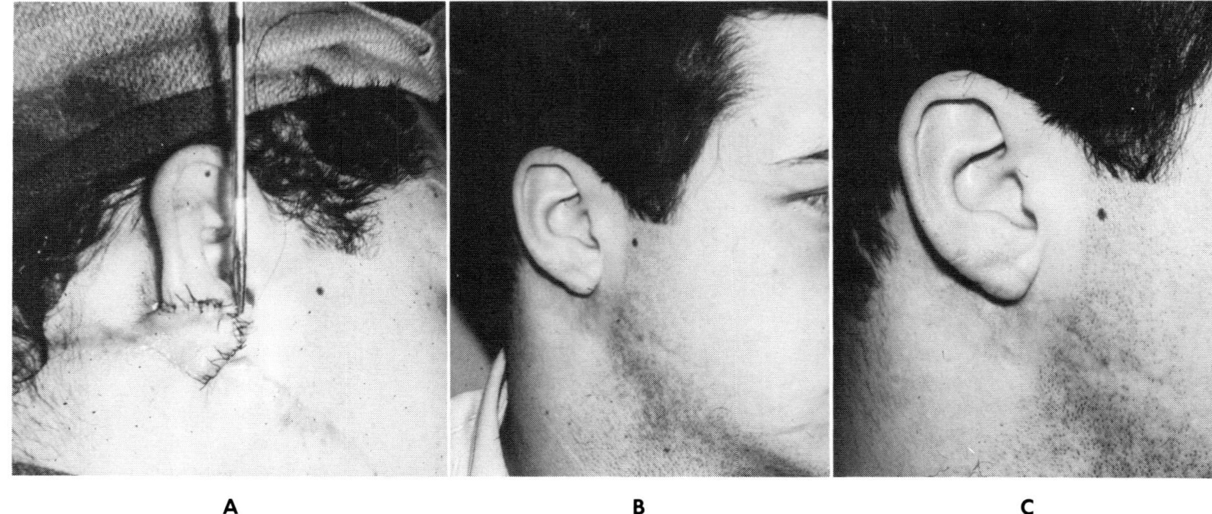

Fig. 28-5. A, Dieffenbach pedicle neck flap repair of traumatic loss of ear lobule of patient D. M. **B,** Postoperative photograph of completed Dieffenbach procedure second stage with definitive result. **C,** Close-up photograph of completed second stage Dieffenbach procedure.

Fig. 28-6. Diagram of bilobed flap procedure of Gavello.

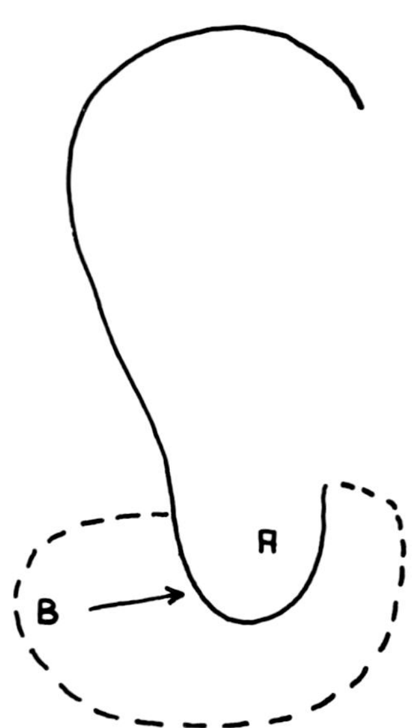

Fig. 28-7. Diagram of Wynn's procedure of one-stage ear lobule reconstruction by imbricated neck flap technique.

surface of the lobule, which was then lined with a full-thickness skin graft for the posterior surface (Fig. 28-8).[7] As the graft contracted, a natural rolling of the lobule margins occurred that produced a normal appearing lobule.

Musgrave in 1967 reported his technique of ear lobe reconstruction in a case of total ear avulsion in a 3-year-old Caucasian male.[15] He designed

a double turn-in "diaper" flap with fat-bearing tissue and a central bridge at the central inferior area of the flap (Fig. 28-9). This bridge is retained for vascular supply and restrains contracture of the flap during the first stage. The bridge was secondarily detached to provide a very adequate appearing lobule.

Subba Rao of Gunter, India, reported seven

Fig. 28-8. Diagram of ear lobule reconstruction technique of Converse. *A*, Anterior surface of ear lobule; *B*, full-thickness graft lining of posterior surface of lobule.

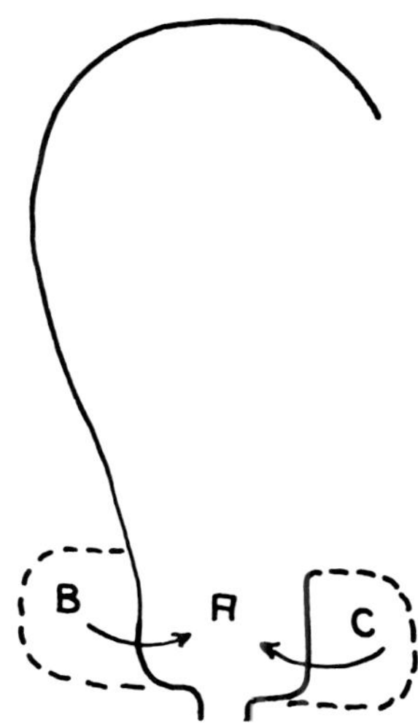

Fig. 28-9. Diagram of double turn-in "diaper" flap of Musgrave. *A* represents central portion of flap. Flaps *B* and *C* folded to form posterior surface of lobule.

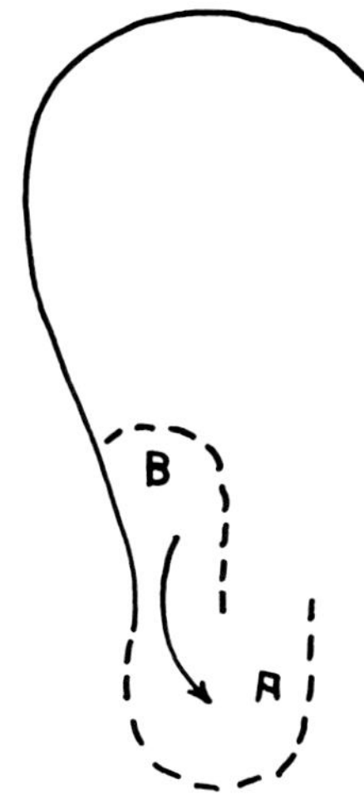

Fig. 28-10. Diagram of posterior auricular surface pedicle flap of Subba Rao. Flap forms anterior surface of lobule, which is lined with full-thickness skin graft for the posterior surface.

cases of his one-stage method of lobule reconstruction that employs an auricular flap based on the posterior surface of the auricle (Fig. 28-10). This flap is turned down to form the anterior surface of the lobule, which is then lined with a full-thickness skin graft.[16] He states that his method avoids neck scars and is well suited to the economic and behavioral needs of his patients.

Finally, Alanis of Mexico City reported his technique of lobule repair that he stated produced the anatomic characteristics and features of a normal ear.[17] The anterior flap is designed after the technique of Converse. An adjoining vertical posterior flap is then used as lining to form the posterior surface.

CORRECTION OF THE PROTRUDING EAR LOBULE

In the usual congenital protruding ear deformity there is also protrusion of the ear lobule. The protrusion of the lobule is usually concurrently corrected with most of the surgical techniques in which posterior auricular skin is excised and carried down into the lobule as in the Luckett and Converse types of procedure. Mustardé in his paper in 1963 entitled "The Correction of Prominent Ears Using Simple Mattress Sutures" stated: "Occasionally a small amount of tissue from the antihelix or a thick wedge from the posterior surface of the lobule has had to be excised when these structures were found to project too far forward after correction of the prominent ear."[18]

Webster has demonstrated that vertical elliptical excisions of some 60% to 70% of the posterior auricular skin will produce long-term results in the correction of auricular protrusion and the associated lobular protrusion as well.[19] Spira and

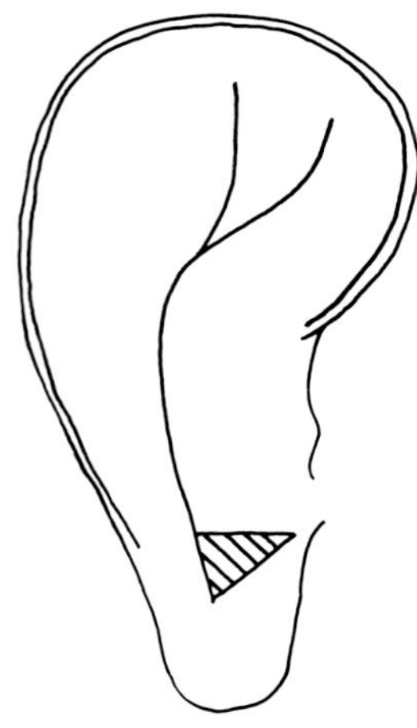

Fig. 28-11. Diagram of triangular wedge resection of inferior portion of conchal cartilage.

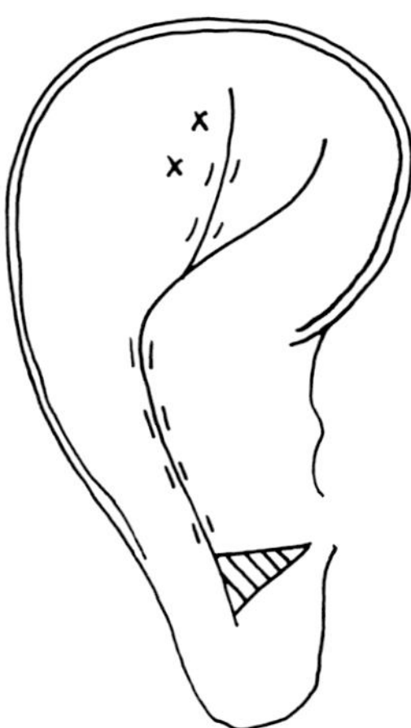

Fig. 28-12. Diagram in which *X* indicates site of "ear-to-mastoid periosteum" sutures for fixation of superior pole of auricle.

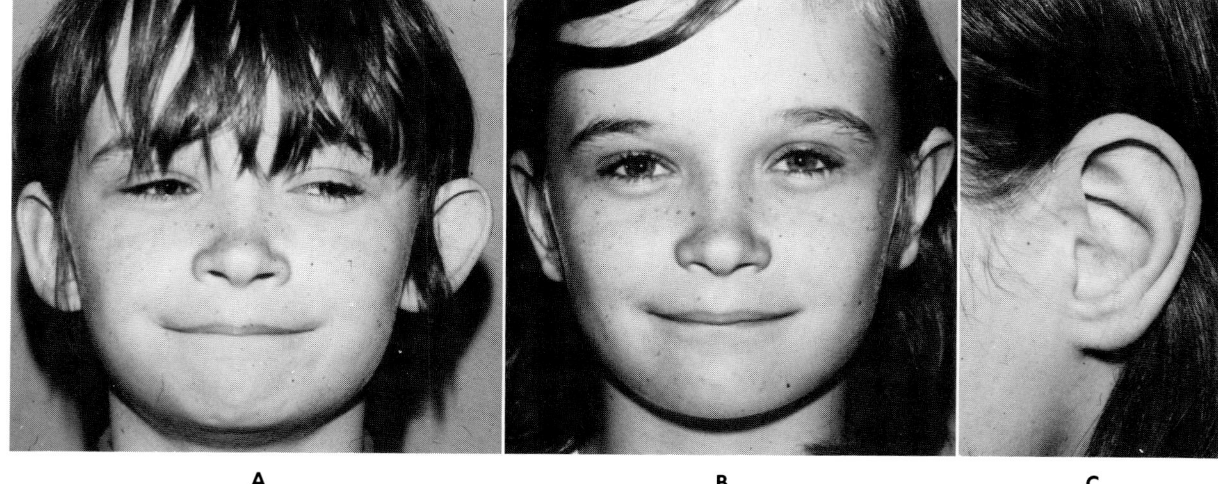

Fig. 28-13. Case 1. **A,** Preoperative photograph of congenital ear protrusion deformity. **B,** Postoperative photograph of corrective otoplasty as a combined procedure of Mustardé, antihelix incision of lower half of cartilage, and triangular conchal wedge resection. **C,** Postoperative view of corrective otoplasty.

associates in 1967 stated: "The ear lobule may be enlarged and/or protruding, thus contributing to the deformity" (of the protruding ear).[16] Spira listed wedge resection, Z-plasty, excision of the tail of the helix, and other procedures as available techniques for the correction of the ear lobule protrusion. He, however, uses excision of excess skin from the posterior lobule surface and a "dermis to periosteum suture" to reposition the lobule.[20]

I excise a triangular wedge of cartilage, combined with excision of excess skin, from the inferior portion of the concha with the base of the triangular excision in line with the antihelix fold cartilage extension incision (Fig. 28-11). This procedure achieves delicacy and repositioning of the ear lobule in conjunction with the corrective otoplasty procedure. I almost routinely use the wedge resection technique to produce an aesthetically well-

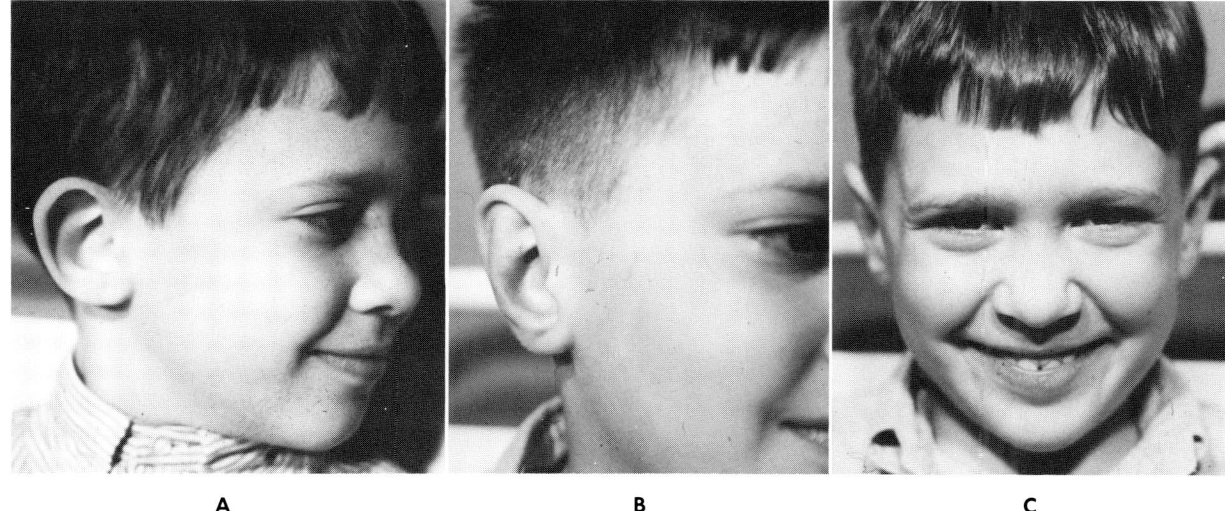

A   B   C

Fig. 28-14. Case 2. **A,** Preoperative photograph of congenital ear protrusion. **B,** Postoperative otoplasty photograph of combined technique of Mustardé, antihelix cartilage incision, and triangular wedge resection of conchal cartilage. **C,** Postoperative otoplasty, front view.

balanced ear lobule that is in symmetry with the corrective otoplasty produced by a combined Mustardé and antihelix cartilage incision procedure for the lower half of the auricle with ear-to-periosteum fixation sutures for the upper and middle third areas. Such sutures stabilize the superior pole and assist in achieving bilateral symmetry of the ears (Fig. 28-12). In this procedure, the lobular position and size are aesthetically improved without using incisions on the anterior surface of the lobule (Figs. 28-13 and 28-14).

DISCUSSION

The auricular lobule is a dependent appendage and component of the auricle. The auricular lobule is characterized as aesthetically essential to the symmetry and balance of the auricle and to the facial area as well. Anatomically, the lobule consists of a heavier and more loosely structured dermis and stroma than other skin areas. It has a rich capillary structure and an abundance of sebaceous glands. Congenital, developmental, traumatic, neoplastic, and aesthetic abnormalities are clinically encountered in problems of the auricular lobule. The foregoing problems are for the most part correctable by well-documented surgical procedures. A review of the literature indicates that the amount of ear lobule surgery performed is statistically not very extensive and that a wide variety of procedures is available. In most instances, corrective surgery

of the lobule provides an excellent or acceptable result.

SUMMARY

Surgical procedures available of problems of the auricular lobule are reviewed and several representative cases are presented.

REFERENCES

1. Gray, H.: Anatomy of the human body (edited by W. H. Lewis), Philadelphia, 1946, Lea & Febiger.
2. Hollinghead, W. H.: Anatomy for surgeons, vol. 1: New York, 1954, Harper & Row, Publishers.
3. Hunt, H. L.: Plastic surgery of the head, face and neck, Philadelphia, 1926, Lea & Febiger.
4. His, W.: Die Formentwickelung des ausseren Ohres: Anatomie menschlicher Embryonen, part III, pp. 211-221, Leipzig, 1885, F. C. W. Vogel.
5. Gradenigo, G.: Die Formentwickelung der Ohrmuschel, mit Rucksicht auf die Morphologie und Teratologie derselben, Zbl. Med. Wiss. **26:**82-86, 1888.
6. Streeter, G. L.: Contributions to embryology, Carnegie Inst. Washington **14:**113-138, 1922.
7. Converse, J. M.: Reconstructive plastic surgery, vol. III, Philadelphia, 1964, W. B. Saunders Co.
8. Padgett, E. C., and Stephenson, K. L.: Plastic and reconstructive surgery, Springfield, Ill., 1948, Charles C Thomas, Publisher.
9. Santos, J. G.: Correction of hypertrophic earlobes in leprosy, Plast. Reconstr. Surg. **46:**381, 1970.
10. Miller, C. C.: Cosmetic surgery: the correction of featural imperfections, Philadelphia, 1924, F. A. Davis Co.

11. Joseph, J.: Handbuch d Speciellen Chirurgische Katz, Preysing, 1912.
12. Dieffenbach, J. F.: Chirurgische Erfahungen, Berlin, 1829, T. C. F. Enslin.
13. Nelaton, C., and Ombredanne, L.: Les autoplasties, Paris, 1907, G. Steinheil.
14. Wynn, S. K.: One stage ear reconstruction, Plast. Reconstr. Surg. 4:105, 1949.
15. Musgrave, R. H., and Garret, W. S.: Management of avulsion injuries of the external ear, Plast. Reconstr. Surg. 40:534, 1967.
16. Subba Rao, Y. V., and Venkateswara Rao, P.: A quick technique for ear lobe reconstruction, Plast. Reconstr. Surg. 41:13, 1968.
17. Alanis, S. Z.: A new method for earlobe reconstruction, Plast. Reconstr. Surg. 45:254, 1970.
18. Mustardé, J. C.: The correction of prominent ears using simple mattress sutures, Brit. J. Plast. Surg. 16:170, 1963.
19. Webster, R.: Personal communication, 1971.
20. Spira, M., and others: Analysis and treatment of the protruding ear, Transactions of the Fourth International Congress of Plastic and Reconstructive Surgery, Excerpta Medica International Congress Series No. 174, Rome, October, 1967.

Chapter 29

Technique, pitfalls, and complications of otoplasty

Allyn McDowell, M.D.

Four and a half years ago a paper was presented to the American Association of Plastic Surgeons entitled "Goals in Otoplasty."[1] The reason for presenting a paper on that subject was that much of the literature during the past two decades had indicated that the most basic goals of the operation were frequently overlooked or unfulfilled.

Various old and new operations were advocated upon the justification of creating a greater degree of roundness in the new antihelical crest. Yet the accompanying photographs of results often demonstrated that regardless of this secondary benefit, these same procedures were failing to first meet the more important criteria of ears with good gross shapes, good gross positions, and good matching corrections. The most common complications or pitfalls in results of otoplasty were categorized as follows: (1) disappearing helix, (2) ears still hanging out at top, (3) "telephone ears," and (4) generally overcorrected ears.

Accordingly it was recommended that the primary goals of otoplasty for protruding ears should include accomplishment of the following:

1. All trace of protrusion in the upper third of the ear should be corrected.
2. From the front view, the helix of both ears should be seen beyond the antihelix.
3. The helix should have a smooth and regular line throughout.
4. The postauricular sulcus should not be markedly decreased or distorted.
5. The ear should not be placed too close to the head, especially in boys.
6. The position of the two ears should match

fairly closely—to within 3 mm. at any given point.

If the surgery can generally fulfill these goals, the patient will be happy with his result regardless of the degree of sharpness of the new antihelical crest, whereas the opposite does not hold true. As various authors have noted, protrusion of an ear is caused by Factor A, underdevelopment or absence of the antihelical fold, or by Factor B, excessive height of the concha, or by both.

Otoplasty procedures that rely upon scoring or weakening the cartilage in the antihelical regions from an anterior approach[2, 3] do provide a simple and most logical way of folding that cartilage and are seemingly quite appropriate in those ears that require only a correction of Factor A, that is, an increase in the folding of the antihelix. When correction of Factor B (conchal height) is needed, some supplemental maneuver is required.

An otoplasty relying upon sutures alone, as advocated by Mustardé,[4] obviously cannot accomplish much toward correcting Factor B when the suture placement is limited to the location of the antihelical fold. Likewise the conchal-mastoidal suture procedure described by Furnas[5] cannot of itself correct Factor A. Two rows of sutures, one in each location, would be necessary (as acknowledged by Furnas and by Spira and associates[6]) when correction of both Factors A and B is needed. Many ears do require both corrections, and the necessary placement and tying of two rows of such sutures in these ears would seem to be as complicated and time-consuming as most other otoplasty operations. Roundness of the new antihelical crest would seem to be guaranteed by these new suture methods. However, the reported difficulties in permanently maintaining, through sutures alone, all of the apparent initial correction is another factor to consider.[7] Certain large deep and severely protruding ears seem to present an undeniable excess of cartilage so that a procedure that merely folds the existing cartilage can hardly suffice. Instead, excision of cartilage in considerable amounts appears necessary in these cases in order to produce an ear of even grossly normal size and grossly normal shape.

The Luckett[8] procedure, by first dividing the cartilage all along the existing or potential concho scaphal junction, provides the operator with an easy access to correct either Factor A or Factor B or to make both of these basic correction maneuvers to whatever proportionate degree he desires for the best aesthetic result. Details of the operative technique have been fully described, step by step, in the earlier paper[1] and remain the same.

In addition to the correction of Factors A and B many ears, especially those with severe protrusion, also require correction of an excessive cephaloauricular angle where the helix meets the scalp. This excessive angle, hereinafter referred to as Factor C, is associated with a "hanging out" of the upper pole of the ear that can persist after an otoplasty and detract markedly from the result. It is of course not corrected by the usual methods for decreasing conchal height nor is it corrected by folding the ear along the normal line of the antihelix and its superior crus, because that line forms an axis of folding that is essentially vertical. Correction of excessive protrusion in the upper pole of the ear appears to require a hinging of that part of the ear along an axis that instead is horizontal. A modification of the Luckett procedure enables one to extend the line of the new superior crus into a horizontal direction behind cover of the helical rim and then on into the crevice between the helix and the scalp. The cartilage can be divided along this line and enough removed to partially collapse the excessive angle and to fold the upper pole of the ear along a horizontal axis, the needed and unnatural direction of which is fortunately concealed. A similar folding of the upper ear along this same extended horizontal axis would

seem difficult but theoretically possible in other methods of otoplasty, but it has not been indicated in published texts or illustrations.

A modified Luckett procedure thus permits correction of Factors A, B, and C to whatever varying degrees the individual ear requires, and it allows for all three corrections to be made with a single cartilage incision. Because it is relatively simple the operator can usually stay out of trouble. Because it is versatile he can usually convert any protruding ear into an ear with a reasonable gross position and gross shape. Lack of sufficient roundness in the antihelical crest in the Luckett procedure may be a subject for criticism for plastic surgeons, but the patients and their peers practically never notice it, much less object to it. The same holds true for a slight subluxation of the scapha medially upon the concha, and providing this is of only a slight degree it may actually enhance the overall correction. For all of these reasons the modified Luckett procedure is utilized most commonly, and yet each new patient must be approached with enough imagination and enough ingenuity to occasionally choose some other procedure that can be expected to solve that patient's problem more easily or more effectively.

REFERENCES

1. McDowell, A. J.: Goals in otoplasty for protruding ears, Plast. Reconstr. Surg. **41:**17, 1968.
2. Ju, D. M. C., Li, C., and Crikelair, G. F.: The surgical correction of protruding ears, Plast. Reconstr. Surg. **32:**283, 1963.
3. Gonzalez-Ulloa, M.: An easy method to correct prominent ears, Brit. J. Plast. Surg. **4:**207, 1952.
4. Mustardé, J. C.: The correction of prominent ears using simple mattress sutures, Brit. J. Plast. Surg. **16:** 170, 1963.
5. Furnas, D. W.: Correction of prominent ears by concha-mastoid sutures, Plast. Reconstr. Surg. **42:**208, 1968.
6. Spira, M., and others: Correction of the principal deformities causing protruding ears, Plast. Reconstr. Surg. **44:**150, 1969.
7. Davenport, G., and Bernard, F. D.: Experience with the mattress suture technique in correction of prominent ears, Plast. Reconstr. Surg. **36:**91, 1965.
8. Luckett, W. H.: A new operation for prominent ears based on the anatomical deformity, Surg. Gynec. Obstet. **10:**635, 1910.

Chapter 30

Treatment of the cauliflower ear

Ross H. Musgrave, M.D.
James E. Conklin, M.D.

Cauliflower ear is seen mostly in those geographic areas where wrestling and boxing are popular sports. The term *cauliflower ear* properly refers to the long-term end-product of fibrosis following hematoma and an attendant chondritis in the adjacent cartilage layers. Treatment of such a traumatized ear is complicated and unsatisfactory if the process has been allowed to progress to a mature fibrosis. However, the prevention of the distorted ear by adequate early treatment is usually quite successful and relatively simple. Therefore, this chapter should probably be titled "The Prevention of the Cauliflower Ear."

A number of surgical approaches varying from opening a "cartilage window" [1,2] and the use of ultraviolet rays,[3] to various forms of incisions, stab wounds, and aspiration methods, have been suggested by various authors.[4-11] Kelleher,[8] who has done the most thorough study of the pathologic process involved, is of the opinion that in the acutely traumatized ear, seen within a few days of injury, there is a central liquid hematoma that is mainly intracartilaginous. Stuteville[9] has confirmed this opinion in some laboratory studies. Davis[2] states that with the shearing force of acute trauma, the anterior layer of auricular skin moves with the perichondrium, tearing an irregular plane through the anterior portion of the cartilage and lacerating some of the small vessels that penetrate the auriculochondral skeleton.

Regardless of the method of treatment employed, it is essential that the hematoma be totally evacuated and constant pressure be maintained during and after the operation to prevent further accumulation. While the dressing must be firm enough to prevent such a recurrence, it should be pliable enough to maintain normal contour of the flexible external ear. As with any bandage, there should be as little interference as possible between the time of its application and its removal.

The vast majority of this sort of injury is seen in a high school or college wrestler, anxious to get back to competition (Fig. 30-1, *A*). The method of treatment here described has worked satisfactorily for these young men as an out-patient or office procedure.

Aspiration of the hematoma or serosanguineous fluid is accomplished with a small hypodermic needle and a 5-ml. syringe (Fig. 30-1, *B*). The needle is inserted at the dependent portion of the collection of fluid, but it may have to be reinserted to pick up any small extra loculations of fluid. Xylocaine 0.5% is first injected in a subcutaneous wheal with a small hypodermic needle, and the same needle and syringe is used to aspirate the serosanguineous fluid. Careful pressure is asserted by the examiner's fingers while he puts on a dressing consisting of strips of 1-inch gauze soaked in *nonflexible* collodion. The gauze must be fine meshed and of good quality cotton. These saturated gauze strips are applied in shingle-like, overlapping fashion (Fig. 30-1, *C*). It takes but a few minutes for this contoured splint to become quite hard. Slight pressure is maintained as the collodion envelope becomes firm. In this method, the through-and-through sutures that are used in most other procedures described in the literature* are avoided. Since in the early postinjury patient we are deal-

____
*See references 2, 4, 6-8, and 11.

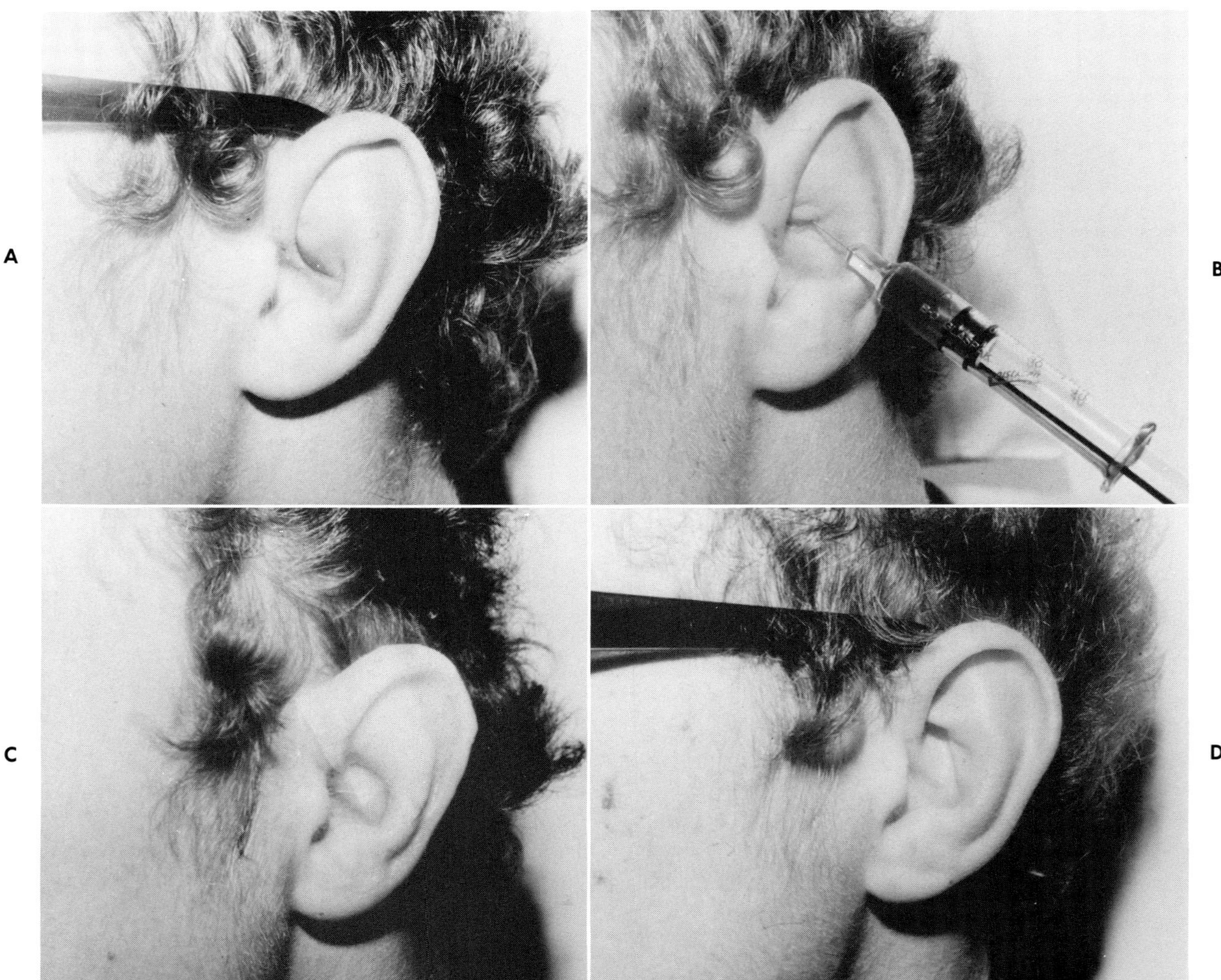

Fig. 30-1. A, Otohematoma obscures contour of upper concha, triangular fossa, and antihelix. **B,** Aspiration of serosanguineous fluid via small hypodermic needle. **C,** After fluid has been withdrawn, 1-inch gauze strips saturated with nonflexible collodion have been applied in shingle-like, overlapping fashion. The surgeon applies light pressure with his finger while the collodion splint becomes hard. **D,** Subsequent appearance several weeks later.

ing with a hematoma or seroma in a bruised and engorged ear, the use of trochars or large stay sutures or large-gauge needles can be a two-edged sword, each modality potentially contributing to further bleeding and hematoma formation.

It is our experience that for this method of treatment to be successful, it must be done within a week of the injury, while the fluid can still be aspirated by a small hypodermic needle. The best results are obtained if the fluid is serum or straw colored rather than bright red. There is an optimum time of between 3 and 6 days for this aspiration. Occasionally, when the athlete is seen on the day of injury, it has been our practice to send the young man away (untreated) for 2 days and have him apply warm compresses, for too early an aspiration (before the small sheared-off vessels have sealed themselves) will only create new bleeding and the optimum result will not be obtained.

The key to this conservative method is the dressing. Many physicians are unaware that there is such a thing as *nonflexible* collodion, and it is essential that this be the material used with the strips of gauze. The dressing does not encompass the entire ear but only overlaps the areas where fluid had accumulated. The patient is instructed to

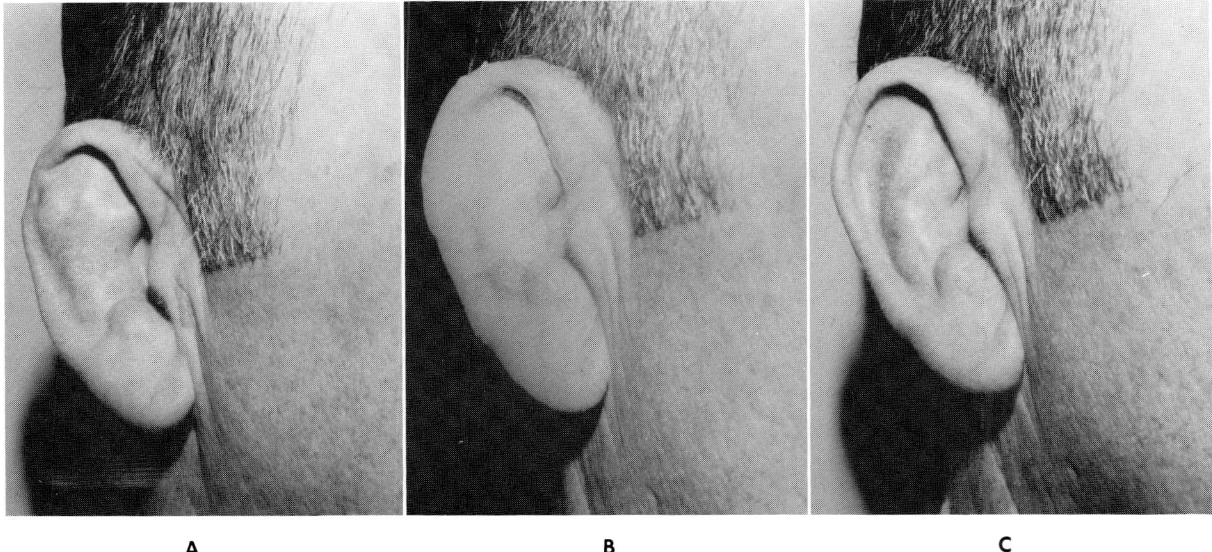

A   B   C

Fig. 30-2. A, More extensive accumulation of hematomatous fluid. **B,** Similar collodion-gauze dressing in place. **C,** Postoperative appearance. This patient has had numerous such collections, all treated in this same conservative manner.

remove the dressing while showering and final removal is done when he notices that it is beginning to become loose, which is usually a matter of 7 to 10 days (Fig. 30-1, *D*).

As noted previously, this sort of injury is almost always seen in a collegiate wrestler. These dedicated athletes, well trained and well motivated, are most anxious to get back to competition. We have found that with this particular dressing, supplemented by appropriate cotton padding in front of and behind the ear, the patient can comfortably wear a wrestler's headcap. Many of these athletes go on to compete in the next meet with such protection in place. If further accumulation of fluid occurs, the wound is reaspirated and the treatment process is repeated (Fig. 30-2). Obviously, this is a calculated risk and the athlete, his coach, and his parents are informed of the possibility of such recurrence.

There have been times when the traumatized ear has not been seen early and the hematoma may already be organizing. In these instances, **the patient is instructed to apply warm compresses and return the next day.** We have no physiologic basis for the empirical findings, but clinically the aspiration seems to be a little easier the day following such warm compresses.

If organization of the hematoma has advanced to the point where aspiration is unsuccessful, then small stab wounds are made and a combination of suction and scooping out of the visible clot is necessary. A small drain may be left in place for 2 or 3 days and a complete head bandage with moist fluff gauze is applied. At the time of the first dressing, the drain is removed and any further fluid is suctioned. A similar dressing is applied for another 4 days. With this form of treatment, we recommend no further wrestling for 3 or 4 weeks.

For the late cases and in those instances when the patient is through with his wrestling career, the aspiration-collodion envelope method is not even attempted. We have instead resorted to through-and-through sutures tied over a firm cotton bolus in a manner similar to that described by Kelleher.[8] In some patients small stab wounds may be inefficient, and it sometimes becomes necessary to make a curved incision within the helix rim or just inside the overhanging ledge of the antihelix fold in order to extirpate the fibrotic cartilage mass by gouging or curetting. After suitable sculpture work, the skin flap is replaced and sutured. Through-and-through sutures are placed and the ear immobilized with tie-over molds of damp cotton sponges.

Kelleher[8] has observed laminations of cartilage and fibrous tissue within the mass, postulating that each stratum represents an episode of trauma sufficiently severe to cause a separation of a layer of cartilage with attendant hemorrhage. He further reported finding occasional calcification and new bone formation.

There is surprisingly little surgery requested for the fully organized and scarred cauliflower ear, some ex-athletes probably wearing this as a "badge of courage." Perhaps this is advantageous since such late surgery is tedious, difficult, and frequently unrewarding.

SUMMARY

An office method of treatment for the cauliflower ear in the still-active athlete has been presented. Aspiration of the intracartilaginous serosanguineous fluid is followed by the application of a contoured nonflexible collodion envelope to prevent further fluid accumulation. These patients, while permitted to continue wrestling (with a protective headcap), must be made aware that this is a calculated risk and that further accumulation of fluid may occur and further treatment may become necessary.

REFERENCES

1. Howard, R. C.: The window operation for hematoma auris and perichondritis with effusion, Laryngoscope **45:**81, 1935.
2. Davis, P. K. B.: An operation for hematoma auris, Brit. J. Plast. Surg. **24:**277, 1971.
3. Lele, D. N.: Ultraviolet in hematoma and perichondritis of the auricle, Arch. Otolaryng. **79:**33, 1964.
4. Stark, R. B.: Plastic surgery, New York, 1962, Harper and Row, Publishers, p. 257.
5. Barsky, A. H., Kahn, S., and Simon, B. E.: Principles and practice of plastic surgery, ed. 2, New York, 1964, McGraw-Hill Book Company, p. 309.
6. Converse, J. M.: Acquired deformities of the auricle. In Converse, J. M., editor: Reconstructive plastic surgery, vol. III, Philadelphia, 1964, W. B. Saunders Co., p. 1107.
7. Konuralp, H. Z.: A new method of surgical treatment for cauliflower ear, Transactions of the International Society of Plastic Surgeons, First Congress, Baltimore, 1955, The Williams & Wilkins Co., p. 336.
8. Kelleher, J. C., and others: The wrestler's ear, Plast. Reconstr. Surg. **40:**540, 1967.
9. Stuteville, O. H., Janda, C., and Pandya, N. J.: Treating the injured ear to prevent a "cauliflower ear," Plast. Reconstr. Surg. **44:**310, 1969.
10. Shambaugh, G. E.: Surgery of the ear, ed. 2, Philadelphia, 1967, W. B. Saunders Co.
11. Eade, G. G.: Preventing cauliflower ears, Northwest Med. **63:**99, 1964.

Aesthetic surgery of the chin and nonsurgical considerations of surgery of nose, ears, and chin

Chapter 31

Chin augmentation for microgenia

Simon Fredricks, M.D., F.A.C.S.

True adult micrognathia, or mandibular retrusion, is the result of mandibular atresia, which in turn is caused by congenital or acquired disturbances of the mandibular growth centers. Acquired micrognathia is usually caused by trauma, infection, or x-ray in infancy or early childhood. It may be unilateral, with compensatory overgrowth of the opposite side, or bilateral, giving the patient a bird-face or "Andy Gump" appearance. These jaws are frequently edentulous, or if teeth are present they are retrognathic, with or without cross-bite. These extensive mandibular underdevelopments require careful preoperative planning, followed by surgical reconstruction by osteotomy, with or without bone graft, as indicated by the particular case. This subject has been discussed by Converse in 1963, and his published report remains a classic in the field.[1]

Microgenia, however, is a condition in which the chin eminence is diminished but the remaining mandible is satisfactory and the dentoocclusal relationship acceptable. It is true that the lower incisor teeth are usually lingually displaced, with an overbite of varying severity, resulting in a vertical reduction of the lower third of the face, but orthodontia can usually deal with this adequately. Chin augmentation, or mentoplasty, is the treatment of choice in microgenia, where increase of the chin eminence is the desired result. The chin may be augmented extraorally, by a classic, small, horizontal submental incision or intraorally as implied by Converse and described by Millard[2] through the lower buccal sulcus. The disadvantages of the extraoral approach are a visible scar, the danger of inferior displacement of the implant below the inframandibular ridge, and inadvertent in-

traoral perforation during pocket dissection. Its advantages are simplicity of approach and reduced hazard of secondary implant exposure or secondary infection. There is no visible scar with the intraoral route, and it facilitates supporting the implant in a position above the inframandibular ridge. Its disadvantages are an implied higher risk of secondary extrusion or infection, and it requires precise placement of incision and meticulous closure of the wound. This must be followed by an increased restriction of oral intake and motion during the period of wound healing. It produces lower buccal sulcus scarring and partial obliteration of the depth of the sulcus.

The most frequently used materials are:

1. Preformed Silastic rubber
2. Soft-carved Silastic sponges
3. Preformed acrylic
4. A variety of other materials
   a. Free silicone injection
   b. Preserved cadaver cartilage
   c. Kiel bone (denatured bovine bone)
   d. Compressed Ivalon and/or Etheron
   e. Autogenous material
      (1) Bone from the iliac crest
      (2) Cartilage from the ribs
      (3) Nasal hump
   f. Cervical fat flaps
   g. Dermafat grafts

In 1966, Safian traced historically the implant material used for chin implantation and benefited us with his more than 40 years of experience with these materials.[3] He concluded with a report of the use of Silastic rubber implant of Dow Corning. These were modeled after the ivory forms used by the late professor, Jacques Joseph, in 1920. In 1963 Blocksma and Spiers reported on the fate of

Silastic sponges following implantation.[4] It is my impression that these two Silastic materials are the most widely used for chin augmentation in this country.

Gonzalez-Ulloa has long advocated the use of preformed acrylic in a buttocks or mustache shape,[5] and more recently Pitanguy reported its use. In this country, acrylic implants appear to be predominately favored by oral surgeons.

Free silicone injections have also been widely used outside of this country. There is no question that the material subsequently satellites into the surrounding tissues, producing numerous micro-encapsulations of the disseminating silicone. Furthermore, late dependent migration marks the end result.

Preserved cadaver cartilage, as reported by Pierce and O'Connor,[6] was an important landmark in chin augmentation, but a high incidence of late resorption of the material was disappointing and led to its abandonment. The use of Kiel bone, as reported by Penn, also demonstrated a high resorption rate. Autogenous bone and cartilage grafts achieved the only long-term predictable results available prior to the advent of newer alloplastic materials. Their use, however, carried an increased morbidity and visible scarring associated with obtaining the material.

Nasal hump is a generally unsatisfactory material for this purpose. The advocates of compressed Ivalon and Etheron are tenacious in their loyalty to the material, but the majority of sponges seem to undergo subsequent distortion and changes of consistency and size with time. Dermafat grafts have proved to be generally unsatisfactory for the usual reasons of late encystments and variations in implanted volume. Transposed submental fat flaps, as reported by Robertson,[7] have a very limited use in association with submental lipectomy, and again the late shape is variable.

Experiences with closed cell, fine Silastic self-carved sponge has not been satisfactory, since initially satisfactory results have been marred by late distortion of shape, secondary to scar contracture. One case of infection 5½ months following successful implantation, with percutaneous liberation of gross purulent material, occurred (Fig. 31-1).

If any sponge material becomes infected, it should be promptly removed and the resultant wound packed open and irrigated until the pocket is obliterated. The subsequent scarring frequently produces sufficient protrusion of the chin so that

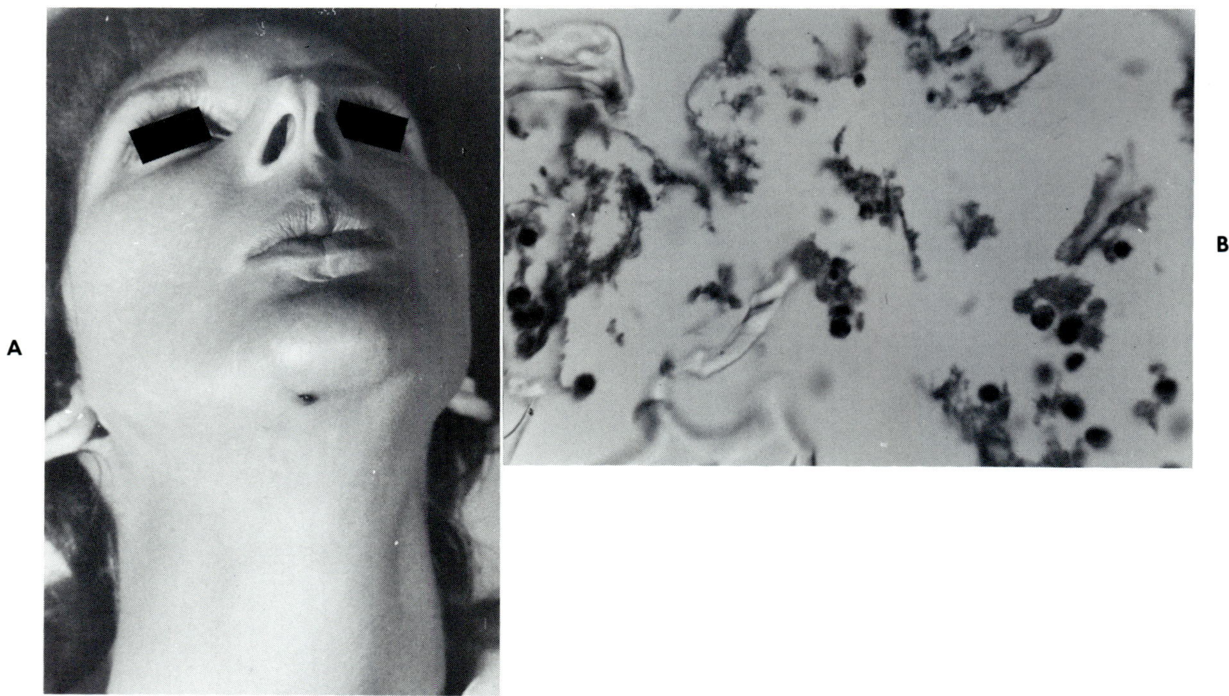

Fig. 31-1. A, Percutaneous liberation of gross purulence 5½ months following Silastic sponge implant. **B,** Histologic section of infected Silastic sponge. Note profusion of macrophages.

reimplantation is unnecessary. One cannot help but wonder if the elaborate decussating lymphatics coursing across the chin, draining the periodontal region, do not leave these sponges forever exposed to infection.

The use of the preformed Silastic rubber implants, which have afforded consistently satisfactory long-term results by both extraoral and intraoral implantation, have proved most satisfactory. The major disadvantage of the Silastic implant is the preformed shape, which may not be ideal for a particular patient. It is also limited by its lack of lateral extension, since ideal reconstruction frequently requires a larger lateral volume than the preformed Silastic implant allows.

To handle this problem, Carlin, Gurdin,[8] and

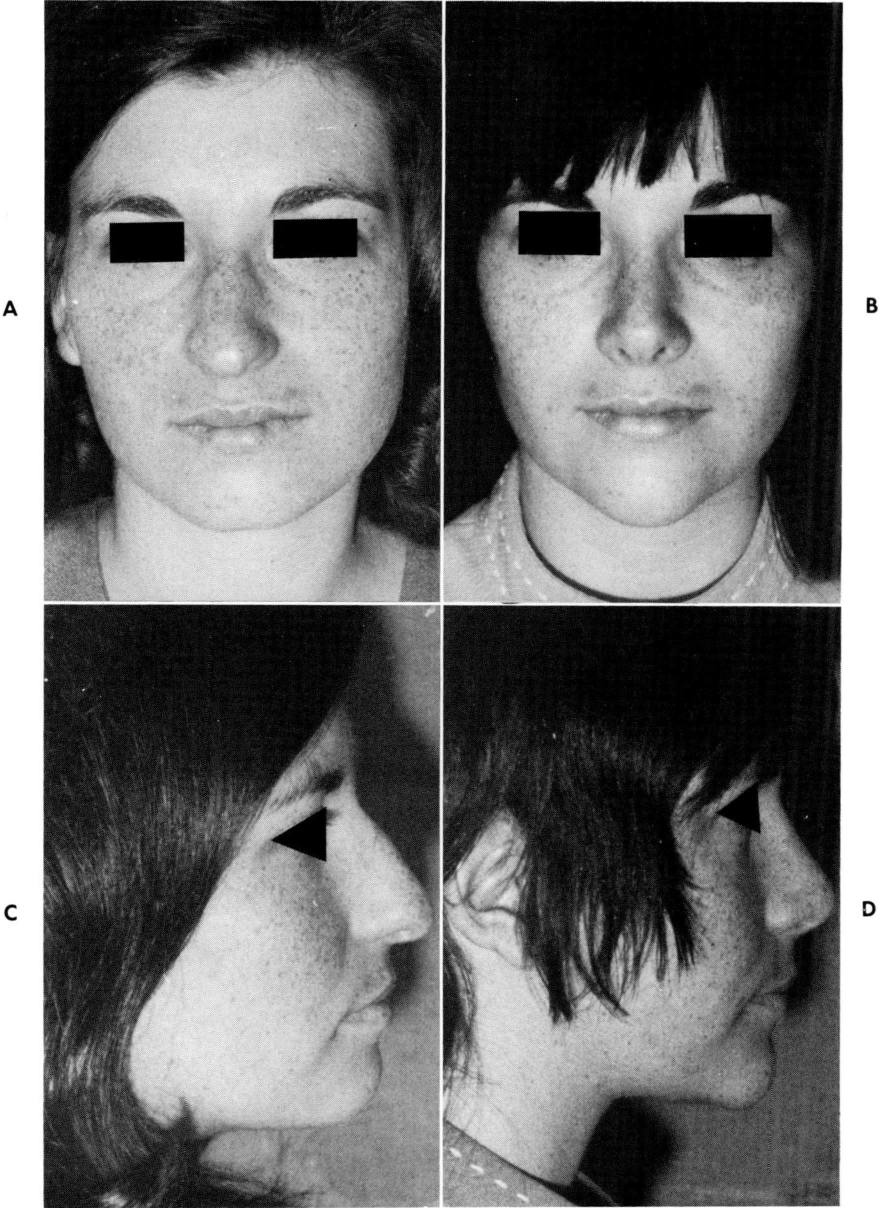

Fig. 31-2. Rhinoplasty-mentoplasty combinations. **A,** Front view. **B,** Postoperative front view after using soft Silastic rubber implant. **C,** Side view. **D,** Postoperative view.

others have molded their own Silastic implants based on moulages of the particular patient from the self-curing Silastic material available from Dow Corning. This, of course, requires the formation of preoperative molds and more careful preoperative planning.

The only matter of concern in the use of silicone rubber implants is a report in 1969 of Robinson and Shuken[9] covering some thirty-three silicone and acrylic implants subperiosteally placed over an 8-year period in which the absorption of mandibular bone beneath the implant was demonstrated. I have been unable to demonstrate such radiographic evidence of resorption but have been

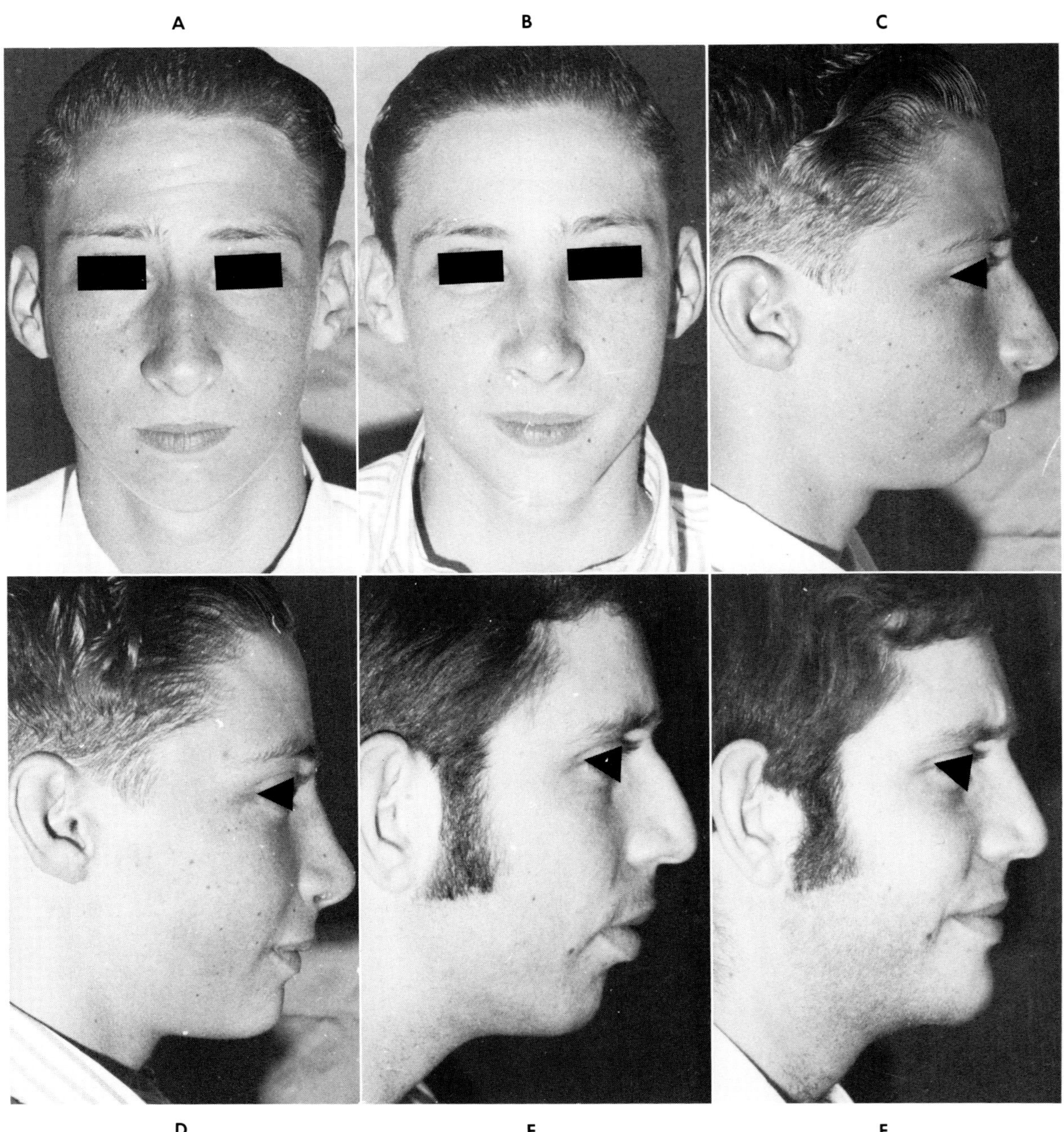

Fig. 31-3. Rhinoplasty-mentoplasty combinations. **A,** Preoperative front view. **B,** Postoperative view after using soft Silastic rubber implant. **C,** Preoperative side view. **D,** Postoperative view. **E,** Preoperative side view. **F,** Postoperative side view.

afforded the benefit of undisputed radiographic evidence in the cases of other surgeons. Although the etiology of such bony resorption is still obscure, it is believed to be caused by either a compression phenomenon or the stripping away of the periosteum beneath the implant.

Successful implantation via the intraoral route is dependent upon producing a satisfactory cuff of buccal mucosa on the side of the incision, suffi-

ciently wide and far away from the alveolus that a good eversion of the wound edge by mattress sutures is easily accomplished, producing a watertight closure. Beneath the mucosal closure, a secondary layer of deeper muscle-to-muscle apposition for support and implant retention is required.

After the horizontal incision is made, the pocket may be produced with the use of a Joseph's elevator. A guarding finger is held at the inframan-

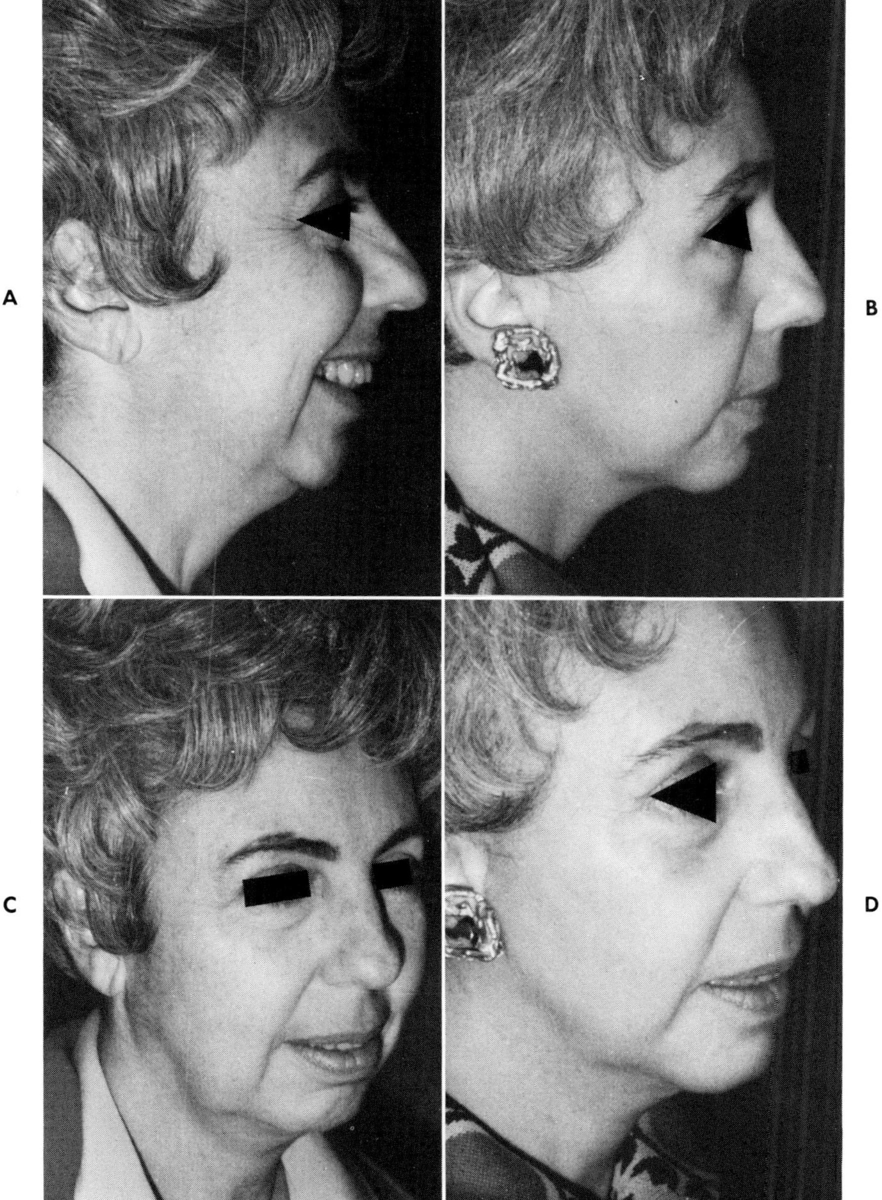

Fig. 31-4. Rhytidectomy-mentoplasty combinations. **A,** Side view. **B,** Postoperative view after using soft Silastic rubber implant. **C,** Three-quarters view. **D,** Postoperative view.

dibular ridge so that the pocket will not be carried beyond this point. The central incisors are utilized to orient the midline and the pocket is made equidistant laterally, with care being exercised to avoid injury to the mental nerves. Dissection is carried down to, but not beneath, the periosteum overlying the mandible. The pocket should be of sufficient size to accept the implant without undue tension of the overlying tissues, but under no circumstances

should the pocket be large enough to permit free and easy movement of the implant. Otherwise, late displacement of the implant will result.

It should be pointed out that normally the most prominent point of the chin lies at a point equidistant between the depth of the buccal sulcus and the inframandibular line. When this orientation is appreciated, proper placement of the maximum chin eminence between this horizontal plane in the

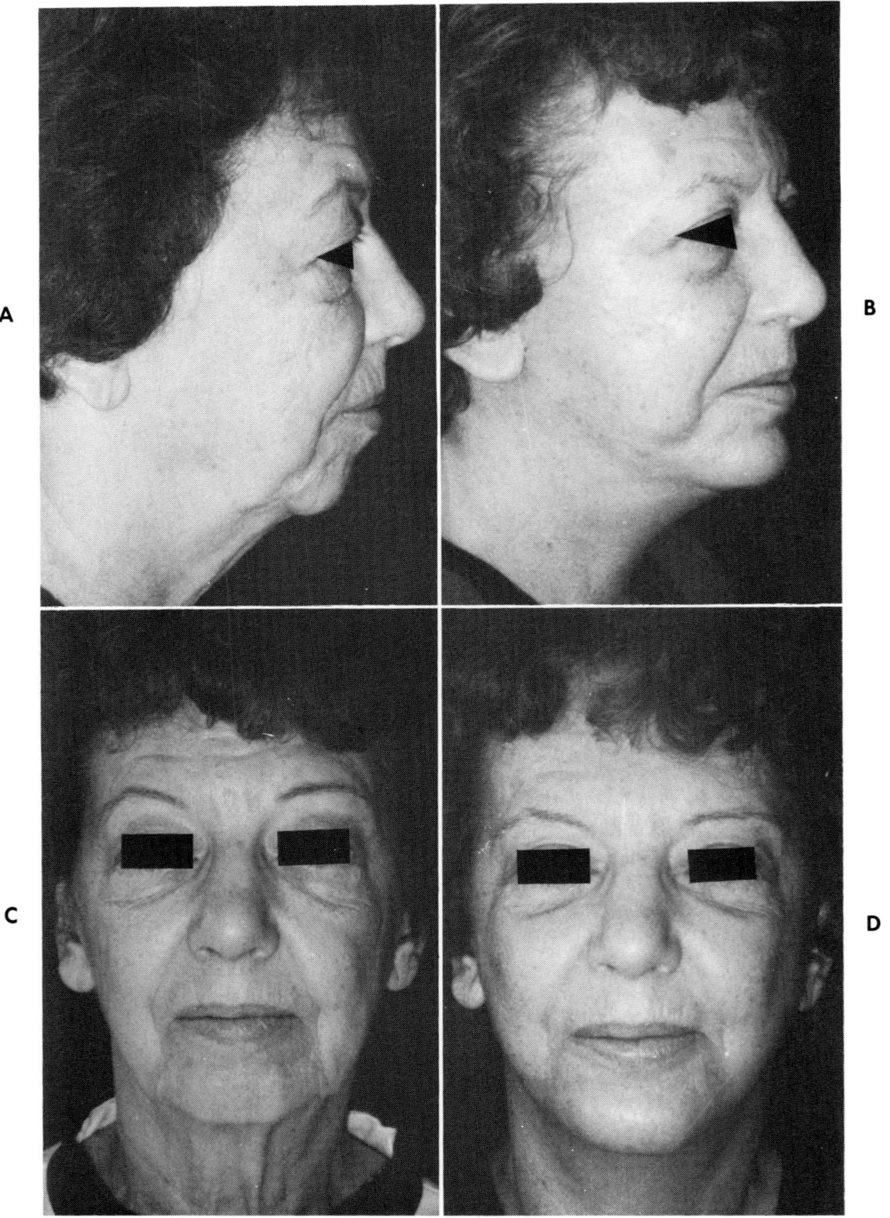

Fig. 31-5. Rhytidectomy-mentoplasty combinations. **A,** Side view. **B,** Postoperative side view after using soft Silastic rubber implant. **C,** Preoperative front view. **D,** Postoperative front view.

line of the central incisor will facilitate proper placement of the implant.

Scrupulous hemostasis is obtained before implantation, either by packing with moist gauze or pinpoint electrocautery, as indicated. The implants are cleansed with ether prior to sterilization, and immediately following sterilization they are placed into saline so that dirt and lint will not fall upon them prior to implantation.

When the implant is in position, it is difficult to determine whether it is centered. To facilitate this, a suture needle may be passed through the upper border of the implant, coated with methylene blue, and then withdrawn. A small blue dot will then be visible when the implant is in place, thereby locating the exact center of the implant. This may be lined up with the space between the two central incisors. A curvilinear strip of ½-inch Elastoplast

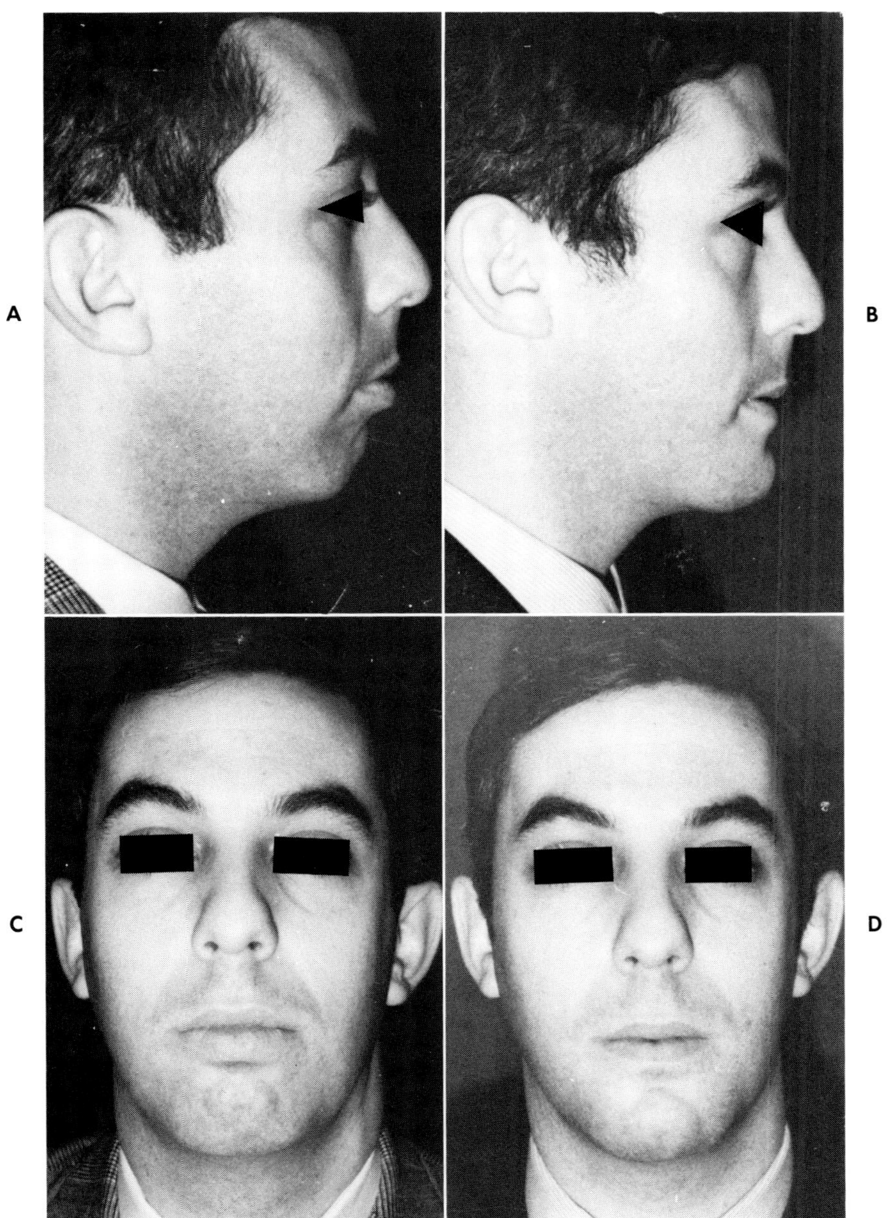

Fig. 31-6. A, Side view before mentoplasty. **B,** Postoperative side view after mentoplasty, using soft Silastic rubber implant. **C,** Preoperative front view. **D,** Postoperative front view.

is loosely applied above the implant to obliterate any dead space in the intraoral wound, and a similar curvilinear ½-inch strip of Elastoplast is placed beneath the implant to prevent its descent beneath the mandibular ridge. No dressing is applied over the center of the implant at the point of maximum tension to avoid the possibility of necrosis of the overlying skin.

Preoperative scrupulous dental hygiene and toilet are mandatory, and undue motion is discouraged postoperatively.

This operation is most frequently performed in combination with rhinoplasty in order to produce an overall balanced profile. However, only 10% to 15% of patients require this combined procedure (Figs. 31-2 and 31-3). It is also frequently performed in combination with rhytidectomy where microgenia is present, enhancing the overall balance of profile prolongation of the intramandibular line and borrowing some of the redundant tissue from the submental region (Figs. 31-4 and 31-5). Here again, no more than 10% to 15% of patients require this combined procedure. The smallest group of patients seeking mentoplasty were those whose only complaint and requirement was isolated correction of a weak chin (Fig. 31-6).

REFERENCES

1. Converse, J. M.: Micrognathia, Brit. J. Plast. Surg. **16:**197, 1963.
2. Millard, R. D.: Buccal incision for chin implants, Southern Med. J. **52:**1371, 1959.
3. Safian, J.: Progress in nasal and chin augmentation, Plast. Reconstr. Surg. **37:**446, 1966.
4. Spiers, A. C., and Blocksma, R.: New implantable silicone rubbers, Plast. Reconstr. Surg. **31:**166, 1963.
5. Gonzalez-Ulloa, M.: The role of chin correction in profileplasty, Plast. Reconstr. Surg. **41:**477-486, 1968.
6. Pierce, G. W., and O'Connor, G. B.: Refrigerated cartilage isografts, Surg. Gynec. Obstet. **67:**796-798, 1938.
7. Robertson, J. G.: Chin augmentation by means of rotation of double chin fat flaps, Plast. Reconstr. Surg. **36:**471, 1965.
8. Carlin, G., and Gurdin, M. M.: Personal communication.
9. Robinson, M., and Suken, R.: Bone resorption under plastic chin implants, J. Oral Surg. **27:**116-118, 1969.
10. Stark, R. B.: Bone graft to chin. In Stark, R. B., editor: Plastic surgery, New York, 1962, Harper & Row, Publishers.
11. Brown, J. B., Fryer, M. P., and Li, M.: Polyvinyl and silicone compounds as subcutaneous prosthesis, Arch. Surg. **68:**744, 1954.
12. Berman, W. E.: Use of silicones in rhinoplasty and chin implant, presented at a meeting of the American Academy of Ophthalmology and Otolaryngology, New York, October, 1963.
13. Conway, H., and Goulian, D.: Experience with an injectable silastic RTV as a subcutaneous prosthetic material, Plast. Reconstr. Surg. **32:**294, 1963.
14. Brown, J. B., and others: Silicones in plastic surgery, Plast. Reconstr. Surg. **12:**374, 1953.
15. Brown, J. B.: Studies of silicones and Teflon as subcutaneous prosthesis (editorial), Plast. Reconstr. Surg. **28:**86, 1961.
16. Brown, J. B., and others: Silicone and Teflon prosthesis, including full jaw substitution: laboratory and clinical studies of Etheron, Ann. Surg. **157:**932, 1963.
17. Brown, J. B., and others: Dimethylsilaxane and halogenated carbons as subcutaneous prostheses, Am. Surg. **28:**146, 1963.
18. Vinas, J. C.: Surgical experience with silicone implants, presented at the Fifth Joint Meeting of Plastic Surgeons of Uruguay and Argentina, 1962.
19. Parkes, M. L.: Chin implants with a newer plastic compound, Arch. Otolaryng. **75:**429, 1962.

Chapter 32

The silicones—their uses in nose, chin, and ear

Silas Braley

The use of silicones for reconstruction of the human body was initiated in 1950 when De Nicola published an article on an artificial urethra.[1]

Since 1959, when active participation was begun by Dow Corning, the field has advanced rapidly, so that certain medical grades of this large family of synthetic polymers are now widely accepted by the medical fraternity. Basic information on the application of these materials to surgery and on their chemistry and physical properties is available upon request.[2, 3]

Each application of these materials to the many requirements of reconstructive surgery needs its own set of rules, so to speak, and this paper is an attempt to evaluate the present position that the silicones have in reconstructive surgery of the ear, nose, and chin. The advantages and disadvantages of their use for these applications, both from experience and from the literature, are discussed, and extensive references are included for those who desire more detailed information. In addition, various special techniques that could aid the physician in this surgery, and which are not widely known, are included.

THE SILICONES IN RECONSTRUCTIVE SURGERY OF THE EXTERNAL EAR

The cosmetically satisfactory reconstruction of the human ear has always been one of the more difficult procedures asked of the plastic surgeon. The many procedures and materials tried bear mute testimony to the fact that no one concept is satisfactory in all hands. Not only is the surgery itself difficult, but a completely faultless material to substitute for the missing cartilage is yet to be found. Cronin[4] lists no less than twenty materials or forms of materials that have been tried. Each of these materials has had advantages and disadvantages: The use of autogenous cartilage requires an added assault upon the patient to obtain the needed tissue, and the spectre of infection and absorption is omnipresent. Heterogenous and homogenous cartilage have failed primarily because of foreign body reaction, infection, and/or absorption. Nonreactive allopathic materials have the allure of easy accessibility and fine architecture, but their use in the ear is always followed by the possibility of extrusion. The surgeon is thus in the uncomfortable position of deciding which advantages outweigh which disadvantages.

Because of the softness and lack of tissue reaction of medical grade silicone rubber and with full knowledge of the generally dismal history of other allopathic materials used for ear reconstruction, it was felt that a silicone rubber prosthesis might be of value, and work was begun early in the last decade. Brown, Fryer, and Ohlwiler[5] reported in 1960 on a hand-carved silicone rubber ear prosthesis for replacement of a rejected paternal homotransplant in a child. Cronin[4] reported on the use of the molded Silastic Otoplasty Prosthesis[6] in 1966. The major problem, as expected, was that of extrusion, but he felt that cautious use and further investigation was justified, especially in patients not subjected to trauma.

This early work also showed some basic factors

that, while widely recognized to be of value in other areas of implant surgery, seemed to be of peculiar importance to ear surgery:

1. The tissue over the implant should be under as little stress as possible.
2. The implant should be as soft as is consistent with the desired rigidity.
3. There should be no sharp edges on the implant.

The extreme inertness of the medical grade silicone rubber is, in this case, actually somewhat of a detriment, since it does not cause the development of a tough fibrous tissue that would serve to protect from erosion.

It was also noticed that, should erosion occur, disaster does not necessarily automatically follow: The trimming of a portion of the implant followed by covering with adjacent skin can be successful. In cases of infection, antibiotics can be used effectively. This same successful repair of cases of exposure of medical grade silicone rubber has since been reported by others in various areas of the body. Every exposure, of course, does not end happily, but the chances of it doing so are well worth the avoidance of automatic removal of the implant.

In a follow-up study in 1968, Cronin, Greenberg, and Brauer[7] reported on nineteen patients with congenital microtias and eleven patients with complete traumatic auricular losses repaired by the use of the Silastic Otoplasty Prosthesis. In half the cases they used a prosthesis to the helix of which Dacron cloth had been attached. The purpose of the cloth was to serve as an artificial stroma into which fibrous tissue would grow.*

Of the fifteen patients with the cloth attached to the prosthesis, there were three exposures, while in the fifteen patients without the cloth there were five. All the exposures (except one patient who had not returned) were subsequently repaired either by closure or by removal and subsequent

replacement. The authors felt that the cosmetic results were excellent and that a minimum of procedures was needed.

Since that report there have been several other articles published on this use of the silicone rubber otoplasty prosthesis. Brown, Fryer, and Morgan[8] in 1969 reported the use of another specially carved Silastic ear prosthesis, with the now familiar warnings concerning use in patients of an age and sex that can be expected to receive ear trauma. Also in 1969, Edgerton[9] reported sixteen successes out of eighteen uses of a silicone rubber ear prosthesis. He also reported both success and failure in the use of Teflon and polyethylene prostheses. In 1969, Curtin, Bader, and Ellenby[10] reported the use of the Silastic Otoplasty Prosthesis to which Dacron mesh had been attached. They reported only three failures out of forty-two patients and felt that the device should be a definite consideration in many cases of reconstruction of the external ear, with caution used in the selection of each candidate and in each surgical state.

In 1971, Monroe[11] reported on the use of the Silastic Otoplasty Prosthesis in ten cases of congenital microtias, five ears reconstructed after burns, and two after trauma. Fourteen of these prostheses had been in place for $1\frac{1}{2}$ to 5 years and three were thus far incomplete. Exposure again is the main complication, one device having been exposed for 3 months. All, including the latter, were either snipped and closed or removed and later replaced successfully. Monroe felt that a wait of 2 to 3 months before bringing out the ear contributed to success.

Lynch, Pousti, and Doyle[12] reported on forty-six patients in whom fifty-six ears had been rebuilt with a silicone rubber prosthesis. Again exposure was the chief complication, with 25% of the congenital implants, 50% of the posttraumatic implants, and 70% of the postburn implants requiring removal. He stated that while the procedure is not condemned, it should not be routinely employed in posttraumatic patients and not used as frequently as cartilage in congenital deformities.

In summary, the consensus of opinion at this writing seems to be that in selected patients, particularly those in whom the possibility of future trauma is low, a silicone rubber ear armature has the following advantages over autogenous cartilage:

1. Easy availability
2. Good architecture

---

*Those interested in this use of an artificial stroma of Dacron are advised of the following points: Be certain that the Dacron cloth is attached to the prosthesis in such a manner that the interstices of the cloth are open to permit tissue ingrowth. If Dacron tricot is used, it can simply be sewed on to the helix, or it can be "spot-welded" in place with Silastic Medical Adhesive Silicone Type A. Dacron felt has also been used, with attachment by means of the Silicone Adhesive Type A. Both types of textile are available upon request from the Dow Corning Center for Aid to Medical Research, Midland, Mich. 48640.

3. Easy sterilization
4. Reduction in number of operative procedures
5. No absorption
6. Reasonably good management of most complications

The major disadvantage has been exposure, with its concomitant problems. Exposure of a silicone implant, however, is not inevitably followed by the problems experienced with other implant materials, not even with autogenous transplants, and efforts at recovery have proved to be surprisingly effective. In some cases the rubber can be snipped away and then covered again with adjacent skin. Infection has been successfully handled with antibiotics. Should these procedures fail, the removal and subsequent replacement of the prosthesis has been successful and is worthy of consideration.

An interesting use of silicones in conjunction with ear surgery was reported by King in 1965.[13] A room-temperature-vulcanizing silicone rubber (Silastic 386 Foam Elastomer) was catalyzed and allowed to foam up around an ear following reduction of a hematoma, thus forming a custom-built splint and providing mild, fairly equal pressure to all portions of such an unsymmetrical organ. The same procedure has been used to hold newly placed skin grafts onto the penis[14] and for supportive collars after release and grafting of contracted neck burn scars.[15] Such a procedure may be of value in surgical reconstruction of the ear, whether an autogenous or allopathic armature is used.

THE USE OF SILICONES IN NASAL RECONSTRUCTION

The first report of the use of a silicone for nasal implant was by Brown, Fryer, and Ohlwiler in 1960.[5, 16] They reported the use of Silastic sheeting following traumatic septal loss. Other papers followed in rapid succession covering many aspects of repair.[17-38]

The advantages of silicone rubber for nasal repair are the same as those in other parts of the body:
1. Easy availability
2. Easy shaping
3. Easy sterilization
4. Reduction of number of operative procedures
5. No absorption

The disadvantages are likewise those encountered in other implantations:
1. Exposure
2. Infection

Again, as in the ear prosthesis, the lack of tissue reaction can be a disadvantage when the silicone is in an area of stress, since a protective layer of heavy fibrous tissue is not laid down. The more delicate subdermal tissues are squeezed against the implant and thus can be destroyed, leading to exposure of the implant.

This exposure has been noted in tight skin over the dorsum and at the tip of the nose, particularly when efforts are made to form a more pointed nose from a flat one, usually by trying to lengthen the columella.

Some surgeons in attempting to do this have used an **L**-shaped prosthesis, others have used two separate implants, one along the dorsum and one in the columella. In either case, a piece of Dacron felt* or tricot* attached to the prosthesis directly under the tip of the nose—combined, of course, with care to minimize the stresses as much as possible—will help to reduce this erosion.

Perhaps a better approach to the lifting of the tip would be to place a prosthesis under the base of the columella, so that the area at the tip of the nose would not be in contact with the implant and the area that is in contact with the implant would be as broad as possible, thus better distributing the stress.

Special prostheses for problem noses can be either carved from silicone rubber block or custom-fabricated.[38, 40, 41]

THE USE OF SILICONES IN MENTOPLASTY

Probably more materials have been successfully used for augmentation mentoplasty than have been successful in almost any other part of the body. This is attributable to the facts that erosion through the skin is not the problem that it can be elsewhere, softness is not necessary, and the skin covering is usually fairly generous.

Teflon[5, 16, 42] has been reported to be successful for mental augmentation. Rubin reports an astonishing 21-year follow-up in the use of polyethylene,[43] while Rish,[44-45] Weinstein,[46] and Pitanguy[47] and

---

*Available upon request from Dow Corning Center for Aid to Medical Research, Midland, Mich. 48640.

others have used the acrylics. Silicone rubbers have been reported on widely.*

The advantages of silicone rubber over other materials for chin augmentation seem to be primarily:

1. Its ease of carving to adjust to individual variations
2. Its easy sterilization
3. The ready availability of a grade manufactured specifically for implantation purposes

Both the buccal and the submental approaches have been used, and each has its advocates and denigrators.

The attachment of Dacron tricot† or felt† serves to supply an artificial stroma for tissue ingrowth and reduces motion of the implant. It has been suggested that holes punched through the device will also serve this purpose.[60]

The use of Silastic Fine Cell Sponge is felt by some to give more desirable results than a soft solid silicone rubber implant. However, the air contained within the pores of the sponge can pass through the rubber and escape, so that if constant pressure is placed on the prosthesis, either intrinsically, such as by the retraction of scar tissue, or extrinsically, as in the case of a "chin-leaner," the sponge will ultimately become smaller and the intended augmentation be lost. (At this writing, the Silastic Fine Cell Sponge is no longer available because of potential explosion hazards encountered during its manufacture.)

Robinson and Shuken[61] reported in 1969 of bone resorption under both acrylic and Silastic chin implants. Since that time other scattered instances have been reported. One even told of a silicone implant sunken into the chin and covered over with bone. As far as can now be ascertained, this sinking of the prosthesis into the bone probably results from the placing of the implant between the bone and the periosteum. Bone is constantly being resorbed and replaced. When an implant is placed between the bone and the periosteum, resorption continues unabated, but replacement is prevented, resulting in the observed phenomenon. At this time no study has been made on a comparison between sub- and supraperiosteal positioning. Such a study would be of value.

In addition to silicone rubber onlays for chin augmentation, other work with the same materials in relation to mandibular replacement is of interest.

Two papers[62, 63] report on the successful implantation of medical grade silicone rubber under the soft tissue and over the bone of the jaw to reconstruct the alveolar ridge. The patients were able subsequently to wear dentures on the oral mucosa overlying the implant.

Silicone rubber has been used to replace the missing mandible in cases of partial or even complete mandibulectomies.[64-66]

SUMMARY

The silicones offer several advantages over other materials for the reconstruction of ears, noses, and chins, but the possibility of erosion is always present, so that, as in all implant and transplant surgery, each surgeon must decide for himself which material in his hands is the best for his patient.

REFERENCES

1. De Nicola, R. R.: Permanent artificial (silicone) urethra, J. Urol. 63(1):168, 1959.
2. Braley, S. A.: The chemistry and properties of the medical-grade silicones, J. Macromol. Sci. Chem. A4 (3):529-544, 1970.
3. Braley, S. A.: Acceptable plastic implants. In Simpson, D. C., editor: Modern trends in biomechanics, London, 1970, Butterworth and Co. (Publishers) Ltd.
4. Cronin, T. D.: Use of a Silastic frame for total and subtotal reconstruction of the external ear: preliminary report, Plast. Reconstr. Surg. 37(5):399-405, 1966.
5. Brown, J. B., Fryer, M. P., and Ohlwiler, D. A.: Study and use of synthetic materials, such as silicones and Teflon, as subcutaneous prostheses, Plast. Reconstr. Surg. 26(3):264-79, 1960.
6. Braley, S. A.: Silicone rubber prosthetic ear frame, U. S. Patent 3,257,668, issued June 28, 1966.
7. Cronin, T. D., Greenberg, R. L., and Brauer, R. O.: Follow-up study of Silastic frame for reconstruction of external ear, Plast. Reconstr. Surg. 42:6, 1968.
8. Brown, J. B., Fryer, M. P., and Morgan, L. R.: Problems in reconstruction of the auricle, Plast. Reconstr. Surg. 43(6):597, 1969.
9. Edgerton, M. T.: Ear construction in children with congenital atresia and stenosis, Plast. Reconstr. Surg. 43:373, 1969.
10. Curtin, J. W., Bader, K. F., and Ellenby, J. E.: Improved techniques for the successful silicone reconstruction of the external ear, Plast. Reconstr. Surg. 44 (4):372-77, 1969.
11. Monroe, C. W.: Experience with the silicone ear prosthesis: a report of 17 ears on 15 patients, unpublished paper presented at American Society of Plastic and Reconstructive Surgeons Meeting at Montreal, October, 1971.

---

*See references 5, 16, 29, 45, and 47-59.

†Available upon request from Dow Corning Center for Aid to Medical Research, Midland, Mich. 48640.

12. Lynch, J. B., Pousti, A., and Doyle, J. E.: Experience with Silastic ear implants, unpublished paper presented at American Society of Plastic and Reconstructive Surgeons Meeting at Montreal, October, 1971.

13. King, J. A.: Silicone rubber molds as an adjunct in the treatment of hematomas of the ears, J. Sports Med. Phys. Fitness 5(1):13-14, 1956.

14. Schneider, P.: Foam dressing on skin grafts, Bull. Dow Corning Center for Aid to Medical Research 7:1, 1965.

15. Blocksma, R.: Personal communication.

16. Brown, J. B., Ohlwiler, D. A., and Fryer, M. P.: Investigation of and use of dimethylsiloxanes, halogenated carbons and polyvinyl alcohol as subcutaneous prostheses, Ann. Surg. 152(3):534-547, 1960.

17. Farrior, R. T.: Implant materials in restoration of facial contour, Laryngoscope 76:934-954, 1966.

18. Farrior, R. T.: Synthetics in head and neck surgery: with particular reference to silicone, Arch. Otolaryng. 84:82-90, 1966.

19. Vinas, J.: Surgical experiences with silicone rubber implants, presented at Meeting of Uruguayan Society of Plastic Surgeons, October 12, 1962.

20. Khoo, B. C.: Some aspects of plastic (cosmetic) surgery in Orientals, Brit. J. Plast. Surg. 22:60, 1969.

21. Khoo, B. C.: Augmentation rhinoplasty in the Orientals, Plast. Reconstr. Surg. 34(1):81, 1964.

22. Uchida, J. I.: A design for a prosthesis in augmentative rhinoplasty: the projection method, Plast. Reconstr. Surg. 33(5):485-90, 1964.

23. Beekhuis, G. J.: Silicone rubber implants in nasal reconstructive surgery, Laryngoscope 74(10):1405-1419, 1964.

24. Beekhuis, G. J.: Use of silicone rubber in nasal reconstructive surgery, Arch. Otolaryng. 86:114, 1967.

25. Barry, W. B.: Experimental observations upon reconstruction of the nasal dorsum, Laryngoscope 75:1320-1333, 1965.

26. Freeman, B., Biggs, T., and Beall, A.: Injectable Silastic in deformities of the facial skeleton, Arch. Surg. 90:166, 1965.

27. Rise, E., Huffman, W., and VerVais, P.: Silicone nasal implants: a pilot study on dogs, Plast. Reconstr. Surg. 36(2):231, 1965.

28. Morani, A. D.: Surgery of facial trauma: review of principles of treatment, Int. Surg. 46:391-398, 1966.

29. Safian, J.: Progress in nasal and chin augmentation, Plast. Reconstr. Surg. 37(5):446-452, 1966.

30. Patterson, C. N.: Reconstructive septonasoplasty, Arch. Otolaryng. 84:457-463, 1966.

31. Riggs, R., and Wilson, J.: Relapsing polychondritis: report of a case, Laryngoscope 77(4):534, 1967.

32. Millard, D. R.: Alar margin sculpturing, Plast. Reconstr. Surg. 40(4):337-342, 1967.

33. Stellmach, R.: Reconstruction of the facial contours with Silastic, Fortschr. Kiefer. Gesichtschir. 12:141-144, 1967.

34. Marshall, A., and Withers, B. T.: Silastic as dorsal transplant for saddle nose deformity: a preliminary report, Eye, Ear, Nose, Throat Monthly 47:70, 1968.

35. Cipcic, J. A.: Silicone implant correction of facial deformities, Laryngoscope 78:4, 1968.

36. Rees, T. D.: Nasal plastic surgery in the Negro, Plast. Reconstr. Surg. 43(1):13-18, 1969.

37. Rhodes, R. D., III: Restoration of facial defects with individually prefabricated silicone prostheses, Plast. Reconstr. Surg. 43:201, 1968.

38. Aufricht, G.: Rhinoplasty and the face, Plast. Reconstr. Surg. 43:219, 1969.

39. Stark, R. B., and Frileck, S. P.: Conchal cartilage grafts in augmentation rhinoplasty and orbital floor fracture, Plast. Reconstr. Surg. 43(6):591-596, 1969.

40. Laub, D. R., and others: Accurate preoperative fabrication of subcutaneous and bony implants with Silastic MDX 4-4515 and MDX 4-4516: technique and results, Bull. Prosth. Res., pp. 206-213, Spring, 1970.

41. Lash, H., Zimmerman, D. C., and Loeffler, R. A.: Silicone implantation: inlay method, Plast. Reconstr. Surg. 34(1):75-79, 1964.

42. Brown, J. B., and others: Dimethylsiloxane and halogenated carbons as subcutaneous prostheses, Amer. Surg. 28(3):146, 1962.

43. Rubin, L. R., Bromberg, B. E., and Walden, R. H.: Long term human reaction to synthetic plastic, Surg. Gynec. Obstet. pp. 603-608, April, 1971.

44. Rish, B. B.: Alloplastic materials in the creation of facial contour, Arch. Otolaryng. 72:212, 1960.

45. Rish, B. B.: The esthetic-plastic profile, Proc. Penn. Acad. Ophthal. Otolaryng., p. 79, Fall, 1965.

46. Weinstein, I. R., and Viener, A.: Chin implant, Dent. Surg., pp. 45-51, February, 1968.

47. Pitanguy, I.: Augmentation mentoplasty, Plast. Reconstr. Surg. 42:5, 1968.

48. Mohnac, A.: Maxillary osteotomy in the management of occlusal deformities, J. Oral Surg. 24(4):305-317, 1966.

49. Mohnac, A. M.: Surgical correction of maxillomandibular deformities, J. Oral Surg. 23(5):393-407, 1965.

50. De Moura Andrews, J.: Inclusoes de silicone: estudo experimental e clinico, Sao Paulo, 1966, Escola Paulista de Medicina.

51. Serson, D.: A new technique of chin augmentation, Plast. Reconstr. Surg. 46(4):406, 1970.

52. Scheunemann, H.: Kinnkonstruktion mit Silastik in combination mit Einem Fromgebenden Kirschner-Draht, Zahnaerztl. Welt. 79(3):100-104, 1970.

53. Bell, W. H.: Correction of the contour-deficient chin, J. Oral Surg. 27:108, 1969.

54. Lund, B. A., and Sather, A. H.: Correction of maxillomandibular discrepancies and occlusal abnormalities, Mayo Clin. Proc. 44:102, 1969.

55. Meyer, R. A., and others: Restoring facial contour with implanted silicone rubber, Oral Surg. 24(5):598-603, 1967.

56. Bayne, H. R.: Pseudoretrognathia: evaluation and treatment, Dent. Digest 72(3):112-117, 1966.

57. Parkes, M. L.: Chin implants with a newer plastic compound, Arch. Otolaryng. 75:429-436, 1962.

58. Calderwood, R. G.: Polydimethylsiloxane implants in oral surgery, J. Oral Surg. 26(1):33-40, 1968.

59. Junghans, J. A.: Profile reconstruction with Silastic

chin implants, U.S.A.F. Med. Service Digest **18:**10-13, 1967.

60. Vinas, J.: Private communication.
61. Robinson, M., and Shuken, R.: Bone resorption under plastic chin implants, J. Oral Surg. **27:**116, 1969.
62. Boucher, L. J.: A new method of ridge extension and tissue protection, Milwaukee Dent. Bull. **30:**4-7, 1964.
63. Gatewood, J. B.: Reconstruction of the alveolar ridge with silicone-Dacron implants: a pilot study, J. Oral Surg. **26:**441-448, 1968.
64. Schecter, L. M., and Pories, W. J.: Partial mandibular replacement with a silicone structure, Am. J. Surg. **113**(6):846-849, 1967.
65. Murray, J. E.: Personal communication.
66. Salem, J. E.: Personal communication.

Chapter 33

Psychiatric and medical-legal implications of rhinoplasty, mentoplasty, and otoplasty

Mark Gorney, M.D., F.A.C.S.

Of all the problems within the plastic surgeon's province, none presents more pitfalls to the inexperienced operator than elective aesthetic surgery of the face. Like an iceberg, the visible part may hide beneath it dangers far greater than the obvious ones. No matter how great the expertise or how deft the hands, nothing can prevent problems better than accurate patient selection and medical-legal awareness. That is the name of the game.

A warm body and money in the bank are not indications for cosmetic surgery. Proper patient selection must be based on two major criteria: anatomic and psychiatric suitability. The first of these has been dealt with in detail by other authors. The criteria that are used in the second category have been developed over the years, from experience. In no way must they be judged as the standard of care but rather as suggestive guidelines that have proved to be effective prophylaxis against trouble.

PSYCHIATRIC CRITERIA

The desire to appear normal or aesthetically pleasing is older than plastic surgery. The Puritan ethic that has, until recently, dominated our culture disapproves of narcissism. Thus, there is still a certain stigma in many patients' minds attached to seeking aesthetic improvement. This may add the element of guilt to a preexisting distorted body image. Edgerton and Jacobson[1] neatly liken the body image to a gyroscope. When it is functioning well we do not notice it. However, though it does not decide the course or steer the ship, the ship is vulnerable to storms and steering becomes difficult if it is not functioning.

Obviously every patient seeking rhinoplasty cannot be referred for psychiatric evaluation, nor is this necessary. A patient seeking a plastic surgeon feels that he is trying to do something positive about his problem and if he has to see the psychiatrist, he may consider himself a failure. It is vital, however, that the young plastic surgeon quickly develop psychiatric awareness and apply it in patient selection.

There are no objective criteria in this gray zone. They are not only subjective but totally different to patient and doctor. The patient has an idea of what she wants; the surgeon knows more or less what he can do for her. The problem is to get together as accurately as possible beforehand. It is much easier than looking sadly back in retrospect.

There is a significant difference in the psychodynamics of the male and female cosmetic patient, as Edgerton and Jacobson[1] have pointed out. Both sexes have positive psychologic expectations and a conscious wish for attractiveness. Women, however, more obviously wish to change themselves and feel more attractive. Men, on the other hand, seem more interested in changing others' attitudes toward them.[1, 2] This is an important difference, since the male seeking cosmetic surgery moves against strong social pressure in our society. The female is less likely to project her concern onto the attitudes and responses of others. The male, again according to Edgerton and Jacobson,[1] often shows:

1. Family or cultural background conflicting with his present life
2. Difficulties in heterosexual adjustment

3. Self-deprecating attitude and feelings of inadequacy

The ear patient, usually much less of an emotional problem, often shows different personality characteristics. He is much more likely to be hostile, agressive, with a "chip on the shoulder." Most often he is brought in by a parent from whom he inherited his deformity and who himself suffered ridicule in childhood. Consider, for example, the difference in the image of a Clark Gable and a Jimmy Durante. There is a significant difference in parental attitude toward nasal and ear revision. The nose patient frequently shows a guilt-tinged second-generation rejection of his ethnic background. Often it is not so much the desire to abandon his ethnic group but to be viewed as an individual and to rid himself of class attributes.

Motivation, rather than specific psychodynamics, should be the plastic surgeon's overriding concern. Is it a pragmatic desire to improve appearance, or is it a pathologic focus of subconscious problems on a physical fault? Contrast these two commonly heard statements:

"I don't like my nose. It's too big for my face, and I don't photograph well."

"I've always been terribly self-conscious about my nose. My father's family all have noses like this. I hate it!"

Many patients can say anything convincingly, but the second statement should trip a red flag and invite further inquiries into the patient's real motivations.

Strength of motivation is important and has a startlingly close relationship with patient satisfaction. A strongly motivated patient will have less pain, a better postoperative course, and a significantly higher index of satisfaction regardless of result.

Let us now explore some of the more nearly objective criteria of patient selection and liability potential in order of descending importance.

Objective deformity versus patient concern. Fig. 33-1 shows graphically a plot of the patient's objective deformity, as judged by the surgeon, versus the patient's degree of concern with that deformity. According to R. Mills,[3] the two opposite extremes of patient selection are as follows: (1) The patient with a major deformity but minimal concern and (2) the patient with a minor deformity causing extreme concern. The latter represents the poorest candidate. The broad group of patients will fall somewhere in between. Decision to accept or reject must be the surgeon's ultimate respon-

sibility. We have found it very useful to note somewhere in the initial visit chart our coded impression. Thus, when the patient whom the plastic surgeon dimly recalls phones and expresses interest in going ahead, a simple notation in the form of a "tic-tac-toe" chart with an x in the appropriate square should put him on guard.

Great expectations. Increasing experience invariably teaches the plastic surgeon to avoid the patient who expects his surgery to change his whole life. If he operates on someone with a large crooked nose and large hang-ups, he is likely to wind up with someone with a smaller straighter nose and large hangups, or worse. Certainly a reasonable degree of positive change is expected and usually occurs. However, aesthetic surgery, regardless of excellence, is dubious therapy for severe personality disturbances.

The demanding patient. The individual who brings pictures, drawings, and exact specifications of what he wants should be refused as a general rule. Such a person has little insight into the realities of reconstructive surgery. More than likely this type of patient is very explicit and very fussy about tiny imperfections and will not understand the fact that the surgeon is working with human tissue, not clay.

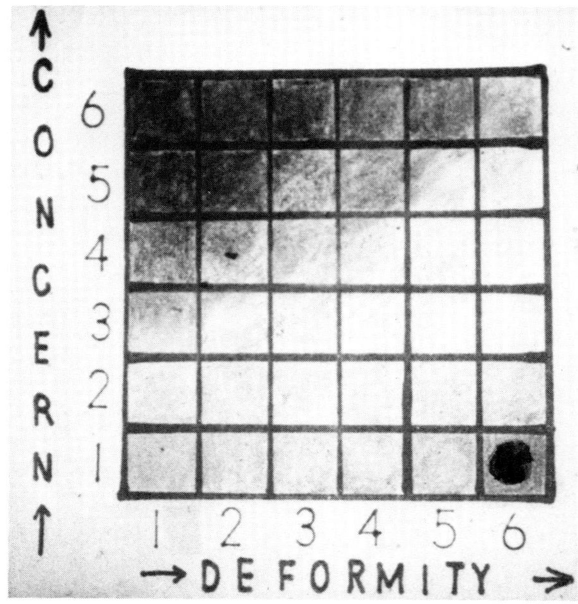

Fig. 33-1. The "Mills coefficient" of patient selection: a plot of objective deformity against patient concern. The patient in the lower right corner represents the best candidate while that in the upper left represents the worst. (After Dr. Robert, Mills, Millbrae, Calif.)

The indecisive patient. We have already spoken about the relationship between motivation and result. To the question: "Doctor, do you think I ought to have this done?" we usually answer: "This is a decision that I cannot make for you. I cannot encourage or discourage this operation; I can only tell you what I think we can accomplish. If you have thought about it carefully and feel strongly that you would like to have it done, you will probably be satisfied with the results. If you have any doubts at all, I strongly recommend that you think about it further or not have it done at all."

The important patient. Beware of the patient who makes a conscious effort to impress others with his stature in his profession, community, peer group, and the like. Such individuals often suggest that a successful result on them will immediately bring on innumerable referrals and undying fame. They will also turn out to be very difficult patients with a weak ego structure needing constant shoring up. They are difficult to satisfy and are prone to forget their financial obligations.

The secretive patient. Some applicants make a fetish of absolute secrecy about their surgery. Besides the fact that such arrangements are difficult to guarantee, exaggerated concern over this aspect of their operation indicates a suspicious degree of guilt over what they are about to do.

The immature patient. For reasons other than growth and development, one should carefully evaluate the degree of maturity in the young applicant. Often they have excessively romantic and unrealistic expectations of the effects of the change. When confronted with a mirror postoperatively, they sometimes exhibit disconcerting shock reactions and alarming behavior. If they have been talked into it by a relative, it only compounds the problem. Naturally, age does not have a linear relationship to maturity in all cases.

Familial disapproval. We much prefer the immediate family to be in agreement with the proposed operation and often refuse if they are not. Too often failure of communication or an unsatisfactory result produces an automatic "See, I told you so" reaction. This only deepens the guilt, the dissatisfaction, and the associated headaches.

MEDICAL-LEGAL IMPLICATIONS

Informed consent. It has been an unpleasant responsibility over the last few years to warn all plastic surgeons of the vital importance of obtaining the patient's informed consent. Suffice it to say that

the emphasis is on the word *informed*. Our attitude, reinforced by our experience, is embodied in our standard opening statement when we begin talking with the patient:

My responsibility to you is different from that of other doctors. I am not taking a sick patient and making him well; I am taking a healthy patient and and making him sick ostensibly to make him better. For that reason it is my duty and your right to know exactly what is going to happen in the most realistic terms possible. Patients have selective hearing. The good things stay, and the bad things go in one ear and out the other. I want you to listen carefully to what I am going to say, and then ask me anything I have not answered.

In our office each examining room is equipped with an inexpensive small blackboard. On this board we illustrate diagramatically the proposed operation in terms clearly understandable to the patient (Fig. 33-2). Naturally we vary the degree of detail according to each patient. We do not use medical terminology and avoid the use of "painful" words. Instead of *incise, chisel,* or *fracture,* we may say *open, take down,* or *remodel.* We use metaphors that are clearly understandable to the patient.

Example 1: "Doing this operation is like walking into a circus tent and reducing the framework so that the tent will shrink around it. If the tent is made of thin elasticized nylon, it will shrink quite nicely and conform to the reduced framework. If it is made of thick canvas it will not."

Example 2: "The bridge of your nose is like a gabled roof on a house. Down the middle of the attic there is a wall dividing it into two rooms with shelves on the side-

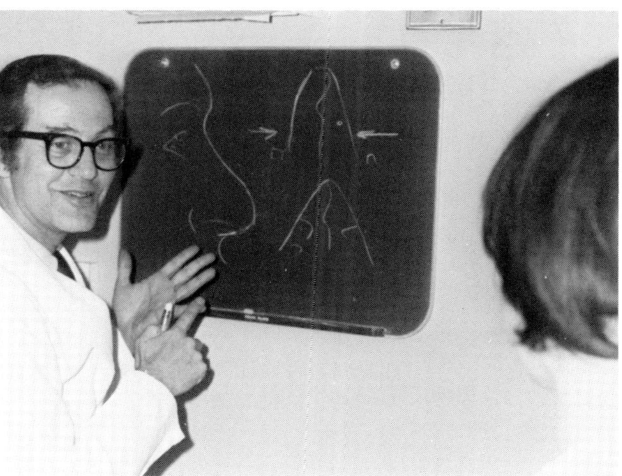

Fig. 33-2. The use of blackboard diagrams to assure full "informed consent." Notes are made in chart emphasizing that patient was informed with aid of diagrams.

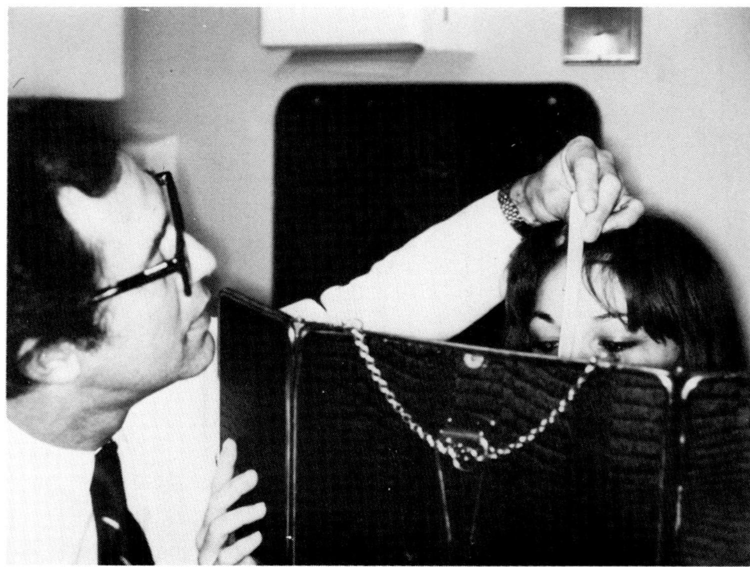

Fig. 33-3. Pointing out to patient preoperative natural asymmetry and mirror exaggeration.

walls. If the wall in the middle is crooked, we must straighten it out so air can get through."

We describe the use of local anesthetic (and the reasons for it), the degree of anticipated pain, the postoperative course, the duration of the disability, and the immediate postoperative sequelae. We discuss the expected changes in balance of facial features. We lay heavy emphasis on the matter of natural asymmetry of the human face (Fig. 33-3). We explain that when we look at another human being we see "a whole me or a whole you." Our mind does not break up the face into its component features, as does the mirror (Fig. 33-4). We point this out on our own faces or the patient's reflection in the mirror. We suggest that postoperatively there might still be a natural degree of asymmetry that, though visible to the patient in the mirror, will not be noticeable to another person.

We tell patients flatly that anyone who guarantees results in this kind of work is a fool or lacks experience.

Potential complications are openly discussed with our patients. How much explanation and in what detail is something that must be decided on balance between the surgeon's feeling about his patients' need to know and the requirements of the law. He need not engage in an orgy of open-minded disclosure. It is not possible to tell all the patients everything that can happen to them and not scare them out of their surgery. Nevertheless, the law

states that if known dangers of a treatment exist, the patient must be told of the "most probable," and preferably in percentage probability. Thus one can discuss them in general terms, while at the same time reminding the patient that he has a statistical probability of falling down and hurting himself that same day.

We tend to lean over backward in discussing with the rhinoplasty patient the outside possibility of the following:

1. Possible asymmetry
2. Postoperative bleeding
3. Airway problems
4. Possible necessity of secondary surgery, if such is foreseen
5. External scars, if any
6. If alloplastic implants are contemplated, possible infection and rejection

With the mentoplasty patient we discuss the following:

1. Rejection
2. Infection
3. Temporary numbness of the lip
4. Location, size, and appearance of scar, if any
5. Possible position shift of implant

With the ear patient we discuss the following:

1. Natural asymmetry
2. Location, size, and possible hypertrophy of scars
3. Temporary numbness

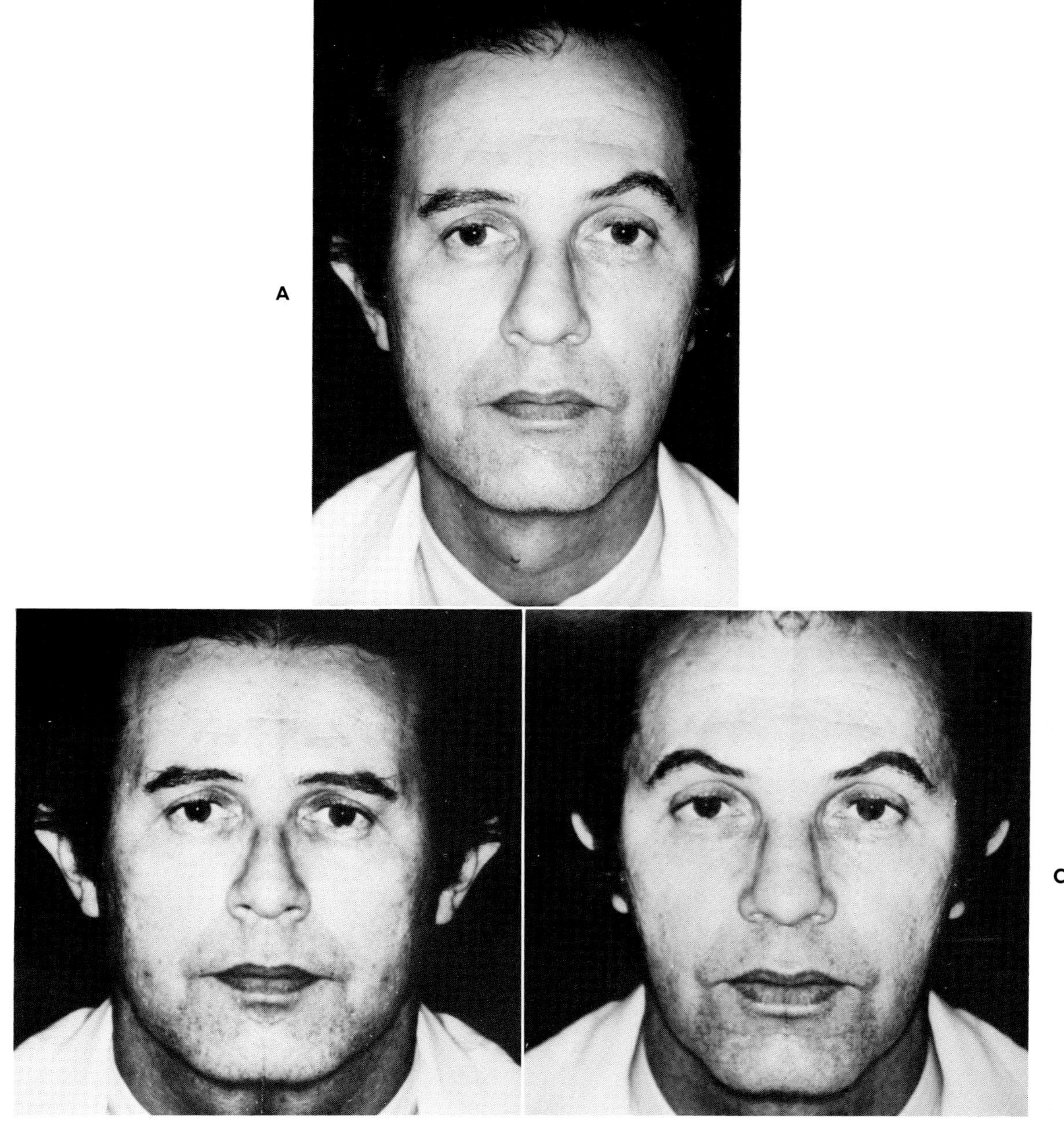

Fig. 33-4. A, Photograph of author (Gorney). **B,** The author's left half of face shown side by side. **C,** The author's right half of face shown side by side.

All patients are emphatically told that there is no warranty offered or implied. We reemphasize that this is human tissue, subject to unpredictable effect. It is not wood or clay. We also make it a habit to end this phase of conversation by mentioning the high percentage of satisfaction with these procedures.

Consent forms. Consent forms can vary between worthless pieces of paper and equally worth-less absolutions from everything except strangulation by hand. Some forms we have seen are appalling recitations of every complication this side of leprosy. No one can sign away his constitutional rights. Nevertheless, proper wording of a statement acknowledging unequivocally that the patient has been informed of the most common complications that can occur will, in most cases, dissuade an attorney from filing suit on the basis of breach of

PERMISSION FOR SURGICAL CARE

I hereby authorize and direct____ M.D.,

and/or associates or assistants of his choice to perform the following operation on______________________________,
                                                                                          Person

my ____, as we have agreed upon:
          Relationship

____

____

I further authorize the doctors to do any other procedure that their judgment may dictate to be advisable for my well being. It is understood that the doctors will give their best professional care toward the accomplishments of the desired results. The nature of the operation has been explained to me. I am advised that though good results are expected, complications cannot be anticipated and that
therefore, there can be no guarantee, either expressed or implied, as to the results of the surgery or cure. The doctor has explained to
me the most likely complications that might occur in this operation and I understand them.
I hereby authorize and direct the above named surgeon and/or his associates or assistants to provide such additional services as they
may deem reasonable and necessary including, but not limited to, the administration of any anesthetic agent, the services of the Department of Pathology, Radiology, the laboratories, and I hereby consent thereto.

Date:____   Time:____________________________________a.m./p.m.

____   ____
          Signature Patient   Signature Relative

____   ____
          Witness   Relationship

Fig. 33-5. Patient's consent form for surgical care.

informed consent. The form that we have lived with comfortably for a number of years, and which has proved its worth on at least one occasion, is shown in Fig. 33-5. It does not vary greatly from most forms, except that the patient acknowledges by his signature that he has been informed and that he *understands.* If the patient does not understand English, there should be an additional notation that the form has been translated. The blanks should always be filled in before the patient signs and the signing should be witnessed.

Caveat. The consent can be gravely weakened if it is not backed up by a handwritten note in the chart briefly stating what complications have been discussed. Our note usually reads as follows:

Have discussed in detail with patient proposed surgery, including diagrams. Advised of probable location, size, and appearance of scars. Described possible complications including ________________. Emphasized to patient no warranty, asked to think about it.

If there are any special conditions particular to this case they are also named in the handwritten note.

Photographic permits are usually fairly standard. If, however, the use of a patient's photographs are contemplated in future publications, it should be noted that the photographs must not in any way identify the patient unless he specifically permits it. A simple blocking out of the eyes on a face picture is not sufficient unless the permit so specifies.

Warranty. We have previously warned of the folly of implied warranty. Showing pictures of other patients has been interpreted by the courts as *expressed* warranty. It implies, "This is the kind of work I do and this is what you can expect." We strongly recommend against it. When asked to do so, we refuse, saying the following:

"These pictures are privileged information and part of the patients' records; they cannot be shown."
"If I were to show you pictures, I would obviously show you my best results only."

If the surgeon plans to draw anticipated results on enlarged photographs or moulages, we recommend that he do so in private and not in front of the patient.

As important as what is said is how it is said. Contrast these two answers to the common question: "Doctor, is there any danger in this operation?"

1. "Don't worry about a thing. We do lots of these every year, and there's nothing to it. You'll do just fine!"
2. "Any operation, no matter how minor, carries with it a certain amount of risk. This is an operation that we do quite commonly and the vast majority of patients obtain satisfactory results. I see no reason to think that yours will be different."

MISCELLANEOUS GUIDELINES

There should be a frank discussion of fees and costs, if not by the surgeon then by someone in his office. Experience has shown that payment in full in advance for cosmetic surgery tends to diminish subsequent unhappiness with final results. We handle all requests for extended payments by saying: "I would rather be your doctor than your banker."

It is an ironclad rule in our office that we never accept a patient at the first visit. He is asked to go home, think about what we have told him, and return (at no fee). We ask him to write down any questions and at the second visit we go over the highlights of our original conversation. Only then do we allow the patient to book surgery.

At surgery, it is axiomatic that all patients under local anesthesia should be adequately sedated. No permits should be signed after sedation is administered, since they will be held invalid. Every member of the surgical team should understand clearly that a patient under the influence of narcotics can misinterpret the most innocent words or jokes, and these can come back to haunt him. Under no circumstances should there be any arguments of any kind, even in jest. There should be no swearing, and assistants should be warned to save questionable doubts for later. There is no such word as "oops" in the operating room, whether the surgeon drops a hemostat or comminutes the nasal bones. It helps to talk to the patient and be highly visible at the beginning and the end of the procedure.

At the end of the operation, we make it a point to report to the family immediately. If they are not present we telephone. A visit on the evening of the operation is immensely reassuring. Our discharge instructions are clear, specific, and in writing. Availability during the first few days is essential, especially about the eighth postrhinoplasty day when bleeding is apt to occur.

When dressings come off, there will be numerous questions, all of which require simple reassuring answers. These questions will be less in number and anxiety if they have been answered preoperatively.

It is an article of faith that litigation in plastic surgery has as its common denominator not poor results but poor communication. Underlying all dissatisfaction is a breakdown in rapport between patient and surgeon. This vital relationship, alas, is often shattered by the surgeon's arrogance, hostility, coldness (real or imagined), but mostly the fact that "He didn't care!" There are only two ways to avoid this debacle. One is to make sure that the patient has no reason to feel that way. The other is to learn to avoid the patient who is going to feel that way whatever is done.

REFERENCES

1. Jacobson, W. E., and others: Psychiatric evaluation of male patients seeking cosmetic surgery, Plast. Reconstr. Surg. **26:**356, 1960.
2. Meyer, E., and others: Motivational patterns in patients seeking elective plastic surgery (women who seek rhinoplasty), Psychosom. Med. **22:**193, 1960.
3. Mills, R. L.: Personal communication.
4. MacGregor, F. C., and Shaffner, B.: Screening patients for nasal plastic operations, Psychosom. Med. **12:**277, 1950.
5. Palmer, A., and Blanton, S.: Mental factors in relation to reconstructive surgery of nose and ears, Arch. Otolaryng. **56:**148, 1952.
6. Stern, K., Fournier, G., and LaRiviere, A.: Psychiatric aspects of cosmetic surgery of the nose, Canad. Med. J. **76:**469, 1957.

Index

"